Learning with Understanding in the Chemistry Classroom

Iztok Devetak · Saša Aleksij Glažar
Editors

Learning with Understanding in the Chemistry Classroom

Springer

Editors
Iztok Devetak
Saša Aleksij Glažar
University of Ljubljana
Ljubljana
Slovenia

ISBN 978-94-007-4365-6 ISBN 978-94-007-4366-3 (eBook)
DOI 10.1007/978-94-007-4366-3
Springer Dordrecht Heidelberg New York London

Library of Congress Control Number: 2013953612

Printed on acid-free paper

Springer is part of Springer Science+Business Media (www.springer.com)

General Preface

The main goal of chemistry education research is to understand and improve chemistry learning and teaching. Research studies show the range of research design strategies and results that have contributed to an increased understanding of learning in chemistry. Practitioners, however, are seldom acquainted with the findings of education research and as a consequence they are not applied into school practice. The challenge is how to link together findings of research and effective practice and study their influence on curriculum, on teaching methods, and on assessment. This will require more effective communication between researchers and practitioners to bridge the gap between chemistry and education disciplines.

This publication's aim is to offer an additional stone in the mosaic of efforts toward changing chemistry teaching and learning from incidental and rote learning to learning with understanding and meaningful knowledge. All contributions in the publication try to follow this goal.

Authors from 12 countries, despite cultural differences and economics of schooling emphasize the same trends, which stem from human physiology and psychology that underline learning and teaching chemistry in 18 chapters.

On the basis of a content analysis of the papers published in selected science education journals for a period of 5 years it was found that research in the field of chemical education could be divided into nine categories: (1) teacher education; (2) teaching; (3) learning—students' conceptions and conceptual change; (4) learning—classroom contexts and learner characteristics; (5) goals and policy, curriculum, evaluation, and assessment; (6) cultural, social, and gender issues; (7) history, philosophy, epistemology, and nature of science; (8) educational technology; (9) informal learning. These science education fields are also illustrated from different perspectives in the present book. This book is according to its content divided into three sections: Section I Teaching and learning chemistry; Section II Approaches in chemistry teaching and learning with understanding; and Section III Curriculum reform and teachers.

The first section "Teaching and learning chemistry" focuses on the general aspects of chemical education research and practice. In this section the teaching and learning of chemical concepts are discussed. This section comprises two parts;

the first part "Understanding Chemistry Concepts Teaching Strategies" deals with learning chemical concepts that results in understanding chemical phenomena; and the second part "Students' characteristics on chemistry learning" describes and analyzes students' characteristics that can foster chemical concepts learning with a low rate of misconceptions.

The first part of this section focuses on learning chemical concepts, and it has been established that chemical concepts can pose different levels of demand on students' working memory. This means that especially abstract concepts demonstrating chemical change should be presented to the students in different ways. But before that teachers should understand concepts and should be able to move easily between all three representations of concepts (e.g. macro-, submicro- and symbolic level). Chemical concepts are because of this characteristic specific and even more demanding in terms of understanding compared to those that can be presented only on the macro level for example. Students' learning chemical concepts with understanding should be stimulated by the teacher. These stimuli should trigger students' mental activities, so that learning would occur. Without students being mentally (and also manually) active during learning, meaningful learning with understanding will not happen. The concepts describing active learning are frequently discussed in the chemistry education literature but a more in-depth analysis should be provided.

The second part of this section comprises two chapters dealing with students' characteristics that can significantly influence chemistry teaching and learning. Students' attributes such as motivation and interest for learning chemistry, different mental abilities (i.e. intelligence, visualization abilities, working memory capacity, formal reasoning ability), social skills, and others, should be considered when the teacher organizes their school lessons, authors design the teaching material, policy makers prepare national curriculums, and teacher educators conduct pre- and in-service teacher education programs.

Section II entitled "Approaches in chemistry teaching and learning with understanding" comprises two parts; the first part "Cooperative and collaborative learning" presents three chapters and the second part "Teaching Strategies" comprises six chapters.

The first part focuses on cooperative and collaborative learning in the science classroom to promote students' learning with understanding. The first part deals with different aspects influencing science learning as students' cultural, racial, ethnic, and social backgrounds can influence collaborative and cooperative learning. The authors explain the development of cooperative learning methods and the integration of these approaches into science education to stimulate peer-to-peer teaching and learning hoping that these approaches will enhance students' academic achievements and stimulate interest for science learning and future careers in science and technology are presented. The differences or similarities between cooperative and collaborative learning are explained by the different authors. Both approaches are sometimes used for the same thing, e.g., small-group activities in the classroom where learning takes place, but differences can be found in the organization of the specific learning approach. Collaborative learning can have fewer roles assigned, the teacher is not the center of authority, group tasks are usually more open-ended, and complex, so collaborative learning is less structurally defined as cooperative learning.

The second part deals with teaching strategies or approaches that support students' engagement in mental activities in science learning. If learning would take place, students should think about the content presented by the teacher, textbook, online or otherwise. Some of these aspects are presented in Part II (Approaches in chemistry teaching for learning with understanding). The most important problem that science teachers face is how to motivate students to learn for their future lives as active citizens. It is difficult to explain to students the fact that they are not learning just to pass the exams, but to become scientifically literate adults, who will make important and correct decisions. To achieve this, teachers and science education researchers try to find ways to make students learn science concepts with understanding and for life. This usually involves experimental work, using different pictorial material, context-based approaches, and multimedia environments.

The last section of this book entitled "Curriculum reform and teachers" deals with the chemistry curriculum and changes influence the chemistry teacher's education. It is mentioned that chemistry curriculums have changed over the decades from traditionally oriented chemistry teaching emphasizing symbolic and mathematical components of the chemical concepts to more context-based enquiry learning-oriented teaching supported by different applications of the informational-communicational technology. It is emphasized that it is important to develop students' scientific/chemical literacy, so that they will be able to use their science knowledge in different real-life situations. On the other hand, teachers should be adequately educated so that they can efficiently implement curriculum innovations. This means that teachers should in pre-service/university level education develop their sense of permanent in-service education, so that they can instantly and effectively apply those innovations that appear in the curriculum into their teaching. It is stressed that teachers are aware of their possibilities to upgrade their teaching with outside school activities for students. Chemistry presented in museums, industry, agriculture, medicine, science centers, forensic TV shows, etc., can influence students' interest to learn chemistry at a formal level. Teachers should for that matter use the informal ways of showing the importance of chemistry in human society to their advantage.

The editors would like to thank Dr. Leopoldina Plut Pregelj (University of Maryland, USA) for numerous prudent suggestions that have helped to make the book as it is today.

Iztok Devetak
Saša Aleksij Glažar

Contents

Section II Approaches in Chemistry Teaching for Learning with Understanding

Cooperative and Collaborative Learning

Teaching Strategies

Contributors

Sevil Akaygun Boğaziçi University, Istanbul, Turkey

Miriam Barak Department of Education in Science and Technology, Institute of Technology, Haifa, Israel

Katharina Beck KGS Kirchweyhe (comprehensive school), Brinkum, Germany

George M. Bodner Department of Chemistry, Purdue University, West Lafayette, USA

Diane M. Bunce Department of Chemistry, The Catholic University of America, Washington, D.C., USA

Liberato Cardellini Università Politecnica delle Marche, Ancona, Italy

Miriam Carmi Department of Education in Science and Technology, The Hebrew Gymnasium "Herzeliya", Institute of Technology, Tel Aviv, Israel

Kirsten Lowrey Casey Department of Chemistry, Purdue University, West Lafayette, USA

Sanoe Chairam Ubon Ratchathani University, Ubon Ratchathani, Thailand

A. L. Chandrasegaran Science and Mathematics Education Centre, Curtin University, Perth Western, Australia

Gail Chittleborough Deakin University, Melbourne, Australia

Richard K. Coll University of Waikato, Hamilton, New Zealand

Chanyah Dahsah Srinakharinwirot University, Bangkok, Thailand

Iztok Devetak Faculty of Education, University of Ljubljana, Ljubljana, Slovenia

Yehudit Judy Dori Division of Continuing Education and External Studies, Department of Education in Science and Technology, Technion—Israel Institute of Technology, Haifa, Israel

Nina Dunker University of Bremen, Bremen, Germany

Ingo Eilks Department of Biology and Chemistry, Institute for Didactics of Science (IDN), University of Bremen, Bremen, Germany

Wiebke Endres University of Oldenburg, Oldenburg, Germany

Saša A. Glažar Faculty of Education, University of Ljubljana, Ljubljana, Slovenia

Katarina S. Wissiak Grm Faculty of Natural Sciences and Engineering, University of Ljubljana, Ljubljana, Slovenia

Ninna Jansoon Thaksin University, Songkhla, Thailand

Loretta L. Jones University of Northern Colorado, Greeley, USA

Mojca Juriševič Faculty of Education, University of Ljubljana, Ljubljana, Slovenia

Stephan Kienast Ursulinen-gymnasium (grammar school), Werl, Germany

Paweł Kościelniak Department of Analytical Chemistry, Institute of Forensic Research, Jagiellonian University, Kraków, Poland

Iwona Maciejowska Department of Chemical Education, Jagiellonian University, Kraków, Poland

Patricia A. Metz Department of Chemistry, Purdue University, West Lafayette, USA

Bettina Most Konrad-Adenauer-Realschule (middle school), Hamm, Germany

Zuzi Mthembu Science and Mathematics Education Centre, Curtin University, Perth Western, Australia

Ilka Parchmann IPN—Leibniz Institute for Science and Mathematics Education, University of Kiel, Kiel, Germani

Renata Wietecha-Posłuszny Department of Analytical Chemistry, Jagiellonian University, Kraków, Poland

Norman Reid University of Glasgow, Glasgow, UK

Vesna Ferk Savec Faculty of Natural Sciences and Engineering, University of Ljubljana, Ljubljana, Slovenia

Kristina Sodja Faculty of Natural Sciences and Engineering, University of Ljubljana, Ljubljana, Slovenia

Keith S. Taber Faculty of Education, Science Education Centre, University of Cambridge, Cambridge, UK

David F. Treagust Science and Mathematics Education Centre, Curtin University, Perth Western, Australia

Georgios Tsaparlis Department of Chemistry, University of Ioannina, Ioannina, Greece

Margareta Vrtačnik Faculty of Natural Sciences and Engineering, University of Ljubljana, Ljubljana, Slovenia

Vickie M. Williamson Texas A&M University, College Station, USA

Torsten Witteck Westfalen-Kolleg (grammar school), Bielefeld, Germany

Michał Woźniakiewicz Department of Analytical Chemistry, Jagiellonian University, Kraków, Poland

Reviewers

Maija Aksela
Bette Davidowitz
Philip Johnson
Rosaria Justi
Barica Marentič Požarnik
Primož Šegedin

Section I
Teaching and Learning Chemistry
Understanding Chemistry Concepts

Section I of the book focuses on more general aspects of chemical education research and practice. In this section, teaching and learning of chemical concepts are discussed. This section comprises two parts; Part I deals with learning chemical concepts that result in understanding chemical phenomena, while Part II describes and analyses those students' characteristics that can foster chemical concepts learning with a low rate of misconceptions.

As mentioned above, this part focuses on learning chemical concepts, and it is well known from studies that chemical concepts can pose different levels of demand on students' working memory. This means that especially abstract concepts demonstrating chemical change should be presented to the students in different ways. But before that teachers should understand concepts and should be able to move easily between all three representations of concepts (e.g. macro-, submicro- and symbolic level). Chemical concepts are because of this characteristic specific and even more demanding for understanding as those that can be presented only on a macro level for example. Students' learning chemical concepts with understanding should be stimulated by the teacher. These stimuli should trigger students' mental activities, so that learning will occur. Without students' being mentally (and also manually) active during learning, meaningful learning with understanding will not happen. The concepts describing active learning are frequently discussed in the chemistry education literature but a more in-depth analysis should be provided.

For that reason, Taber in Chap. 1 entitled "Constructing active learning in chemistry: concepts, cognition and conceptions" argues that all meaningful learning is 'active' in the sense that the learner actively (although not necessarily consciously) links new learning with, and interprets teaching through, existing ways of making sense of the world. It follows then that conceptual learning in chemistry is iterative. Sound foundations in the subject support progression in understanding; but, equally, alternative conceptions (ideas at odds with the scientific models) support the misconceiving of teaching. Teaching can be misunderstood when the learner's existing understanding does not match the prerequisite knowledge assumed in the teacher's presentation. A range of different categories of 'learning impediment' may result, when learners either fail to make the intended links with prior learning, or form idiosyncratic links with existing ideas that seem relevant from the student's perspective. An engaging chemistry teacher, who provides students with a range of relevant learning activities, will inevitably produce active learning in the sense of the mental construction of new knowledge. The first chapter of this book for these reasons offers an outline of constructivist thinking about learning, and presents a classification of the main types of learning impediments that misdirect learning.

Chapter 2 entitled "The development of theoretical frameworks for understanding the learning of chemistry" by Chittleborough focuses on the importance of the triple nature of chemical concepts presentations that gives, according to the author, chemistry unique characteristics that make it a difficult subject to understand. Drawing on data from a study involving first-year university students learning introductory chemistry, this chapter looks at how these students' understanding of the characteristics of chemistry influences the way they understand and learn chemistry. Two theoretical frameworks to describe how chemical concepts can be presented and understood are developed based on research data: the expanding triangle and the rising iceberg. These interesting ideas about students' learning chemistry on a triple level can further develop the ways of thinking about how students learn chemistry. The author proposed these two frameworks as useful tools for chemistry educators to better understand students' learning, linking chemical education research to practice so as to inform pedagogical content knowledge.

One of the most important ideas about meaningful learning in chemistry—the triple nature of chemical concepts is further developed in Chap. 3 by Tsaparlis. His text entitled "Linking the Macro with the Submicro Levels of Chemistry: Demonstrations and Experiments that Can Contribute to Active/Meaningful/Conceptual Learning" discusses chemistry as a multi-representational structure. Studies have shown that students have great difficulties when trying to grasp concepts at the submicrolevel. In this chapter, a set of demonstrations and experiments is proposed that, if properly used in teaching by means of an active-learning methodology, can contribute to meaningful learning and conceptual understanding of the particulate concepts of matter by properly linking the macro with the submicrolevels. Different laboratory work is presented and the importance of linking different levels of chemical concepts presentations is proposed.

The last chapter in Part I of Section I "Teaching and Learning Chemistry" by Bunce entitled "Challenging Myths about Teaching and Learning Chemistry" argues about students' chemistry learning when and if teaching is active or it comprises different multi-model representational approaches. Some things in education are repeated so often that they become embedded in the collective memory of both students and teachers. We have come to accept as 'truths' such things, for example students' attention during lecture, the use of modern technology will increase students' achievements in chemistry, students just memorise the learning material and do not study for understanding, and students forget most of what they learn in chemistry immediately after completing an exam.

Bunce also discusses the proofs to support the acceptance of these ideas within the academic community and she tries to explore the truth behind these beliefs and some of the intervening variables that affect their measurement and interpretation. The goal of this chapter is to move our knowledge of how students learn from unsubstantiated opinion to a more accurate research-based foundation.

Chapter 1
Constructing Active Learning in Chemistry: Concepts, Cognition and Conceptions

Keith S. Taber

Active Learning and Chemistry Education

This chapter explores the nature of active learning in chemistry in terms of how learners develop their conceptions of chemical concepts through cognition. The term '*active* chemistry learning' may suggest images of busy classrooms, with students moving about undertaking practical work, to find out the 'secrets of nature' for themselves. However, whilst such a classroom certainly can facilitate much chemical learning under certain conditions, it is not necessarily the case. Practical work, unless carefully set up, can engage hands more than minds (Abrahams 2011). Moreover, practical work that does engage minds is often unlikely to lead to the desired learning outcomes (Driver 1983), unless it is very carefully structured and integrated within well-planned teaching sequences. So whilst physical activity is certainly a candidate for a feature of good chemistry teaching, it is not of itself a good sign of active learning. Rather, the focus needs to be on mental activity (Millar 1989).

However, whilst ensuring students are mentally active and have their minds focused on the chemistry being taught in a lesson is likely to bring about learning, even this is not enough to ensure that student learning closely matches the intended learning. This can be appreciated by considering the large number of studies of student thinking in chemistry that have reported 'misconceptions' or 'alternative conceptions' (Kind 2004; Taber 2002). Research has elicited from chemistry learners a wide range of alternative conceptions (or misconceptions), which are inconsistent with the scientific concepts (Duit 2009).

The 'constructivist' perspective, which has dominated thinking about science education internationally for some decades (Taber 2009b), interprets these alternative ideas as the outcomes of active learning processes; but active learning processes that led the student to a somewhat different understanding than that

K. S. Taber (✉)
Faculty of Education, Science Education Centre, University of Cambridge, Cambridge, UK
e-mail: kst24@cam.ac.uk

I. Devetak and S. A. Glažar (eds.), *Learning with Understanding in the Chemistry Classroom*, DOI: 10.1007/978-94-007-4366-3_1,

intended by the teacher. Constructivism has drawn upon psychological models of how conceptual learning is an iterative process, and has highlighted the nature of students' own conceptions in science topics. These 'alternative conceptions' (Gilbert and Watts 1983) reflect how each learner actively constructs their own knowledge, interpreting teaching in terms of their own existing understanding.

Constructivist Premises

Constructivism has been informed both by philosophical arguments about the nature of knowledge, and by studies of learning from psychology and other cognitive sciences (Taber 2009b). Whilst there are many variations in the way constructivism is presented, it is based on some simple premises. In particular, human beings are inherently driven to make sense of the world. This is not something that depends upon a particular motivation, but rather it is hard-wired into our brains as part of our evolutionary heritage. We interpret flashes of light, and short extracts of overheard conversation, instinctively. We feel frustrated when we cannot understand something. We are by nature meaning-makers.

However, because much of this meaning making takes place at pre-conscious levels of cognitive processing, we are usually only aware of the *outcomes* of the process, not the process itself (Smith et al. 1993). We recognise a face, or a snippet of Vivaldi or the Beatles, without being aware how the actual sensations (of patterns of light; of vibrations in the air) became interpreted as something familiar. The same processes are at work when a student watches a chemistry teacher's demonstration or listens to her explanation for some chemical phenomenon. What is presented to consciousness is not raw data to be interpreted by the conscious mind, but the output of automatic processing that has often matched what is seen or heard to some familiar pattern represented at pre-conscious levels in the brain (diSessa 1993; Taber and García Franco 2010).

Whilst is it possible to learn 'non-sense' information by rote, *meaningful* learning (Ausubel 2000) requires the learner to associate what they see and hear with something they already 'know'. So the student makes sense of what they are taught in an internal as well as an external context. The external context is the classroom, in which the teacher talks and demonstrates, and students carry out various activities. This public context is shared by the teacher, and all the students in the class. The internal context is highly personal: it is the mental environment in which new information is interpreted. This environment may be rich and multi-leveled: and as suggested above, includes stages of processing that occur before anything is presented to the conscious mind.

The term 'conceptual ecology'—drawing on Toulmin's (1972) notion of the evolution of concepts in an intellectual ecology—has been used to describe the context in which ideas are understood, and develop, in the human mind. The analogy here with how living things evolve in a particular habitat draws attention to the potential complexity of the mental system in which learning occurs

(Taber 2001b, 2009b). The conceptual ecology is not just the student's existing understanding of a topic, but also includes a range of *meta-conceptual* factors. As one example, explanatory coherence is something that is highly valued in science (Thagard 1992): scientific explanations should be consistent across topics and even disciplines, and explanations that use already well-accepted principles are to be preferred to those that need to introduce new, additional premises. Any student who shares such values is primed with certain expectations regarding the scientific explanations met in class, and so is biased to interpret them in certain ways. Any student who has not adopted these values may not appreciate the unspoken assumptions of much teacher exposition, and so may miss much of the motivation for certain scientific ideas (Taber 2008a).

Three Broad Classes of Learning Outcome

Learning is perhaps best understood as a change in the potential for behaviour: that is, learning has taken place if there is some change in the learner such that after learning they can behave differently in some possible situation than had been the case before learning (Taber 2009b). This is a general description, but commonly the type of behaviour we are most interested in is responses to questions and other such set tasks. If a learner undergoes some experience such that she is able to provide an answer to the question 'is carbon a metal' that was not part of her repertoire before, then she has learnt something. We need to note that such a general definition has implications: learning brings about a change in potential that may only be realised in specific situations; *and* learning that does take place in classrooms is not necessarily desirable from the educational perspective.

So for example let us consider a hypothetical student called Hilda. If she was asked the question 'is carbon a metal', she would answer 'no'. However, Hilda then attends a chemistry lesson on electrolysis, where she undertakes practical work using graphite rods as electrodes. Hilda has existing knowledge that graphite is a form of carbon, and that metals conduct electricity. During the lesson Hilda makes sense of the use of carbon electrodes in terms of her belief that metals (and only metals) act as conductors. Hilda comes to think of carbon as a conductor, and so a metal. As a result of this learning experience, there are physical changes in the structure of Hilda's brain, such that the knowledge represented there is altered. We might say there have been changes in her 'cognitive structure'. If Hilda were now asked the question 'is carbon a metal', she would answer 'yes'. However, as Hilda is given no reason to demonstrate her new thinking in the lesson, the teacher does not detect this learning.

A week later, in a subsequent lesson, the chemistry teacher might ask the class if anyone remembers what material was used for the electrodes. Hilda is able to reply 'carbon, graphite'. Her active processing of the information that the electrodes were made of graphite, and her linking that into her knowledge about carbon, and about metals as conductors, supports her in remembering this as meaningful information (Taber 2003b). The teacher is pleased with Hilda's learning. Although Hilda now thinks carbon is a metal, this is not elicited by the teacher's question, and a misconception remains unidentified. This is just a hypothetical case (some real examples are discussed later in the chapter), but illustrates both (a) how learning may be real, but not actively demonstrated unless elicited by a specific set of circumstances; and (b) how learning does not necessarily shift understanding in the intended direction.

If Hilda's teacher was committed to helping students form links between their scientific knowledge when opportunities arise, she might think to follow-up her question about the electrodes by asking something like 'why might we be surprised that we can use carbon as a component of an electrical circuit?', providing an opportunity to explore how carbon is generally considered a non-metal, but that the graphite allotrope has some properties that are unusual in this regard. We might even conjecture that despite (or perhaps because of) her earlier false assumptions, Hilda—a student actively looking to link her knowledge together—would be especially primed to learn from this aside. In this hypothetical case we might consider that Hilda held a particular epistemological commitment to the nature of scientific knowledge that was an active factor in her conceptual ecology (Hammer and Elby 2003).

In principle, then, it is possible to identify three possible general classes of outcome when a student is exposed to teaching (see Table 1.1).

One possibility is that no learning takes place. Whilst this is a theoretical possibility, it is seldom going to be the case in absolute terms. Any experience we have will activate some cognitive process (i.e., remind us of something) and is likely to forge some new links in cognitive structure (without necessarily being related to target knowledge: e.g., 'the colour of the teacher's tie is the same as the shirts worn by Manchester United footballers'). Unless we are comatose, we cannot avoid some level of learning from our experiences. However, if a student can make little sense of a lesson, and has no motivation to pay attention, it is feasible that any learning related to chemistry will be fairly minimal, and we might for practical purposes consider there to have been no significant learning.

Rote Learning

The second possibility is that some rote learning will take place (see Table 1.1). Rote learning concerns the learning of material that has no inherent meaning. An example might be a telephone number, where there is no automatic link between the pattern of numerals and the person who can be called on the number. Such information is not easy to learn, unless one spots some pattern to latch onto. For example, the number 19141918 may be a burden to remember, but becomes easier to recall if recognised as the dates of the 'first world war'. Of course even a number which does *not* suggest such a pattern has been 'made-sense of' compared with the raw perceptual data (the sensory impression of the pattern made by ink on

Table 1.1 A caricature of three levels of learning from teaching

Level	Description	Notes
No learning	A student who pays no attention to a lesson may in principle undergo no learning	In practice, we can learn incidentally even without consciously focusing on our surroundings. However such learning is unlikely to be effective in terms of academic progression
Rote learning	Material may be learnt by repetition—e.g., mentally repeating it *verbatim* until it can be recalled	Accessing such material in memory tends to rapidly become more difficult, unless there is medium- and long-term reinforcement. Isolated material learnt this way tends to only be useful for low-level tasks (i.e., being able to *recall* that Kekulé proposed structures for benzene; but not for explaining the significance of the structures he proposed)
Meaningful learning	Material that is actively processed by being explored in terms of existing thinking can be learnt meaningfully	Meaningful learning is integrated into the learner's existing conceptual structure, which makes it easier to access later, and allows it to be used more flexibility in higher level tasks (such as forming and critiquing explanations). Meaningful learning can be just as effective at representing incorrect understandings of chemistry as correct understandings

the page of the telephone directory): the numerals are themselves familiar, as is the process of constructing a telephone number from a string of numerals.

Interestingly, a good deal of early research into human memory was undertaken with this type of target—for example lists of nonsense letter triads to be recalled. Humans certainly can memorise such material, but it usually takes some effort. This is especially so if recall is required not later in the same test session, but some days, weeks or months later. Motivation is clearly important here. Learning something by rote usually requires time and effort that is unlikely to be invested without good reason. Indeed, the ability to effortlessly learn a large amount of such meaningless material is not only rare, but seems to be pathological (Luria 1987).

This is highly relevant to education. If much course material has to be learnt by rote, then the students' task becomes both substantial and tedious. Meaningful learning is both easier and more interesting. It also offers flexibility in application as material learnt by rote can be regurgitated when, and only when, we recognise it is an appropriate response. However, not understanding the significance of learnt material means that it can only be presented 'as is', as so much mental ballast. Chemistry, as a science, is not primarily about isolated facts (the formula of ammonia, the electronic configuration of sodium, the molecular mass of sulphur dioxide—such facts are of little significance in isolation), but about *concepts that can be used to build extensive theoretical frameworks that offer explanatory value.*

Concept Learning as Meaningful

Concepts are inherently meaningful. A student may learn a concept label by rote, and even an associated definition, but if that is done without understanding then the student has not learnt the concept. There is certainly a good deal of rote learning in classrooms around the world, and sadly some approaches to chemistry teaching may indeed encourage such an approach. Yet students in such classes are learning facts, and NOT learning science. Although there is considerable discussion on how to best understand the nature of concepts (Gilbert and Watts 1983), they may be most easily understood as categories. A student can be considered to have acquired a concept of 'metal', 'methane', 'molecule', 'metallic bond' or 'molecular formula' if they are able to make discriminations that allow them to decide when something is or is not a metal, some methane, a molecule, a metallic bond or a molecular formula. If they can make such discriminations, then they have a concept with that concept label: although this does not necessarily mean they make the same discriminations as the chemistry teacher would, and so have the 'same' concept. Hilda's concept of metal included carbon as an example, whereas her teacher's did not. Concepts tend to be understood in terms of the links they have with other concepts: metals *conduct electricity*, *copper* is a metal, metals have *metallic bonding*, metals are *ductile*, metals form *cations*, metals are a type of *material*, etc.

So the third main category of learning, then, is meaningful learning, where new information is understood in terms of existing conceptual frameworks, and new concepts are incorporated into those frameworks to extend them (see Table 1.1). This type of learning is educationally more valuable, offering flexible, applicable knowledge; is more interesting for the student; and involves the development of the type of knowledge that science itself seeks—knowledge that is coherent, integrated, systematic and so forth.

An irony, perhaps, in the context of a discussion of active learning, is that meaningful learning requires less effort than rote learning. Learning by rote requires deliberate focused acts of concentration. Meaningful learning just builds upon the brain's evolved ability to make sense of new information, which is automatic. Indeed a student who is intrinsically motivated by interest in a topic, and who is working at a level where new concepts are being met, or existing ones being developed, at a pace and level that matches their existing level of understandings, may experience a mental state of 'flow' (Csikszentmihalyi 1988) where sustained concentration seems effortless.

So the kind of active learning we should seek is not that where we encourage students to be active in terms of either physical manipulation or hard mental effort; but rather that where the match between current knowledge and new experiences allows engagement in the subject matter that best activates the natural cognitive processes associated with accessing existing knowledge, exploring how new material fits with old, and looking for new links and ways to incorporate new ideas into existing understanding.

Of course, student study experiences are seldom *explicitly* perceived this way—unless they are undertaking activities designed to make concept linking explicit, such as concept mapping (Taber 1994b). This type of mental activation can *sometimes* be achieved when a skilled teacher demonstrates and explains ideas to motivated students—although in general students taking notes from lectures will not fit the bill. Practical work can sometimes be effective, but not practical work for its own sake (Abrahams 2011; Millar 2004). Discussion tasks, where students have to explain and justify their reasoning in groups, can be very effective. For that matter, written exercises can sometimes support effective learning. In all these cases, the key is to structure the activity so that the student is thinking about the new in terms of their existing understanding, something that is only possible if there is good matching so that the new material does not seem trite, and is not pitched at a level too high for the students to make sense of it.

Indeed, the general principles here are no different in teaching chemistry than in effective teaching of history or geography or many other subjects. However, what chemical education research has revealed over recent decades is just how challenging the task of matching the new to the old is for chemistry teachers. In this regard, a key problem of chemistry education is NOT how to find ways of making learning meaningful for students, but rather *how to channel students towards the particular meanings the chemistry teacher is charged with teaching.*

When Active Learning Goes Wrong

Extensive research shows that whilst students do indeed commonly make sense of their chemistry lessons in terms of their existing understandings, it is often in ways rather different from that expected by their teachers (Kind 2004; Taber 2002). One way of thinking about this is in terms of the teacher's role in bringing about learning. When the teacher presents a chemical topic, the learners will each interpret her words in term of their existing knowledge. Unfortunately, as learning is an iterative process, when students come to classes with alternative understandings of chemical phenomena, it is very likely that they will go on to further misinterpret the teacher's intended message. New alternative conceptions that the student finds useful for making sense of chemistry will be reinforced, and can in time be well integrated into the students' understanding of the subject. Such robust learning—whether matching scientific models or not—has potential to act as the foundations for further later learning (Taber 2005).

The teacher then needs to present the material to be learnt in such a way that it can be understood as intended in terms of the learners' existing knowledge of the subject. The justification for studying learners' conceptions in chemical topics is that knowledge of how students understand chemical topics can inform teachers so that they can better support learners in acquiring scientifically acceptable models. As we have learnt more about the nature of learners' ideas it has become clearer that this is by no means a straightforward matter (Taber 2009b).

The chemistry teacher clearly expects and intends their teaching to be understood correctly, and so (whether through careful planning, or simply the implicit assumptions behind any attempt at communication) presents the information on the basis of a personal mental model of the learner's existing understanding. As an extreme example, a teacher taking an introductory chemistry class in a school is not going to base her explanations on explicit solutions of the Schrödinger equation, as she will know that the pupils will not be in a position to understand the chemistry in these terms. Whilst this is obvious, it is often much less clear exactly what level of prior understanding can be assumed when planning teaching. Certainly, an assumption that the class will understand correctly all the science that has been studied prior to the new lesson is likely to be rather optimistic given the catalogue of common alternative conceptions reported in the literature. For learning to be successful, there needs to be a good match between the presentation of material and the conceptual frameworks that pupils can call upon to interpret it, and that means a good match between the actual conceptual structures available to students, and the mental model of those structures used by the teacher to plan teaching.

Learning Impediments

Learning can go wrong when there is a mismatch (Taber 2001a). Such mismatches act as impediments to learning. Sometimes a student makes no sense of the teacher's presentation at all (either because the assumed prior knowledge is lacking, or because the student is not able to make the links the teacher intended). These situations have been referred to as 'null' learning impediments. We might imagine that our junior chemistry teacher using the Schrödinger equation would fall into the former category: a 'deficiency' learning impediment where the expected prior knowledge is lacking. An example of the second type of case, a 'fragmentation' learning impediment could come about when a teacher refers to the 'valence' shell of an atom, but the students have only previously heard this called the 'outer' shell. The students here do have the conceptual knowledge to understand the teacher, but due to the use of a different label do not make the intended links with prior knowledge.

Many cases of learning going wrong in chemistry, however, involve the learner actively making a link with existing knowledge, but an inappropriate one. These 'substantive' learning impediments are again of different kinds. In particular they may either derive from making links with existing alternative conceptions ('grounded' learning impediments), or by making inappropriate links with knowledge that is not relevant ('associative' learning impediments). An example

Table 1.2 A simple typology of bonds in compounds

Type of bond	Found in
Covalent	Compounds formed between non-metallic elements
Ionic	Compounds formed between a metallic element and a non-metallic element or elements

of an inappropriate association would be that of a student inferring that the neutralisation process *necessarily* leads to a neutral product (Schmidt 1991). Although the teacher does not make such a statement, the human brain seeks links and connections, and adopts a linguistic clue from the word 'neutralisation'. Here the active nature of learning is unhelpful from a chemical perspective.

I have found that some students who study biology and chemistry come to understand the term 'hydrogen bond' as meaning a covalent bond to hydrogen. What seems to happen here is that students learn from school chemistry that there are two types of bonds, ionic and covalent, according to the classification rules given in Table 1.2.

Later on in their chemical education they will be taught about metallic bonding, intermolecular bonding, polar bonding and so forth: but the most elementary courses often limit consideration of bonding to the two types shown in Table 1.2.

However, when they start advanced biology classes, students often find teachers referring to hydrogen bonds (which are obviously important in such contexts as proteins and nucleic acids), even though this concept has not yet been taught in their chemistry classes. Rather than realise this is a new class of bond, students may simply assume that these bonds between hydrogen and other non-metals are covalent bonds. So when the teacher uses the term 'hydrogen bond' it is understood to mean a covalent bond to a hydrogen atom. The student misunderstands, but having made a connection that allows the teaching to make sense in terms of prior learning, the student does not realise that they are misunderstanding.

Other associative learning impediments may be based upon drawing inappropriate analogies (something that has been labelled a 'creative' learning impediment). As one example, 17-year-old Alice (a real case, but an assumed name) explained that a balloon that had been rubbed on a jumper would stick to a wall because of a 'relative' difference in charge: although the wall was neutral, this made it charged *relative to* the charged balloon, so they would attract. This seemed to be an argument by analogy with potential difference: an object at zero potential can be a source or sink for charge compared with an object at some other potential, as there is a potential difference. In making this creative link between how to conceptualise charge and potential, Alice missed another potential link that might have helped her. Alice knew that polar molecules can induce dipoles in other molecules leading to intermolecular attraction, but she did not think this might be relevant to the question of why a charged balloon would stay attached to a neutral wall (Taber 2008a).

A related category of problem concerns what has been labelled 'epistemological' learning impediments, where the student fails to appreciate the role and nature of models and such devices as metaphors when they are used in science teaching. Models have limited ranges of application (Gilbert and Osborne 1980), but may well appear to students to be intended as accounts of how things actually are. Metaphors are only intended to give a flavour of how things are—but can be taken literally (Lakoff and Johnson 1980). A classic example of this is the delay before chemists managed to form compounds of the inert gases. The description of these elements as 'noble' came to be taken as an absolute description, so that few chemists would have thought of trying to react them with other substances. It is not just students who may find that the brain's tendency for active meaning-making sometimes leads us astray.

Grounded Learning Impediments

So students may fail to learn because of lacking prior knowledge, or because they do not spot the intended connections; and they may learn something other than what was intended because they make unexpected and unintended connections. The other category of problem suggested above was grounded learning impediments. Here the student does recognise the area of prior knowledge relevant to teaching (the general area of prior learning targeted by the teacher), and makes appropriate links, but with existing conceptions that are already at odds with scientific models.

This immediately raises an important question: how do students come to already have alternative conceptions about chemistry, such that these types of situations can arise. This is particularly the case when we acknowledge that many of these alternative conceptions concern chemical concepts that are themselves abstract, and relate to theoretical entities such as molecules and bonds, and the like, that are by-and-large only met by pupils in the context of chemistry classes.

The model of different types of learning impediments I am drawing upon here (Taber 2009b) suggests three types of origins of student ideas which may be important when students develop grounded learning impediments about science topics. These are 'intuitive', 'life-world' and 'pedagogical' learning impediments.

The term intuitive learning impediment refers to those alternative conceptions that pupils appear to develop from their direct experience of the world (rather than being mediated through language for example). In physics education it has been found that a majority of students in most classes have, before receiving physics instruction, developed an intuitive understanding of the relationship between force and motion which somewhat reflects the historical 'impetus' theory (Gilbert and Zylbersztajn 1985). That is, to make something move you give it a push, and as that push gets 'used-up' the object comes to a stop. Now that is not compatible with the account of force and motion presented in school physics, but it *does* describe our everyday experience of moving objects around. It is not too difficult

Table 1.3 Student judgements about the stability of the hypothetical Na^{7-} ion

Study	N	Students judging Na atom less stable than Na^{7-} anion (%)
(Taber 2000)	29	72
(Taber 2009a)—study 1	19	89
(Taber 2009a)—study 3	33	64

to understand how most children acquire an intuitive feel for everyday dynamics (Georgiou 2005), and indeed it took Newton to appreciate and codify the modern scientific understanding.

That can explain children's conceptions of dynamics, but it is not immediately obvious such an explanation can have much relevance to many alternative conceptions in chemistry. For example, most students asked to compare the three chemical species Na^{+}, $Na^{\bullet}$ and Na^{7-} thought that the neutral atom would be a less stable species than the seven-minus sodium anion. This would seem an obscure deduction for most chemists or chemistry teachers. Students should know that metals form cations; that sodium has a valency of one; that highly charged ions are difficult to stabilise and so rare. Sodium compounds met in school and college chemistry inevitably only involve one sodium species, the Na^{+} ion. Whilst the neutral sodium atom is readily ionised, it has no tendency to attract electrons. Yet in a series of small-scale studies, involving 16–18 year-old UK students studying chemistry in a range of schools and colleges, it was found that clear majorities of each sample thought the anion would be more stable than the atom (see Table 1.3). Students appear to be implicitly applying intuitive schemas inappropriately to reach chemically unsound conclusions.

A second source of alternative conceptions has been labelled 'life-world learning impediments' as they relate to what is taken as commonly accepted knowledge in the 'life-world' of everyday discourse (Jegede and Aikenhead 1999)—the way ideas get communicated through culture, whether they are scientifically valid or not (Solomon 1987). So in everyday discourse it is common to think that pure substances are safe, chemicals and radiation are dangerous, that acids burn through objects, and so on. Most of these ideas need some *realignment* to fit with the canonical chemical understandings. It would actually be more appropriate to say that these ideas need *translating*. For it might be better to understand such terms as homonyms for chemical terms (Watts and Gilbert 1983). 'Acid' in the life-world is the label for a different, if overlapping, concept to 'acid' in chemistry. In everyday discourse freshly squeezed orange juice is considered pure because it does not contain any chemicals, especially nasty ones like acids. To the chemist, the orange juice is not pure, contains acids, and must by definition comprise chemical substances. It is understandable that such different usages and meanings cause problems when students cross the cultural border from the life-world to the discourse of the chemistry classroom (Aikenhead 1996).

However, whilst this explains some learning difficulties in chemistry, it again does not seem to offer a viable explanation for many of the reported alternative conceptions that relate to the submicroscopic world of atoms and molecules (Harrison and Treagust 2002). Consider, for example, how students commonly respond to being asked why hydrogen, H_2, reacts with fluorine, F_2. Chemists may think here in terms of thermodynamic considerations. Yet when students who studied this topic at senior high school/college level were asked this question the most common response was that the reaction occurred so that the hydrogen and fluorine atoms could fill their outer electron shells (Taber 2002).

Now the most bizarre thing about this response is that it does not make any sense in its own terms: the atoms concerned already have full shells in the reactants! Yet most of the students were so convinced that reactions occur to allow atoms to complete their electron shells and/or gain 'octets' of electrons, that they did not notice they were offering an answer that was inconsistent with the information given in the question. This raises the question of why students could become so committed to the abstract and unscientific notion that the driving force for chemical change is the need of atoms to complete electron shells. We might explain why school pupils assume gases have no weight in terms of their intuitive learning about the world; and why they may think all polymers are 'plastics' in terms of life-world discourse; but developing an explanatory principle based on electron configurations is hardly the stuff of common experience or everyday conversations.

Pedagogic Learning Impediments

This leads to the final category of grounded learning impediment that can lead to alternative conceptions about chemistry: what pupils have previously been taught. That students commonly form alternative conceptions about the nature of the theoretical submicroscopic entities used as the basis for so many explanations in chemistry—entities such as ions and molecules that they have never directly experienced, and which are seldom the subject of everyday discussion outside of the science classroom—points to teachers ourselves being culpable in misleading students. So sometimes, and perhaps more often than we might wish to acknowledge, students come to classes with existing prior knowledge that is inconsistent with the chemistry they have to learn, and yet derives directly from what they have been taught previously.

Sometimes this is due to limitations in teacher subject knowledge. The experienced chemistry teacher who told me that strong acids *always* have a pH of 1 simply did not understand (or had been teaching at a basic level for so long that he had forgotten) the scientific principles involved. School level textbooks that state unequivocally that the third atomic electron shell is filled with eight electrons would seem to reflect limitations of the authors' own subject knowledge. In both of these cases the statement is wrong, but is unproblematic in the context of the level of teaching being undertaken. However, in both cases, if students learn these 'facts' and then opt to study chemistry at higher levels, they will find that their prior learning interferes with their understanding of later teaching. Such pedagogic learning impediments are unfortunate, and would not happen if teachers (and textbook authors) had perfect subject knowledge. Yet we are all fallible, and most teachers are likely to have subject knowledge with some flaws (Goodwin 2002).

The Octet Alternative Conceptual Framework

However, this cannot be the whole picture. Students do not only acquire isolated alternative conceptions, but extensive conceptual frameworks based around dubious learning. Indeed a number of the examples I have used in this chapter relate to an alternative conceptual framework based around the central idea that chemistry occurs to allow atoms to obtain full shells or octets (Taber 1998). This is clearly the basis for students' explanations of why hydrogen and fluorine react. It is the starting point for students claiming that Na^{7-} will be stable, along with a range of other chemically dubious species (Be^{6-}, C^{4+}, C^{4-}, Cl^{11-}).

Yet it seems unlikely that teachers deliberately teach that the reason chemical reactions occur is to allow atoms to fill their electron shells. Perhaps some do (Taber and Tan 2011), but it seems more likely that the situation is more complex than this. Usually students will have studied several years of basic chemistry before they meet chemical explanations for why reactions occur. Initially students may not think about why some combinations of substances react, but not others. Rather, they will tend to simply make sense of chemical reactions in terms of intuitive knowledge elements that are no more than generalised patterns abstracted from experience: e.g., 'it is just natural for chemicals to react when mixed'; the 'stronger chemical forces the weaker one to react' (Taber and García Franco 2010).

However the 'explanatory vacuum' created by ignoring the driving force for chemical reactions in elementary classes comes to be filled by students' interpretations of what they are taught about the submicroscopic world. Bonding is often presented in terms of the 'needs' of atoms to fill their shells. Strictly, arguments about electronic configuration should only be used to explain valency, not the existence of bonds *per se*. However, the impression often given is that bonding occurs because atoms 'want' to gain full shells. Isolated atoms are seldom important in real chemical processes, but they provide a convenient place to start explaining chemistry, and students readily acquire notions of the atom as the starting point for all chemical processes (Taber 2003a). So when students learn about the two basic classes of bonding found in compounds (Table 1.2), they are often taught that covalent bonding is 'sharing' of electrons (which allows atoms to have full shells) and that the ionic bond can be understood in terms of electron transfer between isolated atoms. That is, they see a hypothetical and often irrelevant electron transfer—which allows atoms to have full shells—as the basis of, or even as, the ionic bond.

It is worth considering the status of the information in Table 1.2, i.e., does this represent sound chemical knowledge? Clearly, Table 1.2 makes no reference to bonding in metals as it only concerns compounds, and it ignores intermolecular bonding. It also includes unrealistic ideal cases. Bonds in compounds can seldom be considered as pure covalent, and never purely ionic. In a sense then, Table 1.2 is not scientifically accurate. However, Table 1.2 presents a level of knowledge often considered suitable for basic level chemistry learning. The most sophisticated scientific knowledge available is seldom suitable as target knowledge in the school curriculum. Rather there is a process of reconceptualising scientific knowledge into something more suitable for the learners (Gilbert et al. 1982; Taber 2008b).

Table 1.2 presents a model of bonding in compounds suitable for introductory learning. If the model in Table 1.2 is taught and learnt as if absolute, factual knowledge then it is inaccurate. If, however, it were to be taught and learnt as a useful model that can often be applied, then it is no longer problematic. After all, this simple classification is often good enough for many purposes in chemistry, and is used by professional chemists all the time.

However, for students, bonding is about atoms filling their shells, and the ionic and covalent models are closely linked to achieving this. This makes sense of why students commonly see ionic substances such as NaCl as pseudo-molecular (Butts and Smith 1987; Taber 1994a; Taber, Tsaparlis and Nakiboğlu 2012). The ionic bond, students deduce, is between specific pairs of ions that have a shared history of having been involved in an electron transfer event. It follows from this way of thinking that the ions in NaCl can only form one bond, as the atoms only had one electron to donate or accept in achieving full outer shells. This also suggests that when NaCl dissolves, these strongly bonded ion pairs will enter solution, having only been attached to other ion pairs by 'just forces', not actual chemical bonds.

This model of ionic bonding does not explain the properties of hard crystalline NaCl that dissolves to form electrolyte solutions, and when students make NaCl by neutralising acid and alkali, and evaporating the water, there are no electron transfer events involved. However, despite the limitations of this way of thinking, it offers an enticing and coherent narrative of chemistry being about atoms needing to fill their shells that seems to be accepted by many students. The brain's tendency to actively seek meanings and patterns latches onto a principle (the desirability of full shells) that can be widely interpreted to make sense of a good deal of chemistry at the submicroscopic level.

Unfortunately, this way of understanding chemistry provides a major learning impediment in more advanced studies. As bonds are not seen as physical interactions between chemical species, students find it difficult to accept that intermolecular interactions can be considered as bonds (as they do not help atoms full their shells); do not appreciate that there can be bonds 'in-between' ionic and covalent; have difficult understanding compounds such as CO, $AlCl_3$, or SF_6 that do not have atoms with 'full shells', and they readily revert to explaining chemical reactions in terms of the need of atoms to fill their shells, even after being taught canonical chemical explanations (Taber 2001b).

Chemical Concepts, Chemical Learning and Correcting Conceptions

To some extent the alternative 'octet' conceptual framework can be considered a *pedagogical* learning impediment. It is an aspect of prior learning, based on school chemistry teaching, which blocks later effective learning of chemistry. Yet that is a simplification, for few chemistry teachers are intentionally teaching this framework. Rather the combination of the abstract and inaccessible nature of the concepts (atoms, bonds, etc.); the delaying of teaching any canonical basis for chemical reactions; the general intuitions about the world that students bring to lessons; the limited epistemological sophistication of learners; and the particular simplifications teachers use in basic chemistry courses, conspire to lead many students to develop the alternative conceptual framework.

The 'explanatory vacuum' provides a niche into which the active learner (automatically seeking connections with prior understanding) interprets what she sees and hears. So she makes sense of the teaching models presented as best she can.

The Limitations of Models and Metaphors

The simple bonding typology represented in Table 1.2 is a teaching model; a simplification that is useful provided it is understood as a model with a limited range of application. That may seem obvious to the teacher—after all, most of what we teach in chemistry can be understood as models in this way. Yet pupils lack the sophistication to appreciate this until we teach them about the nature and role of scientific models. If the teacher does not make the status as model explicit when presenting the bonding typology, then students learn it as a fact, and continue to see bonding as a dichotomy even when taught about polar bonds (e.g., seeing them as no more than a variation on covalent bonding, rather than the most common class of bonds). In terms of a typology of learning impediments, we might better class this as an associative (epistemological) learning impediment, rather than a grounded (pedagogical) one (see Table 1.4).

The topology presented in Table 1.4, like the one in Table 1.2, is a model. The typology is intended to help teachers think about where learning can go wrong, but like all models it has limitations. Probably, in most cases, the octet framework is something of a hybrid of 'epistemological' and 'pedagogic' learning impediments, with traces of some other categories present as well.

The failure to appreciate the nature of models can be very frustrating for students—so when faced with learning an orbital-based model of the atom, some students feel that earlier teaching about electron shells was little more than lies. The loose anthropomorphic metaphors that chemistry teachers commonly use in their classes—'carbon wants to form four bonds', 'metals like to form cations', 'the chlorine atom needs to fill its electron shell'—are not literally true: they are

Table 1.4 How active learning can go wrong: types of substantive learning impediment—after (Taber 2009b)

Main category	Nature	Sub-categories
Grounded learning impediments	Occur because existing understanding is inconsistent with accepted scientific thinking. Such 'alternative conceptions' may derive from various sources	• 'intuitive': …the students' own intuitive interpretation of the way the world seems to be • 'life-world': folk beliefs—common scientifically dubious ideas acquired from friends, family, the media etc. • 'pedagogic': impediments due to limitations of previous teaching, such as over-simplification, use of poor analogies and unhelpful models, etc.
Associative learning impediment	Occur because the student makes an unintended link with prior learning. These may be of various types	• 'linguistic':—taking a cue from a word's 'everyday' usage, or the similarity of a word with the label for an existing concept • 'creative': inappropriate analogies—spotting (creating) an unhelpful analogy between the material being taught and some existing knowledge • 'epistemological': over-interpreting models—or lacking the epistemological sophistication to appreciate the limitations of models, analogies and metaphors used in science teaching, and so interpreting teaching in a too literal and absolute sense

shorthand ways of talking about low energy configurations, and charge interactions, and so forth. But when such language is habitually used, it is little surprise that students who have not yet met the scientific explanations, come to adopt these metaphors as scientific principles (Taber and Watts 1996). The notion of atoms with full shells having a particular special status also seems to appeal intuitively: being whole and complete and symmetrical perhaps suggests desirable, and strong and stable.

Conclusion

This chapter has explored the notion of active learning in chemistry in terms of cognition, the mental activity that leads to the development of conceptual understanding. In general we want learning to be 'active' in this sense. Active learning is more interesting, easier and leads to knowledge that is more readily recalled, better integrated and more flexibly applied. All of this is to be welcomed.

However, the activity of the brain leads to each student interpreting teaching in a unique way in terms of their existing knowledge, and various nuances of how they understand particular terms, and whether they appreciate the nature of the models and metaphors teachers use to communicate abstract and difficult ideas. A key message of this chapter is that active learning can easily go wrong. However the alternative—learning by rote so that what is recalled is an empty facsimile of what was taught—is not a useful one if we are trying to teach a science rather than a chemical catechism.

In some ways this chapter may seem very negative, as it illustrates how a whole range of types of learning impediment can stand in the way of chemistry teachers communicating scientific ideas to learners. However, this could also be seen as demonstrating just what an achievement it is when students do learn the scientific models and become good chemists.

The main message of the chapter is intended to be neither despondent nor celebratory, but rather to be guardedly optimistic. There are considerable challenges in teaching the abstract concepts of chemistry, and much potential for the active learner to misinterpret teaching. Yet the examples discussed here show that we are beginning to move beyond research that reports students' alternative conceptions, to understand what is going on when students develop their alternative understandings, the intuitions they bring to class, and the ways they tend to interpret our teaching models. That is surely an important step towards designing chemical instruction that can draw upon the brain's inherent tendency to meaningful, active learning, rather than so often being thwarted by it.

References

Abrahams, I. (2011). *Practical work in school science: A minds-on approach*. London: Continuum.

Aikenhead, G. S. (1996). Science education: Border crossing into the sub-culture of science. *Studies in Science Education, 27*(1), 1–52.

Ausubel, D. P. (2000). *The acquisition and retention of knowledge: A cognitive view*. Dordrecht: Kluwer.

Butts, B., & Smith, R. (1987). HSC chemistry students' understanding of the structure and properties of molecular and ionic compounds. *Research in Science Education, 17*(1), 192–201.

Csikszentmihalyi, M. (1988). The flow experience and its significance for human psychology. In M. C. Csikszentmihalyi, & I. S. Csikszentmihalyi (Eds.), *Optimal experience: Psychological studies of flow in consciousness* (pp. 15–35). Cambridge: Cambridge University Press.

diSessa, A. A. (1993). Towards an epistemology of physics. *Cognition and Instruction, 10*(2&3), 105–225.

Driver, R. (1983). *The pupil as scientist?*. Milton Keynes: Open University Press.

Duit, R. (2009). *Bibliography: Students' and teachers' conceptions and science education*. Kiel: IPN Lebiniz Institute for Science Education. Retrieved from http://www.ipn.uni-kiel.de/aktuell/stcse/stcse.html

Georgiou, A. K. A. (2005). *Thought experiments in physics learning: On intuition and imagistic simulation.* Mphil Thesis. Cambridge: University of Cambridge. Retrieved from http://people.ds.cam.ac.uk/kst24/ResearchStudents/AndreasGeorgiisou.html

Gilbert, J. K., & Osborne, R. J. (1980). The use of models in science and science teaching. *European Journal of Science Education, 2*(1), 3–13.

Gilbert, J. K., Osborne, R. J., & Fensham, P. J. (1982). Children's science and its consequences for teaching. *Science Education, 66*(4), 623–633.

Gilbert, J. K., & Watts, D. M. (1983). Concepts, misconceptions and alternative conceptions: Changing perspectives in science education. *Studies in Science Education, 10*(1), 61–98.

Gilbert, J. K., & Zylbersztajn, A. (1985). A conceptual framework for science education: The case study of force and movement. *European Journal of Science Education, 7*(2), 107–120.

Goodwin, A. (2002). Teachers' continuing learning of chemistry: Some implications for science teaching. *Chemistry Education: Research and Practice in Europe, 3*(3), 345–359.

Hammer, D., & Elby, A. (2003). Tapping epistemological resources for learning physics. *The Journal of the Learning Sciences, 12*(1), 53–90.

Harrison, A. G., & Treagust, D. F. (2002). The particulate nature of matter: Challenges in understanding the submicroscopic world. In J. K. Gilbert, O. de Jong, R. Justi, D. F. Treagust, & J. H. Van Driel (Eds.), *Chemical education: Towards research-based practice* (pp. 189–212). Dordrecht: Kluwer.

Jegede, O. J., & Aikenhead, G. S. (1999). Transcending cultural borders: Implications for science teaching. *Research in Science and Technological Education, 17*(1), 45–66.

Kind, V. (2004). *Beyond appearances: Students' misconceptions about basic chemical ideas* (2nd ed.). London: Royal Society of Chemistry.

Lakoff, G., & Johnson, M. (1980). Conceptual metaphor in everyday language. *The Journal of Philosophy, 77*(8), 453–486.

Luria, A. R. (1987). *The mind of a mnemonist.* Cambridge, MA: Harvard University Press.

Millar, R. (1989). Constructive criticisms. *International Journal of Science Education, 11*(special issue), 587–596.

Millar, R. (2004, June 3–4). *The role of practical work in the teaching and learning of science.* Paper Presented at the High School Science Laboratories: Role and Vision, National Academy of Sciences, Washington, DC.

Schmidt, H. J. (1991). A label as a hidden persuader: Chemists' neutralization concept. *International Journal of Science Education, 13*(4), 459–471.

Smith, J. P., diSessa, A. A., & Roschelle, J. (1993). Misconceptions reconceived: A constructivist analysis of knowledge in transition. *The Journal of the Learning Sciences, 3*(2), 115–163.

Solomon, J. (1987). Social influences on the construction of pupils' understanding of science. *Studies in Science Education, 14*(1), 63–82.

Taber, K. S. (1994a). Misunderstanding the ionic bond. *Education in Chemistry, 31*(4), 100–103.

Taber, K. S. (1994b). Student reaction on being introduced to concept mapping. *Physics Education, 29*(5), 276–281.

Taber, K. S. (1998). An alternative conceptual framework from chemistry education. *International Journal of Science Education, 20*(5), 597–608.

Taber, K. S. (2000). Case studies and generalisability—grounded theory and research in science education. *International Journal of Science Education, 22*(5), 469–487.

Taber, K. S. (2001a). The mismatch between assumed prior knowledge and the learner's conceptions: A typology of learning impediments. *Educational Studies, 27*(2), 159–171.

Taber, K. S. (2001b). Shifting sands: A case study of conceptual development as competition between alternative conceptions. *International Journal of Science Education, 23*(7), 731–753.

Taber, K. S. (2002). *Chemical misconceptions—prevention, diagnosis and cure: Theoretical background* (Vol. 1). London: Royal Society of Chemistry.

Taber, K. S. (2003a). The atom in the chemistry curriculum: Fundamental concept, teaching model or epistemological obstacle? *Foundations of Chemistry, 5*(1), 43–84.

Taber, K. S. (2003b). Lost without trace or not brought to mind?—a case study of remembering and forgetting of college science. *Chemistry Education: Research and Practice, 4*(3), 249–277.

Taber, K. S. (2005). Learning quanta: Barriers to stimulating transitions in student understanding of orbital ideas. *Science Education, 89*(1), 94–116.

Taber, K. S. (2008a). Exploring conceptual integration in student thinking: Evidence from a case study. *International Journal of Science Education, 30*(14), 1915–1943.

Taber, K. S. (2008b). Towards a curricular model of the nature of science. *Science and Education, 17*(2–3), 179–218.

Taber, K. S. (2009a). College students' conceptions of chemical stability: The widespread adoption of a heuristic rule out of context and beyond its range of application. *International Journal of Science Education, 31*(10), 1333–1358.

Taber, K. S. (2009b). Progressing the constructivist research programme to advance teaching and learning about the nature of science. In I. M. Saleh & M. S. Khine (Eds.), *Fostering scientific habits of mind: Pedagogical knowledge and best practices in science education* (pp. 37–57). Rotterdam: Sense Publishers.

Taber, K. S., & García Franco, A. (2010). Learning processes in chemistry: Drawing upon cognitive resources to learn about the particulate structure of matter. *Journal of the Learning Sciences, 19*(1), 99–142.

Taber, K. S., & Tan, K. C. D. (2011). The insidious nature of 'hard core' alternative conceptions: Implications for the constructivist research programme of patterns in high school students' and pre-service teachers' thinking about ionisation energy. *International Journal of Science Education, 33*(2), 259–297.

Taber, K. S., Tsaparlis, G., & Nakiboğlu, C. (2012). Student conceptions of ionic bonding: Patterns of thinking across three European contexts. *International Journal of Science Education, 34*(18), 2843–2873.

Taber, K. S., & Watts, M. (1996). The secret life of the chemical bond: Students' anthropomorphic and animistic references to bonding. *International Journal of Science Education, 18*(5), 557–568.

Thagard, P. (1992). *Conceptual revolutions*. Oxford: Princeton University Press.

Toulmin, S. (1972). *Human Understanding: The collective use and evolution of concepts*. Princeton: Princeton University Press.

Watts, M., & Gilbert, J. K. (1983). Enigmas in school science: Students' conceptions for scientifically associated words. *Research in Science and Technological Education, 1*(2), 161–171.

Chapter 2
The Development of Theoretical Frameworks for Understanding the Learning of Chemistry

Gail Chittleborough

Introduction

There seems to be a mystique about chemistry. Many students do not recognise the chemistry in their everyday lives, many students consider chemistry to be a challenging and difficult subject beyond their capabilities and many students fail to recognise the value of chemistry in their future careers—even for those students who are majoring in a science and especially those who are not majoring in a science.

Chemical literacy can be defined as those skills and knowledge required for understanding chemistry in a social, democratic, cultural and utilitarian sense (Nuffield Curriculum Projects Centre 2001). But falling chemistry enrolments rates at both school and university level (DEST 2003) will result in fewer people having that basic chemical literacy and chemistry knowledge, and yet chemistry knowledge is expanding to include new processes, attitudes and approaches. This expansion of chemical knowledge is seen in the inclusion of emerging sciences in curricula such as green chemistry, with processes that are environmentally aware and designed to reduce waste production, and nanotechnology, biotechnology and neuroscience.

The lack of connectedness of chemistry with the real world and the lives of the learners is a common criticism (Gabel 1998) founded and reinforced in the traditional chemical content and teaching approaches that are resilient to change. This applies to both "what" is taught—both the conceptual knowledge and the procedural knowledge including operative and cognitive skills, and the pedagogical approach to how it is taught—from memorisation of definitions and solving algorithmic type problems to more student centred, active approaches that use open-ended challenges requiring application and problem solving. Commonly, the teacher practice is secure in the chemistry textbook—interpreting the curriculum

G. Chittleborough (✉)
Deakin University, Melbourne, VIC, Australia
e-mail: gail.chittleborough@deakin.edu.au

I. Devetak and S. A. Glažar (eds.), *Learning with Understanding in the Chemistry Classroom*, DOI: 10.1007/978-94-007-4366-3_2,

for the teacher, providing what is to be learnt, problems, practical activities, simulations, and experiments thereby directing or attending to the curriculum, and the assessment. With a textbook approach, there is the risk that the existing understandings of the students may not be considered and individual needs of the students may not be met and the depth of understanding may not be optimised.

This chapter explores chemical epistemology as a way of interpreting students' understandings of how chemical ideas and concepts develop. Chemical epistemology is an understanding of the knowledge of how chemical ideas and knowledge are built up and an understanding of the way of knowing about chemical processes. This understanding will inform teachers' pedagogical practice as explained by Erduran and Duschl (2004, p. 126): "For chemistry teaching to be effective, prospective teachers will need to be educated about how knowledge is structured in the discipline that they are teaching". The interplay between Subject Matter Knowledge (SMK), the philosophy of chemistry and Pedagogical Content Knowledge (PCK) is examined to help identify opportunities for the chemistry teacher to be better informed about the ways students learn chemistry. This should inform their teaching and their use of the textbook and resources. Data from a research study into first-year university students' understandings and learning approaches of chemistry is used to support the development of the chemical epistemology.

Representation Versus Levels of Representation of Matter

The study of chemistry is essentially about the abstract concept of the atomic theory of matter that can be portrayed at various levels of representation corresponding to the scale and symbol being considered. It is important to distinguish the three levels of representation of matter as described by Johnstone (1982, 1993) from the term "representation" which according to *The Australian Concise Oxford Dictionary* (Hughes 1995) has numerous meanings including: to symbolise; to call up in the mind by description or portrayal or imagination; to place a likeness of before the mind or senses; to serve or be meant as a likeness of; to describe or to depict as. These terms reinforce the metaphorical nature of a representation—providing a description of real phenomena in terms of something else with which the learner is more familiar. Under this broad definition, all representations such as models, analogies, equations, graphs, diagrams, pictures and simulations used in chemistry, can be regarded as metaphors because they are helping to describe an idea—they are not literal interpretations, nor are they the real thing. The metaphorical status and role of the symbolic representations used in chemistry is most important and needs to be understood if the metaphor is to be used successfully (Bhushan and Rosenfeld 1995; Treagust and Chittleborough 2001). Because scientific concepts are foreign to students and difficult for them to understand, metaphors are commonly used to provide links to familiar concepts and provide a foundation on which students can build new ideas. These considerations are in line with a constructivist approach to teaching in which the students' prior knowledge is the foundation on which to build further knowledge (Yager 1991). Johnstone (1993) refers to the *level* of chemical representation of matter, which must not be confused with the term representation commonly used for symbolic representations of chemical phenomena including almost any explanatory tool. Johnstone's hierarchical level is a framework that provides an overview of how chemical data are portrayed and presented whereas the term representation can be used for any chemical depiction that the learner encounters.

Johnstone distinguished three levels of chemical representation of matter which are described as: (1) the macroscopic level—comprising tangible and visible chemicals, which may or may not be part of students' everyday experiences; (2) the sub-microscopic level—comprising the particulate level, which can be used to describe the movement of electrons, molecules, particles or atom; (3) the symbolic level—comprising a large variety of pictorial representations, algebraic and computational forms.

Johnstone (1982) describes the macroscopic as descriptive and functional, and the sub-microscopic level as representational and explanatory. An overview of the three levels of chemical representations of matter, presented diagrammatically in Fig. 2.1 encourages the use of multiple representations, using all three levels simultaneously (Hinton and Nakhleh 1999) and develops an understanding of the importance of the scale that is being represented. Examples of each of the three levels of chemical representation of matter are shown in Fig. 2.2. Harrison and Treagust (2002) point out that for many Grade eight students and even for some Grade 8–10 science teachers, their understanding of the particulate nature of

Fig. 2.1 Three levels of chemical representation of matter (Johnstone 1982)

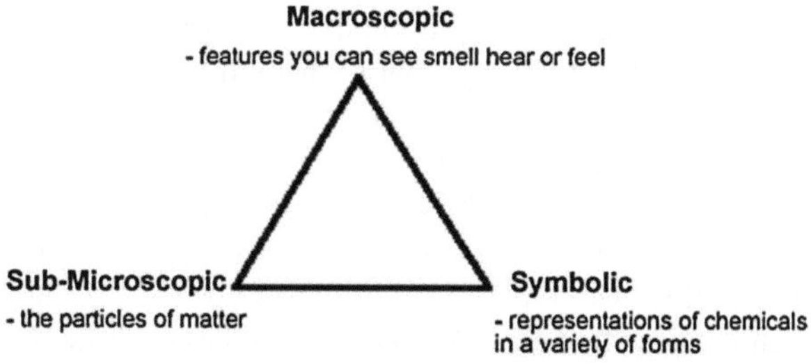

Fig. 2.2 Examples of each of the three levels of chemical representation of matter

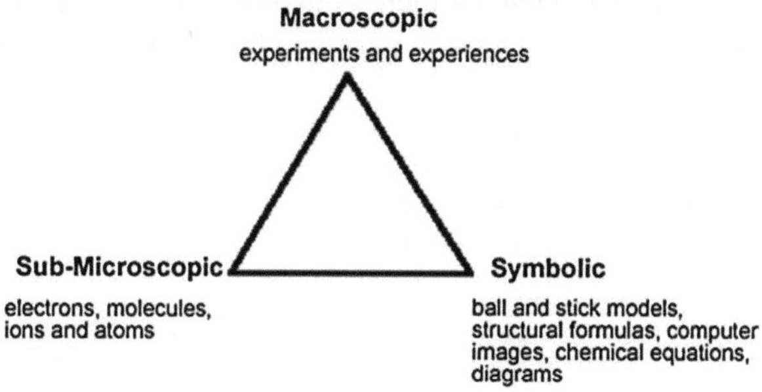

matter, i.e. the sub-microscopic level is poor. The use of the term sub-microscopic refers to levels from the microscopic through to the nanoscopic level and even smaller. Research shows that many secondary school and college students, and even some teachers, have difficulty transferring from one level to another. These findings suggest there is a need to emphasise the difficulty of transferring between different types of representations within each level, as well as transferring from one level to another (Treagust and Chittleborough 2001). At each level many different representations are used in a variety of modes to convey meaning. Johnstone (1997, p. 263) proposes the gradual development of the three inter-connected levels and warns against introducing all three levels simultaneously with novices because the "working space" of our brains cannot handle all three levels simultaneously.

Reality Versus Representation

Inherent in Johnstone's classification scheme is the understanding that the macroscopic and sub-microscopic levels of representation of matter are in fact reality not a representation. The reality of the level of representation is represented in Fig. 2.3 showing the relationship between the three levels of chemical representations and real and represented chemical data. The differences between reality and representations are not often confronted as it is usually assumed that they are understood. However, from discussions with colleagues, it would appear that there is some ambiguity between chemists and educators as to the reality of the sub-microscopic level, with some chemists confident that it is real and some educators believing that it is a representation of a theoretical model—hence the dotted line in Fig. 2.3. The difference between reality and theory needs to be considered here because the sub-microscopic level is based on the atomic theory of matter. The sub-microscopic level is as real as the macroscopic level—it is the scale that distinguishes it, and the fact that the sub-microscopic level cannot be seen easily makes it hard to accept as real. Chemists are now able to observe atoms or molecules, using an electron microscope (but not always in real time), and so they can be classified as real rather than a theory; however, it is not possible to see how the atoms interact, so for this the chemist relies on theories. Theories rely on models—so when we picture an atom we are in fact picturing a model of an atom or a number of pictures of atoms based on various models (Taber 2003).

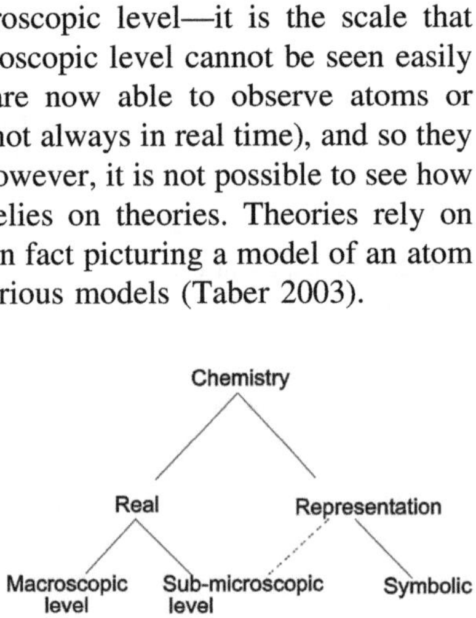

Fig. 2.3 The relationship between the three levels of chemical representations and real and represented chemical data

The three levels of chemical representation of matter are described as follows: (1) the macroscopic level is real and able to be seen; (2) the sub-microscopic level is based on real observations but still needs the theory to explain what is occurring at the molecular level and uses representations of theoretical models; (3) the symbolic level is a representation of the reality.

Johnstone (2000) emphasises the importance of beginning with the macroscopic and symbolic levels because "both corners of the triangle are visualisable and can be made concrete with models" (p. 12). The sub-microscopic level, by far the most difficult (Nelson 2002), is described by the atomic theory of matter, including particles such as electrons, atoms and molecules and is commonly referred to as the molecular level. Johnstone (2000) describes this level simultaneously as the strength and weakness of the subject of chemistry: it provides strength through the intellectual basis for chemical explanations, but it also presents a weakness when beginning students try to learn and understand it. The lack of a mental model of many novice students appears to be a result of the sub-microscopic level being ignored or marginalised when compared to the macroscopic and symbolic levels of representation.

The sub-microscopic level cannot easily be seen directly, and while its principles and components are currently accepted as true and real, it depends on the atomic theory of matter. The scientific definition of a theory can be emphasised here with the picture of the atom constantly being revised. As Silberberg (2000) points out, scientists are "confident about the distribution of electrons but the interactions between protons and neutrons within the nucleus are still on the frontier of discovery" (p. 58). Similarly, the discovery of sub-atomic particles such as the *god* particle, known as the *Higgs boson*, is evidence of the tentative nature of scientific knowledge, and a reminder that it is a construct of changing interpretation. This aspect of scientific knowledge is dynamic and exciting. Appreciating this overview of how scientific ideas are developing may help students to expand their epistemology of science.

The images of the sub-microscopic level available through advances in technology has the potential to provide the visualisation required to teach this level more adequately, even though the projections are still representations (Stevens et al. 2002). Nanotechnology describes research where the dimensions are less than about 1,000 nm (remembering that 1 nm is one-millionth of a millimetre) with descriptions and vision of particular atom.

A teacher's professional expertise demonstrated through their pedagogical content knowledge (PCK) is in scaffolding the learning for students, by selecting the most appropriate form of representation(s) for the concept and for the learner, depending on their experience and background knowledge (Bodner 1986). By acknowledging the level(s) of the representation of matter as proposed by Johnstone, highlighting the weakness and strength of the representation and linking the representation to other forms and other levels—the learner has knowledge of the content, but also distinguishes the content from the features of the explanatory tools being used. The use of multiple representations becomes important in allowing students to distinguish these features (Chittleborough and Treagust 2009).

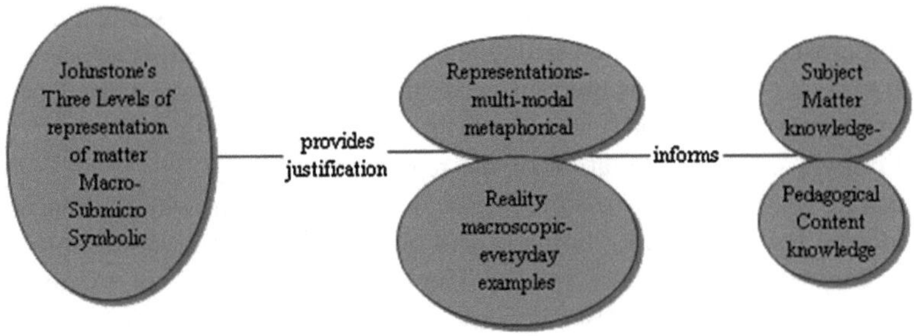

Fig. 2.4 Chemical epistemology ways of thinking about chemical knowledge to inform teaching

This approach empowers the student to appreciate the variety of explanatory tools that are commonly used to help understand abstract concepts.

PCK forges links between the content and how it is best taught. Erduran and Scerri (2002, p. 8) emphasise a philosophical approach to chemistry education and recommend that the "teaching and learning of chemistry can be improved through an understanding of the structure of chemical knowledge". Figure 2.4 represents the links between the structure of chemical knowledge and the tools for teaching. This approach is not inconsistent to Johnstone's ideas that provide students with a means of understanding the nature of chemical knowledge. The emphasis on the philosophy underpinning the knowledge of chemistry has the potential to re-energise the importance of the role of models and representations in the process of science (Treagust et al. 2002). While Johnstone has focused on the representation of the subject matter of chemical knowledge, the pedagogical content knowledge as proposed by Shulman, that connects across contexts, and looks at ways to best help students learn. Carolan et al. (2008) draw on the Peirce's triadic model when discussing representational competence and the way meaning is constructed by learners using representations. This model has the physical object (referent), the meaning (concept) and the symbolic representations—such as verbal, visual, symbol. There is a similarity to Johnstone's three levels of chemical representation of matter with both frameworks designed to add meaning to understanding. When teaching with an emphasis on the role of representations, students began to use the term "representation" in their vocabulary, describing the explanatory tools they are using and demonstrating an appreciation of their role in portraying the abstract (Prain and Tytler 2013).

Explanatory Power of Symbolic and Sub-microscopic Levels of Chemical Representation of Matter

It is the theoretical nature of the sub-microscopic level that is essential for chemical explanations. Symbolic representations of atoms and molecules are usually a snapshot of an instant in time focusing on the single successful reaction only, for example, a reaction mechanism or an equation. By focusing only on the successful reaction, the unsuccessful reactions are forgotten and the probability of success is not represented. There is a risk that the kinetic molecular theory relating to the motion of the sub-atomic particles such as the magnitude of the number of chemical species in the vessel and the constant movement and the many unsuccessful collisions are not appreciated (Krajcik 1991). This omission in understanding the events of the kinetic molecular theory highlights the risk that a representation can be taken out of context and the meaning jeopardised. Explanations of chemical phenomena usually rely on the behaviour of the sub-microscopic particles that are represented symbolically. Consequently, the students' understanding of all three levels is central to the success of any explanation.

As Kozma and Russell (1997) point out, "understanding chemistry relies on making sense of the invisible and the untouchable" (p. 949). Explaining chemical reactions demands that a mental picture or model is developed to represent the sub-microscopic particles in the substances being observed. Observations at the macroscopic level of changes in colour or volume of a reactant, or the evolution of a gas, for example reveal nothing about the sub-microscopic behaviour of the chemicals involved. Yet, explanations are nearly always at the sub-microscopic level—a level that cannot be observed—but is described and explained using symbols by which personal mental models are constructed.

Unfortunately, students often transfer the macroscopic properties of a substance to its sub-microscopic particles, observing, for example that sulphur is yellow, so believing that the atoms of sulphur are yellow also. Indeed, this is not surprising considering the graphical representation of yellow circles in textbooks to represent the atoms (Andersson 1990; Garnett et al. 1995). Colour at the macroscopic level is not a characteristic at the sub-microscopic level. To overcome this problem, Gabel et al. (1992) recommend that teachers provide physical examples or at least descriptions of the chemicals in the problems, in addition to the representations, so that students can establish their own links between the three major levels for portraying the chemical phenomena. In a study into students' understanding of acids and bases, Nakhleh and Krajcik (1994) reported that students' explanations made many more references to the macroscopic level than the sub-microscopic level and more about the sub-microscopic than the symbolic, indicating that they are more confident describing these chemicals at the macroscopic level. Given this not unexpected finding, which is supported by other studies, it is somewhat surprising that so few chemistry curricula emphasise the chemistry of students' everyday experiences (Garnett et al. 1995) or embed the chemical concepts in a familiar or relevant context for the learner. These results lend support to the proposal of placing greater emphasis on the macroscopic level of representation of matter.

Fortunately, there are now exceptions to the atomic structure approach of the 1960s and 1970s that emphasised the abstract symbolic and sub-microscopic levels (Fensham 1994). The use of familiar items in chemistry laboratory work has been used to reinforce the link between chemistry and home, resulting in improved students' perceptions of chemistry (Ramsden 1994; Roberts et al. 1996). In

England, the Salters Chemistry course incorporated a constructivist approach using familiar chemicals as the starting point to motivate students and create a positive classroom climate (Campbell et al. 1994). Nelson (2002) has had positive results with "teaching chemistry progressively" (p. 215) by beginning with student observations at the macroscopic level to provide the examples and foundation to learn the atomic and molecular level, followed by the electronic and nuclear level. Wright (2003) supports the approach of introducing students to atoms and molecules early in their middle years of schooling so that students have a sound foundation before introducing the sub-atomic level. Forgoing content chemistry for contextual chemistry alone is not the solution; moreover, a change in the philosophical approach is needed whereby the unique nature of the structure of chemical knowledge underpins the direction of changes to the curriculum (De Vos et al. 2002; Erduran and Scerri 2002).

Data Source

The progress of first-year university students undertaking an introductory chemistry course was monitored over two consecutive years. The aim of the study was to investigate how first-year university students, who have little or no chemistry knowledge, perceive the role and use of models in science, interpret diagrams of chemical phenomena at the macroscopic and sub-microscopic level, make links between the three levels of chemical representation, develop mental models of chemical phenomena and choose learning strategies. Quantitative and qualitative data were collected from 18 students in year 1 and 19 in year 2 through interviews throughout the academic year, instruments on their use of model and modelling ability, and their work samples were analysed. The data is used here and is also reported elsewhere (Chittleborough and Treagust 2008; Chittleborough and Treagust 2009; Chittleborough et al. 2002, 2005).

The Implications of Johnson's Triangle for Teaching

Because representations are the focus of many chemical explanations, students' understanding of them is critical to their value. Equally important is the students' appreciation of the role of the model and/or representation in the scientific process and their understanding of the concepts of theory, model, fact and reality that are inherent in their epistemological understanding. The research data, from the study with first-year university students undertaking an introductory chemistry course are drawn on here to show that generally most students had a good understanding of the macroscopic and symbolic levels of chemical representation of matter, however, students' understanding of the sub-microscopic level varied, with some students being able to spontaneously envisage the sub-microscopic view while for others their understanding of the sub-microscopic level of chemical representation was lacking. Students with little or no chemical background could not talk seriously about the sub-microscopic level because it was not real to them. Leanne and Kathy, are two such students from the university study whose responds to the following questions:

Int.: Last time I interviewed you, we talked about the mental picture you have of a chemical phenomenon. Can you give me an example of a chemical phenomenon?

Kathy: No not really. If you think of the reaction of photosynthesis—I know the equation; I know what really happens and the equation describes what happens. But I don't picture the little carbon dioxide molecules combining with the water molecules, it just happens; we just know that it does (3.9.1.50).

With experience, the sub-microscopic level becomes real to the learners because they begin to understand its value in explaining why and how the atomic and molecular movements occur. However, Kathy had no need to know any more about the sub-microscopic level than she already knew. Both Kathy and Leanne considered the questions about the sub-microscopic level to be trivial. Similarly, Leanne who left high school the previous year and had not studied chemistry before had no comprehension of the levels of representation. She had studied science to Year 10 level where she was in the non-chemistry group. In the first interview, Leanne applies macroscopic properties to the sub-microscopic nature of matter, displaying a poor modelling ability. There is obvious confusion between the representational nature and the reality of the sub-microscopic level. She was unable to understand the representational nature of the diagrams of various atoms, and molecules, as is shown in the following excerpts.

Int.: If I gave you a sample of copper for example. Can you explain how the copper atoms are arranged?

Leanne: They would be all together.

Int.: What would they be like?

Leanne: No idea.

Int.: Coppers hard we know that but what about the atoms?

Leanne: Coppers hard, then doesn't mean that they are tightly packed. They would be together. (3.8.12.13–19).

Int.: What would sodium chloride atoms look like?

Leanne: It would look like little white things.

Int.: If you get down from the little white things and go down to the atoms what are the atoms going to look like?

Leanne: White.

Int.: OK (3.8.12.24–26).

Leanne's comment demonstrates a common assumption by learners in associating the macroscopic qualities to the sub-microscopic level (Andersson 1990). This misconception arises because the student doesn't understand the differences of the three levels of representation of matter. Learning chemistry requires modelling ability and representational competence—to be able to use the multi-model representational forms that are explanatory tools (Gilbert 2007). Johnstone's (1993) triangle which tries to explain why students find learning chemistry so difficult has become a significant theoretical framework in understanding how chemistry concepts are represented. In considering how and why Johnson's triangle is used, I have proposed two interpretations: the expanding triangle and the rising iceberg.

The Expanding Triangle

Commonly, students are exposed to all three levels of chemical representation of matter simultaneously as part of their chemistry curriculum. A common scenario would be in junior high school for students to perform experiments to observe simple chemical and physical changes; to be taught about the characteristics of the particulate nature of matter and to learn the symbols of atoms—briefly touching on all three levels of chemical representation of matter. Curricula are often arranged as a spiraling concern, consistent with a constructivist philosophy, beginning with basic ideas, returning and repeating what has already been learnt and building on it in a recursive and repetitive manner. In terms of Johnstone's triangle, the students learn some chemistry at all three levels of chemical representation of matter simultaneously and return and learn a bit more at all three levels of chemical representation of matter and so on moving from I to II to III (Fig. 2.5). So the students' depth of knowledge increases and the triangle (representing students' knowledge) grows as the student learning proceeds.

As students continue to understand more chemistry at each of the three levels they can make the connections between the three levels, but this is not guaranteed

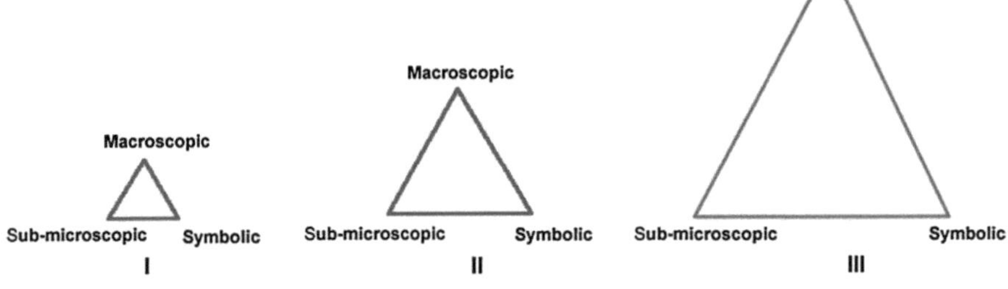

Fig. 2.5 The expanding triangle—a framework for learning chemistry

(Boo 1998; Gabel 1998). The theoretical framework is titled—*the expanding triangle*—because as the students learn more and more at each of the three levels, the triangle expands; however, there is no guarantee that they relate the three levels to each other.

The Rising Iceberg

The three-dimensional image of an iceberg that the title creates—emphasise not only the depth of chemical concepts, but more importantly the sequence of the use of the three levels of representation of matter, with the macroscopic being the central focus in introductory programs and the symbolic and sub-microscopic being introduced subsequently. The shaded triangle—determined by the position of the horizontal line in Fig. 2.6 represents students' growing understanding. It is consistent with Johnstone's (1991) recommendation of starting with the macroscopic and symbolic levels and emphasises using the level(s) of chemical representation of matter that best suits the students' ability level. The macroscopic level of chemical representation of matter at the top corner of the triangle is always included, whereas the sub-microscopic and symbolic levels are only introduced as needed. A horizontal line is drawn across the triangle to indicate the depth of chemistry understanding to be achieved. Obviously the position of this horizontal line depends on the students' abilities, age and stage of chemical knowledge development. The shaded area above the horizontal line is deliverable and achievable for the particular students being considered. As the literature recommends that the macroscopic level is most appropriate for beginning students, so the chemistry should maintain an observable and experimental focus without having to use the particulate nature of matter. When students move to higher levels of understanding then more of the symbolic and sub-microscopic levels can be introduced.

This rising iceberg framework is based on the constructivist philosophy and is consistent with the literature recommendations of starting with the macroscopic, visible and observable chemical occurrences that are often part of students'

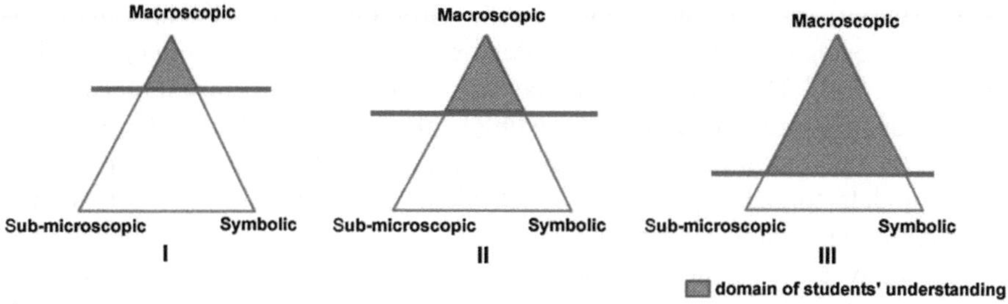

Fig. 2.6 The rising iceberg—a theoretical framework for learning chemistry

everyday experiences and observations, thus providing a contextual learning experience. This was shown to be successful with the Salter approach (Ramsden 1992). The intention of this framework is not to marginalise the sub-microscopic level—especially as it is nearly always the basis of chemical explanations, rather to reassess its role and importance, with evaluation of what detail of the sub-microscopic level is needed to be known in order to understand particular chemical concepts.

The rising iceberg framework is designed to emphasise the importance of the macroscopic level, provide a contextual setting for learning and to critically evaluate how the sub-microscopic level is explaining the chemical phenomena. However, the literature reports that traditionally there is conflict between chemical ideas and everyday ideas; for example, everyday words adopt new and specific meanings in chemical settings; everyday experiences support a continuous nature of matter whereas the more theoretical particulate nature of matter depends on models and representations to help generate mental models; and confusion is evident between the sub-microscopic and macroscopic nature of matter. In order to combat these potential misunderstandings, a constructivist approach is recommended, with the students' understanding as a starting point. The literature emphasises the importance of students' prior knowledge and understanding for their future understanding. The representations at the symbolic level probably contribute most to the students' mental model. Using the rising iceberg framework, initially inexperienced students' mental models are undeveloped corresponding to the small triangle; as they learn more chemistry, then their mental model expands as they focus on the sub-microscopic level.

Johnstone's Triangle Informing the Chemical Epistemology

The triangle can provide insights into chemical epistemology by helping students in their process of knowing about chemical knowledge. The explanatory function of the three levels supports a framework of knowledge that can help the development of a students' epistemology. Through modelling, students are able to gain an understanding of the analogous relationship between the model (analogue) and reality (target) (Gilbert and Boulter 1995). Grosslight et al. (1991) suggest that "different levels of understanding models reflect different epistemological viewpoints" (p. 799). This important link between modelling and epistemology is data from students expressing an understanding of the role of models in the process of science as reported in Chittleborough and Treagust (2007) and also from students using models to make predictions, test ideas and undertake scientific inquiry (Chittleborough and Treagust 2008).

For many students in the study at university year 1, there were dramatic improvements in their epistemology of chemistry; through hard work and application of knowledge, students developed a way of thinking about chemistry. This could be described as an enculturation whereby the sub-microscopic level of matter promotes a chemical way of thinking—a chemical epistemology. For other students, however, the learning experience was driven by the course requirements encouraging a rote-learning regime that did little to improve their epistemology of chemistry. These students circumvented the sub-microscopic level of matter.

Notwithstanding the importance of the framework consisting of the three levels of chemical representation of matter that has been described in the previous section, it is vital to question the relevance of some theoretical and highly mathematical chemistry to all students. There is a need to assess the appropriate depth of chemistry that is required to learn. Johnstone (1982) uses an analogy of the use of a car—for most of the time the car exists at the descriptive and functional level (macroscopic)—detailed explanations of the mechanisms of the car (sub-microscopic level) are not needed or cared about by the general public. In chemistry, even without the sub-microscopic understanding, excellent scientific questions can be posed and experiments tested at the macroscopic level. Johnstone (1982, p. 379) suggests "it would be arrogance ... to assume that chemistry must have all three levels if it is to be respectable". So using this concept, the non-major chemists could still be thinking chemically, in a scientific manner, but have a more practical approach. The analogy of the car is consistent with the rising iceberg framework. Johnstone (1982) is in favour of exploiting the macroscopic level and introducing the sub-microscopic as needed. Through using more macroscopic references the chemistry could be more contextual and help to promote a higher standard of chemical literacy along with giving chemistry a better image.

Pedagogical Implications

Will raising teachers' awareness of how chemical ideas are learnt, impact on their pedagogical practice and translate into effective teaching strategies and deeper student understanding? It is futile if the impact of chemical educational research onto practice is not assured; but still there are many links in the chain to students achieving deeper understanding. The interactions between teachers and students, those learning moments, can be elusive to recognise, but it is these very experiences that inform teachers practice. Shulman describes the situation of the isolated classroom teacher working in "pedagogical solitude" (Shulman 2005, p. 29). The frameworks discussed earlier, SMK, PCK and Philosophy, alongside the three levels of representations and the role of representations provide a foundation understanding of how students learn Chemistry. Alongside are more general theories of learning such as constructivism and behaviourism. Learning is complex, different and unique for individuals, so no single formula or template will guarantee success. In teaching chemistry, teaching approaches focus on many aspects such as modelling ability, multiple representations, visualisation, the process of science (inquiry), contextual focus, etc.

Conclusion

Most significant to the learning of chemistry is the ontological framework that the three levels of chemical representation of matter provide for the learner and the teacher. By providing the learner with basic criteria on which to understand the explanatory tools commonly used in chemistry, then the understanding of the chemical content may be improved. Armed with this understanding, the learner can develop a way of thinking about chemical phenomena—described as the chemical epistemology—an understanding of the knowledge of how chemical ideas are built and an understanding of the way of knowing about chemical processes.

References

Andersson, B. (1990). Pupils' conceptions of matter and its transformations (age 12–16). *Studies in Science Education, 18*(1), 53–85.

Bhushan, N., & Rosenfeld, S. (1995). Metaphorical models in chemistry. *Journal of Chemical Education, 72*(7), 578–582.

Bodner, G. M. (1986). Constructivism: A theory of knowledge. *Journal of Chemical Education, 63*(10), 873–878.

Boo, H. K. (1998). Students' understandings of chemical bonds and the energetics of chemical reactions. *Journal of Research in Science Teaching, 35*(5), 569–581.

Carolan, J., Prain, V., & Waldrip, B. (2008). Using representations for teaching and learning in science. *Teaching Science, 54*(1), 18–23.

Campbell, B., Laxonby, J., Millar, R., Nicolson, P., Ramsden, J., & Waddington, D. (1994). Science: The Salters' approach—a case study of the process of large scale curriculum development. *Science Education, 78*(5), 15–47.

Chittleborough, G. D., & Treagust, D. F. (2007). The modelling ability of non-major chemistry students and their understanding of the sub-microscopic level. *Chemistry Education: Research and Practice, 8*(3), 274–292.

Chittleborough, G. D., & Treagust, D. F. (2008). Correct interpretation of chemical diagrams requires transforming from one level of representation to another. *Research in Science Education, 38*(4), 463–482.

Chittleborough, G. D., & Treagust, D. F. (2009). Why models are advantageous to learning science. *Educación Química, 29*(1), 12–17.

Chittleborough, G. D., Treagust, D. F., Mamiala, T. L., & Mocerino, M. (2005a). Students' perceptions of the role of models in the process of science and in the process of learning. *Research in Science and Technological Education, 23*(2), 195–212.

Chittleborough, G. D., Treagust, D. F., & Mocerino, M. (2002). Constraints to the development of first year university chemistry students' mental models of chemical phenomena. In A. Bunker & G. Swan (Eds.), *Focusing on the student* (pp. 43–50). Perth, WA: Professional Development@Learning Development Services.

Chittleborough, G. D., Treagust, D. F., & Mocerino, M. (2005b). Non-major chemistry students' learning strategies: Explaining their choice and examining the implications for teaching and learning. *Science Education International, 16*(1), 5–21.

DeVos, W., Bulte, A., & Pilot, A. (2002). Chemistry curricula for general education: Analysis and elements of a design. In J. K. Gilbert, O. De Jong, R. Justi, D. F. Treagust, & J. H. Van

Driel (Eds.), *Chemical education: Towards research-based practice* (Vol. 17, pp. 101–124). Dordrecht: Kluwer Academic Publishers.

DEST, Australian Department of Education, S. a. T. C. f. t. r. o., & Education., t. a. t. (2003). *Australia's teachers: Australia's future. Advancing innovation, science, technology and mathematics*. Canberra: DEST, Commonwealth of Australia.

Erduran, S., & Duschl, R. A. (2004). Interdisciplinary characterizations of models and the nature of chemical knowledge in the classroom. *Studies in Science Education, 40*(1), 105–138.

Erduran, S., & Scerri, E. (2002). The nature of chemical knowledge and chemical education. In J. K. Gilbert, O. De Jong, R. Justi, D. F. Treagust, & J. H. Van Driel (Eds.), *Chemical education: Towards research-based practice* (pp. 7–27). Dordrecht: Kluwer Academic Publishers.

Fensham, P. J. (1994). Beginning to teach chemistry. In P. J. Fensham, R. F. Gunstone, & R. T. White (Eds.), *The content of science: A constructivist approach to its teaching and learning* (pp. 14–28). London: The Falmer Press.

Gabel, D. L. (1998). The complexity of chemistry and implications for teaching. In B. J. Fraser & K. G. Tobin (Eds.), *International handbook of science education* (pp. 233–248). Dordrecht, The Netherlands: Kluwer Academic Publishers.

Gabel, D. L., Briner, D., & Haines, D. (1992). Modelling with magnets: A unified approach to chemistry problem solving. *The Science Teacher, 59*(3), 58–63.

Garnett, P. J., Garnett, P. J., & Hackling, M. W. (1995). Students' alternative conceptions in chemistry: A review of research and implications for teaching and learning. *Studies in Science Education, 25*(1), 69–95.

Gilbert, J. K. (Ed.). (2007). *Visualization in science education* (Vol. 1). New York: Springer.

Gilbert, J. K., & Boulter, C. J. (1995, April). Stretching models too far. Paper Presented at the National Association for Research in Science Teaching, San Francisco, CA.

Grosslight, L., Unger, C., Jay, E., & Smith, C. (1991). Understanding models and their use in science: Conceptions of middle and high school students and experts. *Journal of Research in Science Teaching, 28*(9), 799–822.

Harrison, A. G., & Treagust, D. F. (2002). The particulate nature of matter: Challenges in understanding the submicroscopic world. In J. K. Gilbert, O. De Jong, R. Justi, D. F. Treagust, & J. H. Van Driel (Eds.), *Chemical education:Towards research-based practice* (pp. 213–234). Dordrecht: Kluwer Academic Publishers.

Hinton, M. E., & Nakhleh, M. B. (1999). Students' microscopic, macroscopic, and symbolic representations of chemical reactions. *The Chemical Educator, 4*(4), 1–29.

Hughes, J. M., Mitchell, P. A., & Ramson, W. S. (1995). *Australian Concise Oxford Dictionary*. Melbourne: Oxford University Press.

Johnstone, A. H. (1982). Macro- and micro-chemistry. *School Science Review, 64*(227), 377–379.

Johnstone, A. H. (1991). Why is science difficult to learn? Things are seldom what they seem. *Journal of Computer Assisted Learning, 7*(2), 75–83.

Johnstone, A. H. (1993). The development of chemistry teaching: A changing response to changing demand. *Journal of Chemical Education, 70*(9), 701–705.

Johnstone, A. H. (1997). Chemistry teaching—science or alchemy. *Journal of Chemical Education, 74*(3), 262–268.

Johnstone, A. H. (2000). Teaching of chemistry—logical or psychological. *Chemistry Education: Research and Practice in Europe, 1*(1), 9–15.

Kozma, R. B., & Russell, J. (1997). Multimedia and understanding: Expert and novice responses to different representations of chemical phenomena. *Journal of Research in Science Teaching, 34*(9), 949–968.

Krajcik, J. S. (1991). Developing students' understanding of chemical concepts. In S. M. Glynn, R. H. Yeany, & B. K. Britton (Eds.), *The psychology of learning science* (pp. 117–147). Hillsdale, NJ: Lawrence Erlbaum Associates.

Nakhleh, M. B., & Krajcik, J. S. (1994). Influence on levels of information as presented by different technologies on students' understanding of acid, base, and pH concepts. *Journal of Research in Science Teaching, 31*(10), 1077–1096.

Nelson, P. G. (2002). Teaching chemistry progressively: From substances, to atoms and molecules, to electrons and nuclei. *Chemistry Education: Research and Practice in Europe, 3*(2), 215–228.
Nuffield Curriculum Projects Centre. (2001). Chemical literacy: Towards a working definition. Retrieved Nov 7, 2003, from http://www.nuffieldfoundation.org/curriculum/fileLibrary/doc/chemlitaug.doc
Prain, V., & Tytler, R. (2013). Representing and learning science. In R. Tytler, V. Prain, P. Hubber, & B. Waldrip (Eds.), *Constructing representations to learn in science* (pp. 1–14). Rotterdam: Sense Publishers.
Ramsden, J. (1992). Going through both ears and staying there: The Salters' approach to science. *Australian Science Teachers Journal, 38*(2), 13–18.
Ramsden, J. (1994). Context and activity based science in action. *School Science Review, 75*(272), 7–14.
Roberts, J. L., Selco, J. I., & Wacks, D. B. (1996). Mother earth chemistry: A laboratory course for non-majors. *Journal of Chemical Education, 73*(8), 779–781.
Shulman, L. S. (2005). To dignify the profession of the teacher. *Change, 37*(5), 22–29.
Silberberg, M. S. (2000). *Chemistry: The molecular nature of matter and change* (2nd ed.). Boston: McGraw Hill.
Stevens, M. G., Owens, C., & Wuhrer, R. (2002). Nanotechnology in society. *Australian Science Teachers' Journal, 48*(3), 22–27.
Taber, K. S. (2003). The atom in the chemistry curriculum: Fundamental concept, teaching model or epistemological obstacle? *Foundations of Chemistry, 5*(1), 43–84.
Treagust, D. F., & Chittleborough, G. D. (2001). Chemistry: A matter of understanding representations. In J. Brophy (Ed.), *Subject-specific instructional methods and activities* (Vol. 8, pp. 239–267). Oxford: Elsevier Science Ltd.
Treagust, D. F., Chittleborough, G. D., & Mamiala, T. L. (2002). Students' understanding of the role of scientific models in learning science. *International Journal of Science Education, 24*(4), 357–368.
Wright, T. (2003). Images of atoms. *Australian Science Teachers Journal, 49*(1), 18–24.
Yager, R. E. (1991). The constructivist learning model: Towards real reform in science education. *The Science Teacher, 58*(6), 52–57.

Chapter 3
Linking the Macro with the Submicro Levels of Chemistry: Demonstrations and Experiments that can Contribute to Active/Meaningful/Conceptual Learning

Georgios Tsaparlis

Introduction

According to Johnstone (1991, 2000, 2007, 2010), Johnstone and Wham (1982), modern chemistry has three main components: the *macro* and tangible (dealing with experiments and observations of concrete substances), the *symbolic* and mathematical or *representational* (dealing with symbols, equations, and calculations), and the *molecular* and invisible or *submicro* (dealing with molecules, atoms, structure and bonding) (see Fig. 3.1). This *multi-representational* structure (the 'triplet relationship') is very important for understanding chemistry (Gilbert and Treagust 2009).

Once we have embedded this structure in long-term memory, we can use it as a powerful tool for looking at the world. However this is not easy. Dealing with levels other than the macro at the early treatments of school chemistry leads to working-memory overload, hence makes learning difficult or impossible. Johnstone (2007, 2010) maintains that almost all the areas of conceptual difficulties and misconceptions that have been studied by researchers over the past 30 years are attributable to the early introduction of the levels other than the macro.

To avoid this overload, we must keep things tangible, staying with the macro level "until pupils have formed new concepts before we attempt to introduce 'explanations' based on micro considerations" (Jonhstone 2007, p. 9). Laboratory experiences provide direct contact with substances and phenomena, and so "are essential throughout science education." Physical science taught without experiments (this must be the case in many countries) is highly unsatisfactory. In addition, experiments and demonstrations are a powerful tool for linking the three levels of chemistry (Tsaparlis 2009). The new kinds of concept at the submicro

G. Tsaparlis (✉)
Department of Chemistry, University of Ioannina, Ioannina, Greece
e-mail: gtseper@cc.uoi.gr

I. Devetak and S. A. Glažar (eds.), *Learning with Understanding in the Chemistry Classroom*, DOI: 10.1007/978-94-007-4366-3_3,

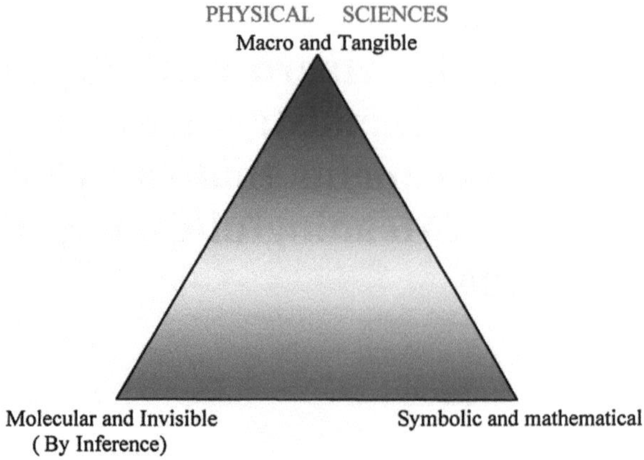

Fig. 3.1 The Johnstone triangle for the three-level representation of chemistry and the physical sciences

level take a long time to develop, making chemistry a complex, difficult, and for many students an unpopular subject. A question then arises: does school chemistry follow an orthodox way to lead students to constructing its multi-representational structure shown in Fig. 3.1?

The actual school chemistry of today, as it is taught and tested all over the world by many if not most teachers, places the emphasis on learning rules and algorithms, which enable conscientious students to respond with success to examination questions, including relatively complicated computational questions. Examples of such 'dexterity' are the placing of electrons in electron shells and subshells or in orbitals, the rote learning of oxidation numbers of the elements, the writing of chemical formulas, the balancing of chemical equations, the calculation of heats of reactions, etc. If we turn however to matters of conceptual understanding, we realize that our students are as a rule ignorant and cannot answer questions such as: why chlorine appears with so many oxidation numbers, why spontaneous endothermic reactions exist, and why reactions lead in general to equilibrium?

Concentrating on the structural concepts, we present to students as absolute truth the foundation of the whole edifice of chemistry. Students have to accept the teacher's word for questions such as: (1) How do we know that molecules and atoms exist? (2) What data forced us to accept that the molecules of several elements are diatomic? (3) How the chemical formulae of compounds are determined? (4) How did we discover the structure of the atom and nucleus? (5) How electric charge and the mass of the electron were measured? (6) How the atomic numbers of the elements were determined? (7) On what experimental evidence the placing of electrons in shells and in orbitals was based? (8) What is an atomic or a molecular orbital? (9) How do we know that atoms in molecules vibrate, and that molecules in gases and in liquids rotate?

The Lack of Deep Understanding is a Real Problem of School Chemistry

In a research project that aimed to examine what students assume that contributes to success in school chemistry, Rop (1999) found that success in chemistry can be defined as "doing the work" and "getting good grades on tests" (p. 221). There were, however, some other students for whom a different definition of success seemed to be in operation. For such students, school chemistry does not require them "to understand molecules, atoms, and the ways things work in the real world": (1) "When I do not understand it, I don't like it [chemistry]"; (2) "[To really understand chemistry,] we have to know that it's there [conceptualization of real atoms and molecules] but I can't grasp it. It doesn't make sense to me that all this stuff [the student points to tables and chairs] would be made of little things"; (3) "... but there must be something awesome out there: [an unseen but wonderful world of moving, acting electrons and atoms—somewhere beyond the constant plodding on of daily life in chemistry class."] (p. 229).

Corpuscular/particulate and structural concepts constitute the corner stone of chemistry. These concepts are highly abstract, lacking both in perceptible examples *and* perceptible attributes, and should be considered formal in the Piagetian sense: hence "it is quite likely that they cannot be totally understood without some formal reasoning" (Herron 1978). Tsaparlis (1997a) employed the following perspectives, and arrived at the same conclusion about pupils' difficulties in learning the atomic and molecular concepts: (1) the Piagetian developmental perspective, (2) the Ausbelian theory of meaningful learning, (3) the information processing theory, and (4) the alternative conceptions movement.

The adoption in teaching of a three-cycle method which separately covers the macro, the representational, and the submicro levels of chemistry should be considered seriously as a good method for introductory chemistry (Georgiadou and Tsaparlis 2000). In the macro cycle, which occupied half of the teaching time, the students became familiar with chemical substances and their properties. Central here was the use of experiments, while chemical notation as well as atoms and molecules were not included. Applying the spiral curriculum, the representational cycle covered the same course material, but added chemical formulas and equations. Finally, the submicro cycle brought atoms and molecules into play. Evaluation of the method, by means of end-of-school-year tests as well as by beginning-of-next year repeat of the same tests, showed that the three-cycle method made the largest single positive effect, compared with a traditional control class and a class in which teaching methods proposed by psychologist R. Case were applied. [According to Case (1978a, b), successful instruction must somehow accomplish the following two objectives: (a) to demonstrate to students that their current strategy must and can be improved upon and (b) to minimize the load on students' working memory.]

Mention should also be made of a freshman chemistry curriculum, in which the topic of atomic structure is delayed until the second semester (Toomey et al. 2001).

In this curriculum, concept development is linked to the observable behavior of matter, while the submicroscopic and symbolic realms are introduced by engaging students in some of the detective work that established the relative atomic masses of the elements and formulas of simple compounds. In this way, students have an opportunity to become familiar with the relationships among facts, definitions, hypotheses, deductions, and predictions, which are central to the enterprise of science. Similarly, Nelson (2002) proposed a way for teaching chemistry progressively, starting with observations at a macroscopic level, interpreting these at an atomic and molecular level, and then at an electronic and nuclear level. Finally, Tsaparlis with colleagues wrote a textbook for eighth-grade chemistry with emphasis given to the macroscopic phenomena and concepts that are treated qualitatively, using constructivist and meaningful-learning teaching methods, while the particulate concepts are delayed (Tsaparlis et al. 2010).

A time then comes, be that early or delayed, that we have to introduce in our chemistry courses the particulate concepts of molecule, atom, electron, etc. It is not then surprising that most chemistry and general science courses introduce these concepts in an almost axiomatic, quasi dogmatic way (see above). Niaz and Rodriguez (2000) defined criteria based on history and philosophy of science, and used them to evaluate presentation of atomic structure in general chemistry textbooks. They found that most of the newer (1970–1992) and older (1929–1967) textbooks not only ignore history and philosophy of science, but also present experimental findings as a 'rhetoric of conclusions.' It was concluded that such presentations are not conducive toward a better understanding of scientific progress.

In this chapter, the aim is to propose a set of demonstrations and experiments that, *if properly used* in teaching by means of active-learning methodology, *can* contribute to *meaningful learning and conceptual understanding* of the particulate concepts of matter. Although essentially all meaningful learning is 'active' in the sense that the learner actively links new learning with his/her pre-existing knowledge and understanding (see Chap. 1), in practice, the application of such teaching methodologies is primarily the job and responsibility of the teacher, and less so of the textbooks. The demonstrations and experiments can be used of course in teacher-centered receptive-learning approaches, but the outcomes in terms of quality of learning might not be the desired ones. Taking into account that secondary school teachers have often a limited knowledge about the findings of educational research (Costa et al. 2000), it is necessary that teachers are aware of active-learning methodology.

Teaching for Active Learning and Conceptual Understanding

Ausubel's Theory of Meaningful Learning

There are various theories of teaching that can contribute to conceptual understanding. One could start with Ausubel's theory of meaningful learning (Ausubel 2000), which concentrates on the influence of *prior knowledge* on *how* learning occurs and is based on the golden rule of educational practice, which states that teaching should be done according to what students already know. Ausubel postulates that *meaningful learning* occurs when the learner's appropriate existing knowledge interacts with the new learning. On the contrary, if such interaction does not occur, the result is *rote learning*. The interaction is realized by means of the so-called *subsumers,* that is, any concept, principle or generalizing idea that the student already knows, and which provides association or anchorage for the components of the new knowledge.

All structural concepts must be built on new ground, that is, the proper subsumers/anchorages for this knowledge should pre-exist in students' minds. In this spirit, we have to admit that chemistry needs basic concepts from physics, such as mass, density, weight, atmospheric pressure, temperature, heat, energy. At higher levels, chemistry is further dependent on physics. According to research findings (Harris 1983), a group of 40 students, who had completed high school chemistry and physics, achieved considerably higher in first-year college general chemistry (79.0 % with standard deviation 9.2) than an equal number (40) who lacked prior physics preparation (63.2 % with s.d. 13.0); that is, physics is deemed an important factor for success in college chemistry. This gives a rationale for physics before chemistry or for chemists to establish physics ideas before chemical ones are attached.

Needless to add that mathematics is also important to physics and chemistry learning, contributing to the complexity of these subjects. Mathematics is essential for the meaningful learning of physics and chemistry, but for this to happen it must be coupled with understanding of the underlying physical concepts. Several studies (Griffith 1985; Hudson and Liberman 1982; Hudson and McIntire 1977; Liberman and Hudson 1979) have attempted to correlate mathematical skill and student reasoning ability with success in physics. It appears that mathematical skills seem to be necessary but not sufficient for success in physics. There are students with marginal mathematical skills, but with well-developed logical and conceptual skills who can be successful in physics. Related to this is the fact that different instructors may place different demands on the students with regard to mathematical ability. Some may be content with the capacity of students to connect physics and chemistry (especially physical chemistry) with mathematics, while others may pay more attention to mathematical operations and calculations.

Constructivism and Active Learning

According to the *Alternative Conceptions Movement,* students form their own models for atoms, molecules, and bonding which are at variance with the scientific views taught in schools. Griffiths (1994) has critically reviewed students' chemistry misconceptions and has enumerated 67 misconceptions about matter and 14 misconceptions about chemical bonding (see also Griffiths and Preston 1992). A similar review has also considered the same concepts (Garnett et al. 1995). These students' concepts are explained by means of the theory of constructivism. It is the duty of teachers, firstly to recognize their students' ideas, and secondly to take them into account in planning and performing their teaching, so that the aim of *conceptual change* is fulfilled. Constructivist teaching and learning (von Glasersfeld 1989) are the banner of modern science education.

Active learning refers to several models of instruction that give the learner the initiative in the learning process (Bonwell and Eison 1991). One form of active learning is discovery learning (guided-discovery learning), the idea of which goes back to John Dewey, but it was fully developed in the 1960s by Bruner (1961). It takes place in problem solving situations. Inquiry-based learning is also a form of active instructional method that developed during the discovery learning movement. In inquiry-based learning, priority is given not to the mere acquisition of knowledge by the student but to the student developing experimental and analytical skills.

To sum up, according to Johnstone (2007, p. 10), "there are a number of messages from research which, if applied, would make our students' experience of science more meaningful, enjoyable and yet intellectually demanding and satisfying. These messages are: (1) What we learn is controlled by what we already know; (2) Learners can process only a limited amount of information at one time; (3) Science concepts exist on more than one intellectual level; (4) Many scientific concepts are of a different kind from everyday concepts; (5) Learners need to start with concepts built from tangible experience and developed later to include inferred concepts."

It is pertinent to emphasize at this point that although Ausubel's meaningful learning and constructivism have been presented here as separate, Ausubel's main ideas are compatible with constructivism. This follows from the fact that, as commented earlier, essentially all meaningful learning is 'active,' and by coupling this with the above argument that constructivist learning is compatible with active learning.

Constructivist and Active Approaches to Teaching Particulate Concepts

A number of years ago, an international seminar (Linse et al. 1990) was dedicated to the relation of macroscopic phenomena to (sub)microscopic particles. Ben-Zvi et al. (1990) confirmed that the root of many difficulties that beginning chemistry students have are due to the deficient understanding of the atomic model and how it is used to explain phenomenology and the laws of chemistry. Appropriate models are essential also to explain the link between energy transfer and temperature change in chemical changes, as well as the link between the molecular model and the energy transfer. Having studied students' relevant views and the problems concerning macro–micro relationships in the area of structure and reaction, the authors proceeded to propose a teaching unit to help overcome students' difficulties. The unit employed a well-known statistical-thermodynamics model, coupled with mechanical models, to explain the energy changes accompanying reactions.

Meheut and Chomat (1990) attempted to make 13–14 year old children build up a particulate model of matter by working out a sequence of experimental facts, starting from properties of gases (compression, diffusion), then moving on to solids, leaving the liquids to last. On the other hand, Millar (1990) placed the emphasis on employing everyday contexts (on the basis of the Salters' approach: Hills et al. 1989), using, e.g., a piece of cloth (which is made of fabrics, made of threads, and made of fibers) to move from the macroscopic to the submicroscopic level (see also Tsaparlis 1989). For Millar, many children need time and experience to appreciate that gases are really matter, so he suggested that it may be wise to start with solids, and postpone consideration of gases until later.

Finally, in a collective volume, Nussbaum (1998), after critically reviewing the various relevant propositions in the 1990 international seminar, dealt with the constructivist teaching of particulate theories, using the history-and-philosophy-of-science approach. Vacuum physics is, according to Nussbaum, the right starting point for particulate physics. Only the existence of a vacuum can justify the noncontinuous nature of matter, hence its particulate nature. In addition, vacuum allows for motion of the particles. Nussbaum bases his introduction of the particulate model on the study of air and other gases, and maintains that the study of the particulate model is a long process of conceptual change, in which students' wrong ideas can play a positive role.

As stated already, in this chapter, the aim is to propose ways that *can* contribute to conceptual understanding and to meaningful/active-learning methodologies for the teaching and learning of the particulate concepts. Though I subscribe to Nussbaum's position that the concept of vacuum is central for a conceptual understanding of particulate concepts, for younger students, I am in favor of starting with liquids, then taking up solids, and leaving (in agreement with Millar) gases to last. Of necessity, some discussion of the properties of gases is essential too. Emphasis will be placed on discussing prerequisite physics concepts and techniques that are deemed essential for realizing the above aim.

Physical science taught without experiments is highly unsatisfactory. Experiments and demonstrations are a powerful tool for providing direct contact with substances and phenomena, as well for linking the three levels of chemistry (Tsaparlis 2009). To carry out and to interpret the experiments, students could work in groups of 2–4. In this way, cooperative learning is encouraged and promoted. If it is not feasible for students themselves to carry out the experiments, the teacher should use demonstrations, where the experiment is carried out in front of the class by a pair of students under the guidance of the teacher.

Needless to add that the psychology of learning requires that, before dipping into the submicro world of particles, students must be fairly familiar with the relevant phenomena at the macro-level. In Ausubel's terminology, it is important to introduce macro anchorages before the submicro concepts are introduced. Again, this makes it imperative that basic physics ideas are established before chemical ones.

Introduction of the Concept of the Molecule

To introduce the concept of the molecule, one needs certain macroscopic concepts and phenomena, such as the phenomenon of diffusion, states of matter, kinetic theory, changes of states of matter, and the concept of temperature. Note that many ideas described below are from an introductory chemistry text aimed at 12 year olds (Johnstone and Morrison 1964).

Diffusion

It is common experience and knowledge that if we open a bottle containing a volatile liquid, e.g., ether, the ether vapor escapes and diffuses into the atmosphere. Similarly, a crystal of potassium permanganate placed on the surface of water is seen soon to dissolve and diffuse into the water. By adding water to the potassium permanganate solution and stirring, we observe that the initially purple color becomes pink and eventually almost disappears. The crystal may have spread itself through water more than a million times its own volume! What has been happening to the crystal as it dissolved? Does the crystal stretches like rubber or has it broken down into minute pieces ('particles'), which disperse themselves through the water?

Is there a limit to the spreading? If one blows some light dust such as fine, dry chalk powder on to the surface of some water contained in a large dish, and then adds one drop of an oily material to the centre of the water surface, the oily material spreads out on the surface. Is there a limit to how far the oil spreads? Again, the oil may have spread out like a rubber sheet, and there is a limit to how far a sheet will stretch. The particle picture is different.

An analogy will help here. Let us allow small wooden balls to float on water, representing the dust on the surface. Then a beaker of wooden balls of another color (perhaps and size), representing particles of oil is poured into the center of the surface. The balls are not on top of each other, but form a single layer. The limit to the spreading comes when there are no more balls piled on top of each other. Again two pictures seem to fit: (1) the oil has spread out like a rubber sheet;

(2) the oil has spread out to give a layer one particle thick, and has pushed the dust to the edge of the bowl. [An experiment based on this principle, using a stearic acid film, can be used for determining the value of the Avogadro constant (Ift and Roberts 1975).]

Collapsing Balloons

Gases can help us resolve the above dilemma with respect to a rubber-like stretching macro material or a particulate structure. Two identical balloons are filled to the same size one with air, the other with helium, and the necks of both are airtight (this can be checked by dipping the necks into water and observing if leaking of gas occurs). The balloons are then left until the next chemistry lesson, when we observe that the balloons have become smaller, with the helium-filled balloon being much smaller. Checking and ruling out the possibility of gas leaking from the necks of the balloons, we are left with the explanation that very small holes must exist in the balloon rubber, which allows the gases to escape. Which of our two models fits now? The particle idea is more appropriate for explaining this experiment: the particles of helium are smaller than those of air, and they must escape through the holes at a higher rate.

Ever-Moving Particles

In the experiment of the crystal of potassium permanganate placed on the surface of water, without any stirring, it is seen that the crystal soon dissolves and diffuses into the water. What is the cause of the observed movement? An analogy can be of help here.

Consider an overcrowded bus. A passenger wanting to alight at a certain station finds it hard to move to the door of the bus, and has to ask and even to push other passengers to make his or her way to the door. The situation in the case of a bus with few passengers is quite different where movement of a passenger is unobtrusive.

Since movement is observed in the case of diffusion, one needs a micro-picture of a gas or a liquid, which allows for empty spaces to be there. Because most gases are colorless and thus invisible, the case of gas is more complicated. A liquid however makes things straightforward: in a liquid, these empty spaces are not directly observable, but instead we sense the continuous presence of the macro-material. The empty spaces make it necessary for the material to be present in distinct lumps. These lumps can be defined as the molecules.

Dissolving salt in water or mixing ethanol and water leads to the fact that the eventual volume of the solution is smaller than the combined volume of salt and water or ethanol and water. A particle model can help explain the observations: fill a beaker with small balls and pour sand to fill the spaces—you do not raise the level of the balls above the top of the beaker.

Half fill a small test tube with a solution of gelatin in water. After the gelatin sets (like the familiar table jellies (Jell-O)), pour some yellow potassium chromate solution on top of it. Take another test tube with gelatin in it, and place a crystal of blue copper sulfate on top.

In the case of solid materials, diffusion is more difficult (or even impossible), and this can be accounted for by assuming that molecules are closely packed in making the crystals, but still leaving empty spaces. Note that the properties of crystals make it imperative for the molecules of a substance to be of the same kind and size (Jones and Childers 1984).

Brownian Motion

One can see movement by watching under a microscope slide on which very fine pieces of blue poster paint are suspended in water. Similarly, observations are possible using some very fine specs of smoke in air, placed in a small box which has two glass windows, with light shining through the side, and observed through the other window by means of a microscope. A model like that helps explain Brownian motion. This kind of movement can be brought about by the bombardment of a large particle with many smaller ones. Similarly, the movement of the smoke and the paint may be due to the bombardment by unseen water or air particles in motion.

From our experiments and our thinking, we have the following picture of the composition of materials: (1) Matter is made up of very small particles, too small to be seen by the human eye; (2) The particles of different substances may have different sizes; (3) The particles are extremely light; (4) The particles are in motion, but different kinds of particles may move at different speeds.

Difference of Properties of a Substance and its Molecule

We can reformulate our definition of molecules as 'the building blocks of substances,' in the same way as bricks are the building blocks of walls, or the rings of a chain, or the threads of a textile. However, many students believe that molecules maintain all (physical and chemical) properties of the macroscopic material, e.g., temperature, physical state, hardness, etc. Thus, a single water molecule is assumed to be like a very tiny droplet of water. This is caused mainly by a faulted definition of a molecule as "the smallest particle of a substance that still retains all [physical and chemical] properties of a mass of the substance." Such a definition is given by some authors (e.g., Fine 1978; Merill 1973) or implied by others (e.g., Sherman and Sherman 1983), and is misleading in suggesting that bulk properties can be attributed to the individual particles (IUPAC 1993). Analogies can be useful in this respect: a wall and the bricks, a chain and its rings, a textile and its fibers (Tsaparlis 1989).

Temperature

Let us consider two visually identical cups of water, one containing water at 5 °C, the other at 30 °C. There is nothing we can see that causes the water in the two cups having a different temperature. We know of course the origin of the temperature difference (for instance cooling or heating the water), but what is there inside water that is responsible for the different temperature? The different rates of diffusion of potassium permanganate in cold and warm water can provide the links to the different rates of motion of particles with temperature change.

Change of Physical State

'Change of physical state' is a topic usually studied within physics. However its relation to chemistry in connection with the concepts of molecules and their varying movement and interaction in the three states is very strong (Meheut and Chomat 1990). An analogy could help here: let us consider a plastic cylinder sitting on a vibrator filled with small polystyrene balls with a card disc sitting on them. This is the solid state. As the vibrator is switched on, on low power, the balls begin to move slightly as a liquid. As the energy input is increased the balls begin to fly about and lift the card. As the energy input increases further, the card rises further showing expansion. This analogy takes in physical state and change of state as well as some idea of energy and temperature change.

The Concept of Energy

In principle, energy is an interdisciplinary scientific concept (Tsaparlis and Kampourakis 2000). However, it is studied more systematically in physics courses. The concept of *energy* (especially chemical energy) as well as the concept of *interaction* are very difficult for young students (Duit 1986; Duit and Häußler 1994). Yet they are essential to many aspects of chemistry and physics. For this reason, the integrated physics and chemistry program proposed by Tsaparlis and Kampourakis (2000) introduces energy from the introductory lesson. Energy is necessary to study changes of state, the concept of temperature, as well as chemical reactions. In addition, it is required as a discriminating factor in distributing electrons to electron shells and to orbitals.

Very helpful for the understanding of the concept of energy is the concept of gravitational energy and especially of gravitational potential energy, that is, is the energy that arises from gravitational force (from the gravitational interaction). In addition to potential energy, we need the principle of energy minimization, as predictor of the most stable (ground-state) electron configuration of atoms and molecules.

In atoms and molecules, potential energy is electrostatic, arising through Coulomb forces. Basic concepts from electricity are essential here. Attach two inflated balloons by a string. Rub both on your hair to give them the same charge and hold the middle of the string. The balloons stand apart as like charges repel. The attraction can be seen by rubbing a balloon on your hair and then taking it away from your head slowly. The hair has a charge opposite to that on the balloon and the hair stands on end, attracted to the balloon.

In contrast to the case of gravitational energy (where we usually set the zero at ground level, so the potential energy at points above the ground assumes positive values), we define the electrostatic potential energy at infinite distance from the nucleus of an atom as zero; hence all energy values at finite distances from the nucleus have a negative sign.

Finally, considering the topics of relative sizes of ions and patterns in ionization energies, Taber (1998) concluded that chemistry teachers base their relevant presentations on the principles of Coulomb electrostatics. However, many students do not have the same background in physics as their teacher, with the result that they apply alternative assumptions in the context of interactions in atoms and molecules.

Vibrational and Rotational Spectroscopies

With more mature students (at upper secondary level and in university general chemistry), vibrational spectra can be used for justifying the concept of vibrating molecules. Rotational (microwave spectra) or the rotational structure of vibrational spectra of gases can be used to rationalize rotation of molecules. It is true that spectra are a theme usually studied in physics, but its strong connections with chemistry should not be overlooked.

The Concept of the Atom

To introduce the concept of atom by means of active/constructivist and meaningful-learning methodology, one certainly needs to dwell on the historical aspects of this concept. Historical experimental evidence for the existence of atoms has been invoked by Jones et al. (1984): the law of definite proportions of Proust, Dalton's atomic theory, Gay Lussac's law, Avogadro's hypothesis, and

Faraday's law of electrolysis. Niaz and Rodriguez (2001) have used examples from the topics of atomic structure, kinetic theory, covalent bonding, and the law of multiple proportions, to illustrate how a History-and-Philosophy-of-Science perspective can facilitate students' conceptual understanding.

Toomey et al. (2001) have shaped a program which follows a historical approach. Observations about gases, liquids, and solids are used to support the atomic theory. Further, the laws of definite and multiple proportions are used to suggest that atoms may be bonding to one another to form molecules when a compound is formed. The gas laws are introduced next, and further connected to the kinetic theory. Students deduce that an oxygen atom should be eight times as massive as a hydrogen atom, and the concept of relative atomic mass unit is introduced. Following that, students are introduced to experimental observations about the volumes of gases that react with each other when the temperature and pressure have the same initial and final values. The law of combining volumes is introduced next, and the fact that *two* volumes of the product gases are produced in various reactions. The Avogadro's hypothesis follows, and students are asked to use the hypothesis and the experimental facts about combining volumes to make various deductions. Returning to the kinetic theory, relative velocities of different gases at the same temperature are compared, and their relative particle masses predicted using Graham's law.

Nelson (2002) suggested that students should be introduced to the following phenomena, which can be demonstrated with suitable experiments (e.g., Fowles 1957; Nelson 1996a; Sienko et al. 1984): law of conservation of mass; phenomenon of constant composition; phenomenon of multiple proportions; phenomenon of proportionate gaseous volumes.

There are several indications that matter may be made up of atoms: Many solids are crystalline; this can be explained in terms of the regular packing of small particles. Gases are much more compressible than liquids or solids, and when they condense there is a large reduction of volume. These observations can be explained if gases comprise separate particles, which come together in the liquid or solid state. These may be atoms or clusters of atoms (molecules). When a small quantity of olive oil is poured on to a large pool of water, the oil only spreads over a limited area of the surface. These considerations, along with the phenomena of the previous paragraph, lead to the following theory of matter, after Dalton and Avogadro (Nelson 2002): (1) Matter is made up of atoms; (2) The atoms of an element are all the same, and differ from those of other elements (provisional statement); (3) Chemical reactions involve changes in which atoms are combined, but not in their number; (4) Atoms of different elements often combine in different ratios; (5) These ratios are often small whole numbers; (6) Avogadro's hypothesis.

This theory explains the law of conservation of mass, constant composition, multiple proportions, and proportionate gaseous volumes, as well as the fact that in the reaction between hydrogen and chlorine to form hydrogen chloride, the volumes are in the ratio 1:1:2. This leads to hydrogen comprising hydrogen molecules H_{2m}, chlorine comprising chlorine molecules Cl_{2n}, and hydrogen chloride comprising hydrogen chloride molecules H_mCl_n (with *m* not necessarily equal to *n*).

Also, postulate six enables the masses of molecules to be compared. For example, the density of hydrogen at STP (standard temperature and pressure) is 0.08988 g/L and of oxygen 1.4290 g/L, thus 1.4290/0.08988 = 15.899. The result is approximate because of the pressure; the limiting value at low pressures is 15.875. If the mass of a hydrogen molecule (μ) is provisionally made the unit of mass for atoms and molecules, the mass of an oxygen molecule is therefore about 16 μ.

To establish the atomic composition of a molecule, a further principle needs to be added to (after Cannizzaro). This is: *The mass of an atom of an element is the smallest mass of the element found in any molecule containing it.*

The conclusion that matter is made up of atoms and molecules is supported by the results of the kinetic theory of gases. Electron tunneling microscopy can be useful at this point for providing images of atom arrangements on metallic surfaces, while mass spectra are useful for the modern way of establishing the relative atomic masses, as well as relative molecular masses. In addition, X-ray diffraction patterns can be used with students at upper secondary level (Tsaparlis 2004).

Electrons and Electron Configurations

In Toomey et al. (2001) approach, electrons are not introduced until week 12, and atomic number is not introduced until week 1 of semester two. On the other hand, topics like quantum numbers and orbitals have been eliminated, and replaced in semester two by presentations that use comparison of ionization energies to suggest the existence of different energy levels in atoms.

The classic experiments that proved the existence of electrons, and determined its charge and mass (Thomson, Millikan) are a must for upper secondary students; similarly Goldsteins's experiment that proved the existence of protons. Models and computer simulations can be very useful here. In addition, atomic spectra, fluorescent tubes with inert gases, as well as the coloring of a flame by metals and salts are useful resources for teaching.

Chemical Bonding

Nelson (1994) distinguished among three "limiting types" of binary compounds: metallic, salt-like (ionic), and nonmetallic. These represent extremes, and most binary compounds fall somewhere in between these extremes. In practice, determining the degree of salt-like character is difficult, since it requires accurate electrochemical measurements at high temperatures on melts. The diagram in Fig. 3.2 places the limiting types at the corners of a triangle. From the triangle, we deduce that there are four intermediate types, A, B, C, and D (see also Table 3.1). The properties of the intermediate types can be inferred from the properties of the limiting types. Thus:

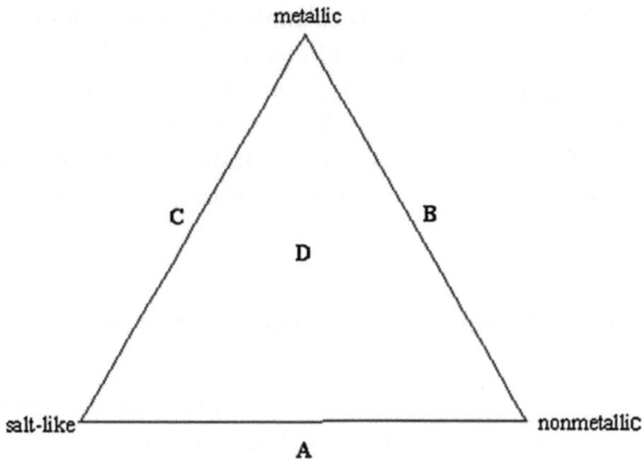

Fig. 3.2 Representation of the three types of binary compounds [metallic, salt-like (ionic), and nonmetallic] as the corners of a triangle, and the four intermediate types

Table 3.1 Types of binary compounds (after Nelson 1994)

Type	Chemical name	Electrical character
Metallic	Metal	Conductor
Nonmetallic	Nonmetal	Insulator
Salt-like	–	Electrolyte
Type A	–	Semi-electrolyte
Type B	Semimetal	Semiconductor
Type C	–	Mixed conductor
Type D	–	Mixed conductor

Type A High transparency, no luster, weak electrolytic conductivity in fused state. Examples: beryllium chloride, $BeCl_2$, and zinc chloride, $ZnCl_2$. These are colorless, and in the fused state they conduct electricity electrolytically. Their conductivities are only a fraction, however, of those of fused $MgCl_2$ and $CaCl_2$ (about 0.5 Ω^{-1} m^{-1}, as compared with about 100 Ω^{-1} m^{-1})

Type B High opacity, possible some luster, semiconducting in solid and liquid. *Example:* iron monoxide FeO, which is black and has a conductivity at room temperature of 2×10^3 Ω^{-1} m^{-1}. Electrolytic conduction in the melt is negligible

Type C High opacity, possibly some luster, both electrolytically conducting and semiconducting (i.e., with a direct current, chemical decomposition takes place, but less than the amount required by Faraday's laws)

Type D Like C, but lower conductance. *Example of type C/D:* Dicopper sulfide, Cu_2S, which is also black. At room temperature, it is a semiconductor, with a conductivity of 3 Ω^1 m^{-1}. At higher temperatures, the conductivity rises, and electrolytic conduction makes a contribution, reaching about 85 % at 400 °C

The distinction between molecular and nonmolecular substances is an important one in chemistry, and can be done without having to appeal to X-ray crystallography, by classifying substances on the basis of volatility and solubility (Nelson 1996b). Finally, the analogy of the covalent chemical bond as an "atomic tug-of-war" (Tsaparlis 1984) is useful for teaching the concept of covalent bond, as well as the distinction between polar and nonpolar covalent bonds.

The Amount of Substance Concept

Central in school chemistry are numerical/stoichiometric calculations that determine the masses (and of volumes in the case of gases) of substances that are consumed and/or produced in chemical reactions (Schmidt 1994). Fundamental here is the amount of substance (Mole) concept. There has been a large literature of the 1970s, the 1980s and the 1990s that focused on the complexity of this concept and the difficulties students and teachers encounter in dealing with and using it (e.g., Bent 1985; Cervellati et al. 1982; Dierks 1981; Duncan and Johnstone 1978; Furió et al. 2000; Ingle and Shayer 1971; Lazonby et al. 1984; Nelson 1991; Novick and Menis 1976; Schmidt 1994; Staver and Lumpe 1993, 1995; Stromdahl et al. 1994; Tullberg et al. 1994). A series of demonstrations will be described below that aim to build the amount of substance concept as a unifying concept in chemistry. They have been used by Johnstone:

- Compare the volumes of equal moles of an organic homologous series (e.g., methanol, ethanol, propanol, butanol, pentanol) to see patterns.
- Compare moles of sugars to "see" mono and disaccharides.
- Compare the volumes of equal moles of finely ground halides with the same cation (e.g., NaCl, NaBr, NaI) to "see" relative halide ion sizes.
- Look at molar heat capacities for partners. Take, for instance, the metals Li, Mg, and Al. Their specific heat capacities are respectively: 3,390, 1,030, and 900 J kg^{-1} K^{-1}. By multiplying each of these values by the corresponding relative atomic mass, we find the following values for the molar heat capacities: 23,730, 24,720, 24,300 J $kmol^{-1}$ K^{-1}, which are very close to each other. Similar values are found for other elements.
- Similarly, let us look at molar gas volumes for patterns. For the gaseous elements H_2, He, N_2, and Ne, the corresponding densities under STP are: 0.09, 0.18, 1.25, and 0.90 gL^{-1}. If we divide the corresponding relative molecular or atomic mass by each of these values, we find the following values for the molar volumes at STP: 22.2, 22.2, 22.4, 22.2 $Lmol^{-1}$. Similar values are found for other species.
- Compare a 20 l drum (approximate volume of one mole water vapour) with one mole (18 mL of liquid water).

Quantum Chemical Concepts

Atomic and molecular orbitals and related concepts are highly abstract, and their introduction in high school course may be problematic (Papaphotis and Tsaparlis 2008a, b; Tsaparlis 1997a, b; Tsaparlis and Papaphotis 2002). Alternative ways that avoid the orbitals at both the high school and the general chemistry level might be preferable. Gillespie maintained that Lewis structures, and the VSEPR model are sufficient for high school, while the electron-domain model is sufficient for general chemistry, with emphasis placed on electron density rather than orbitals (Gillespie 1991, 1992a, b, c; Gillespie et al. 1994, 1996; Gillespie and Mata 2001). For instance, using the physical repulsion of balloons (e.g., two, three, four blown balloons tied together) can lead us to VSEPR, and then we could go a very long way in organic and inorganic chemistry (Johnstone et al. 1981), without the need to discuss orbitals and hybridization.

Concluding Remarks

To have a good conceptual understanding of particulate and structural concepts, students need a firm grasp of the underlying physics concepts, for which the use of the history and philosophy of science is very useful. According to Niaz and Rodriguez (2000), this can be introduced in the classroom not necessarily through formal courses in the history of chemistry or comments and anecdotes, but rather by incorporating the 'heuristic principles' that guided the scientists to elaborate their theories. On the other hand, modern techniques (mass, electronic, vibrational, and rotational spectra, electron tunneling microscopy, X-ray diffraction) can be quite conducive to the active/constructivist, and meaningful learning approach, but these should be mostly reserved for the more advanced students at the upper secondary level and for students in university general chemistry.

It is important to realize that we need more knowledgeable teachers, both with respect to the content of science, and the active/constructivist, and meaningful learning methodologies. In point of fact, according to the chart of the American Association for the Advancement of Science (NCRTL 1994), future teachers must know science and their subject of specialty (physics, chemistry, biology) more deeply than it usually is the case. On the other hand, Gillespie (1997) has suggested that new chemistry textbooks should be written that should aim on the one hand to be interesting for the *vast majority of students*, and on the other hand to be providing them with an *understanding* of chemistry. Finally, when is understanding (e.g., of atomic structure) sufficiently good and complete? This question sets up a paradox: "the more one learns about some aspect of the world, the more aware one is likely to become of the depth of one's ignorance of it. That does not necessarily mean that as a consequence of learning, one's understanding actually decreases, but simply that one's appreciation of the complexity of that aspect of the world is likely to increase—which may be, after all, a better understanding of a fundamental sort" (Rop 1999, p. 233).

Acknowledgments I would like to express my gratitude to Professor *Alex H. Johnstone* for his work and ideas, and the discussions I have had with him, all of which largely influenced this work. I also thank the two anonymous reviewers who made numerous constructive suggestions that contributed greatly to the improvement of this chapter.

References

Ausubel, D. P. (2000). *The acquisition and retention of knowledge: A cognitive view*. Dordrecht: Kluwer Academic.

Bent, H. A. (1985). Should the mole concept be X-rated? *Journal of Chemical Education, 62*(1), 59.

Ben-Zvi, R., Silberstein, J., & Mamlok, R. (1990). Macro-micro relationships: A key to the world of chemistry. In P. L. Lijnse, P. Licht, W. De Vos, & A. J. Waarlo (Eds.), *Relating macroscopic phenomena to microscopic particles* (pp. 183–197). Utrecht: University of Utrecht, Centre for Science and Mathematics Education.

Bonwell, C. C., & Eison, J. A. (1991). *Active learning: Creating excitement in the classroom.* AEHE-ERIC Higher Education Report No. 1. Washington, D.C.: Jossey-Bass.

Bruner, J. S. (1961). The act of discovery. *Harvard Educational Review, 31*(1), 21–32.

Case, R. (1978a). Implications of developmental psychology for the design of instruction. In R. Glaser, A. Lesgold, J. Pellegrino, & J. Fokkema (Eds.), *Cognitive psychology and instruction* (pp. 441–463). New York: Plenum.

Case, R. (1978b). Intellectual development from birth to adulthood: A new-Piagetian interpretation. In R. S. Siegler (Ed.), *Children's thinking: What develop*. New Jersey: Hillsdale, Erlbaum.

Cervellati, R., Montuschi, A., Perugini, D., Grimellini-Tomasini, N., & Pecori Balandi, B. (1982). Investigation of secondary school students' understanding of the mole concept in Italy. *Journal of Chemical Education, 59*(10), 852–856.

Costa, N., Marques, L., & Kempa, R. (2000). Science teachers' awareness of findings from educational research. *Chemistry Education Research and Practice, 1*(1), 31–36.

Dierks, W. (1981). Teaching the mole. *European Journal of Science Education, 3*(2), 145–158.

Duit, R. (1986). In search of an energy concept. In R. Driver & R. Millar (Eds.), *Energy matters* (pp. 67–101). Leeds: University of Leeds.

Duit, R., & Häußler, P. (1994). Learning and teaching energy. In P. J. Fensham, R. F. Gunstone, & R. T. White (Eds.), *The content of science: A constructivist approach to its teaching and learning* (pp. 185–200). London: The Falmer Press.

Duncan, I. M., & Johnstone, A. H. (1978). The mole concept in chemistry. *Education in Chemistry, 10*(6), 213–214.

Fine, L. W. (1978). *Chemistry* (2nd ed.). Baltimore: Williams & Wilkins.

Fowles, G. (1957). *Lecture experiments in chemistry* (4th ed.). London: Bell.

Furió, C., Azcona, R., Guisasola, J., & Ratcliffe, M. (2000). Difficulties in teaching the concepts 'amount of substance' and 'mole'. *International Journal of Science Education, 22*(12), 1285–1304.

Garnett, P. J., Garnett, P. J., & Hackling, M. W. (1995). Refocusing the chemistry lab: A case for laboratory-based investigations. *Australian Science Teachers Journal, 41*(2), 26–32.

Georgiadou, A., & Tsaparlis, G. (2000). Chemistry teaching in lower secondary school with methods based on: a) Psychological theories; b) the macro, representational, and submicro levels of chemistry. *Chemistry Education Research and Practice, 1*(2), 217–226.

Gilbert, J. K., & Treagust, D. F. (Eds.) (2009). *Multiple representations in chemical education.* Dordrecht: Springer.

Gillespie, R. J. (1991). What is wrong with the general chemistry course? *Journal of Chemical Education, 68*(3), 192–194.

Gillespie, R. J. (1992a). The VSEPR model revisited. *Chemical Society Reviews, 21*(1), 59–68.

Gillespie, R. J. (1992b). Multiple bonds and the VSEPR model. *Journal of Chemical Education, 69*(2), 116–121.

Gillespie, R. J. (1992c). Electron densities and the VSEPR model of molecular structure. *Canadian Journal of Chemistry, 70*(3), 742–750.

Gillespie, R. J. (1997). Reforming the general chemistry textbook. *Journal of Chemical Education, 74*(5), 484–485.

Gillespie, R. J., Eaton, D. R., Humphreys, D. A., & Robinson, E. A. (1994). *Atoms, molecules and reactions: An introduction to chemistry*. Englewood Cliffs: Prentice-Hall.

Gillespie, R. J., & Matta, C. F. (2001). Teaching the VSEPR model and electron densities. *Chemistry Education Research and Practice, 2*(2), 73–90.

Gillespie, R. J., Spencer, J. N., & Moog, R. S. (1996). Demystifying introductory chemistry, Parts 1 & 2. *Journal of Chemical Education, 73*(7), 617–626.

Griffith, W. T. (1985). Factors affecting performance in introductory physics courses. *American Journal of Physics, 53*(9), 839–842.

Griffiths, A. K. (1994). A critical analysis and synthesis of research on students' chemistry misconceptions. In H. J. Schmidt (Ed.), *Problem solving and misconceptions in chemistry and physics* (pp. 70–79). Hong Kong: ICASE.

Griffiths, A. K., & Preston, K. R. (1992). Grade-12 students' misconceptions relating to fundamental characteristics of atoms and molecules. *Journal of Research in Science Teaching, 29*(6), 611–628.

Harris, S. P. (1983). Physics, an important factor in the success of general college chemistry students. *Journal of Chemical Education, 60*(9), 739–740.

Herron, J. D. (1978). Piaget in the classroom. *Journal of Chemical Education, 55*(3), 165–170.

Hills, G., Holman, J., Lazonby, J., Raffan, J., & Waddington, D. (1989). *Introducing chemistry: The Salters' approach.* London: Heinemann Educational Books.

Hudson, H. T., & Liberman, D. (1982). The combined effect of mathematics skills and formal operational reasoning on student performance in the general physics course. *American Journal of Physics, 50*(12), 1117–1119.

Hudson, H. T., & McIntire, W. R. (1977). Correlation between mathematical skills and success in physics. *American Journal of Physics, 45*(5), 470–471.

Ift, J. B., & Roberts, J. L, Jr. (1975). *Frantz/Malm's essentials of chemistry in the laboratory.* San Francisco: Freeman.

Ingle, R., & Shayer, M. (1971). Conceptual demand in Nuffield 'O' level chemistry. *Education in Chemistry, 8*(5), 182–183.

IUPAC (1993). Recommendations for language, symbols and representation in chemistry: Atom. *International Newsletter on Chemical Education*, (39), 7–10.

Johnstone, A. H. (1991). Thinking about thinking. *International Newsletter on Chemical Education,* (6), 7–11.

Johnstone, A. H. (2000). The presentation of chemistry—Logical or psychological? *Chemistry Education Research and Practice, 1*(1), 9–15.

Johnstone, A. H. (2007). Science education: We know the answers, let's look at the problems. In *Proceedings of the 5th Greek Conference Science Education and New Technologies in Education* (Vol. 1, pp. 1–13). Retrieved from http://www.kodipheet.chem.uoi.gr/fifth_conf/pdf_synedriou/teyxos_A/1_kentrikes_omilies/1_KO-4-Johnstone.pdf

Johnstone, A. H. (2010). You can't get there from here. *Journal of Chemical Education, 87*(1), 22–27.

Johnstone, A. H., & Morrison, T. I. (1964). *Chemistry takes shape* (Vol. 1). London: Heinemann.

Johnstone, A. H., Morrison, T. I., & Reid, N. (1981). *Chemistry about us*. London: Heinemann.

Johnstone, A. H., & Wham, A. J. B. (1982). The demands of practical work. *Education in Chemistry, 19*(3), 71–73.

Jones, E. R, Jr, & Childers, R. L. (1984). Experimental evidence for the existence of atoms. *The Physics Teacher, 22*(6), 354–360.

Lazonby, J. N., Morris, J. E., & Waddington, D. J. (1984). The muddlesome mole. *Education in Chemistry, 19*(4), 109–111.

Liberman, D., & Hudson, H. T. (1979). Correlation between logical abilities and success in physics. *American Journal of Physics, 47*(9), 784–786.

Lijnse, P. L., Licht, P., DeVos, W., & Warlo, A. J. (Eds.). (1990). *Relating macroscopic phenomena to microscopic particles*. Utrecht: CD-β Press.

Meheut, M., & Chomat, A. (1990). The bounds of children's atomimism: An attempt to make children build up a particulate model of matter. In P. L. Lijnse, P. Licht, W. De Vos, & A. J. Waarlo (Eds.), *Relating macroscopic phenomena to microscopic particles* (pp. 266–282). Utrecht: CD-β Press.

Merrill, M. A. (1973). *Chemistry: Process and prospect*. Columbus: Bell & Howell.

Millar, R. (1990). Making sense: What use are particle ideas to children. In P. L. Lijnse, P. Licht, W. De Vos, & A. J. Waarlo (Eds.), *Relating macroscopic phenomena to microscopic particles* (pp. 283–293). Utrecht: CD-β Press.

NCRTL (National Center for Research on Teaching and Learning) (1994). *A blueprint for the education of project 2061 science teachers*. East Lansing: Michigan State University.

Nelson, P. G. (1991). The elusive mole. *Education in Chemistry, 28*(4), 103–104.

Nelson, P. G. (1994). Classifying substances by electrical character. *Journal of Chemical Education, 71*(1), 24–26.

Nelson, P. G. (1996a). Demonstrating constant composition. *Education in Chemistry, 33*(1), 22.

Nelson, P. G. (1996b). To be a molecule, or not to be? *Education in Chemistry, 33*(5), 129–130.

Nelson, P. G. (2002). Teaching chemistry progressively: From substances, to atoms and molecules, to electrons and nuclei. *Chemistry Education Research and Practice, 3*(2), 215–228.

Niaz, M., & Rodriguez, M. A. (2000). Teaching chemistry as rhetoric of conclusions or heuristic principles—A history and philosophy of science perspective. *Chemistry Education Research and Practice, 1*(3), 315–322.

Niaz, M., & Rodriguez, M. A. (2001). Do we have to introduce history and philosophy of science or is it already 'inside' chemistry? *Chemistry Education Research and Practice, 2*(2), 159–164.

Novick, S., & Menis, J. (1976). A study of student perceptions of the mole concept. *Journal of Chemical Education, 53*(11), 720–722.

Nussbaum, J. (1998). History and philosophy of science and the preparation for constructivist teaching: The case of particle theory. In J. J. Mintzes, J. H. Wandersee, & J. D. Novak (Eds.), *Teaching science for understanding—A human constructivist view* (pp. 165–194). New York: Academic Press.

Papaphotis, G., & Tsaparlis, G. (2008a). Conceptual versus algorithmic learning in high school chemistry: The case of quantum chemical concepts. Part 1, Statisitcal analysis of a quantitative study. *Chemistry Education Research and Practice, 9*(4), 323–331.

Papaphotis, G., & Tsaparlis, G. (2008b). Conceptual versus algorithmic learning in high school chemistry: The case of quantum chemical concepts. Part 2, Students' common errors, misconceptions, and difficulties in understanding. *Chemistry Education Research and Practice, 9*(4), 332–340.

Rop, J. (1999). Student perspectives on success in high school chemistry. *Journal of Research in Science Teaching, 36*(2), 221–237.

Schmidt, H. J. (1994). Stoichiometric problem solving in high school chemistry. *International Journal of Science Education, 16*(2), 191–200.

Sherman, A., & Sherman, S. J. (1983). *Chemistry and our changing world*. Englewood Cliffs: Prentice Hall.

Sienko, M. J., Plane, R. A., & Marcus, S. T. (1984). *Experimental chemistry* (6th ed.). Tokyo: McGraw-Hill.

Staver, J. R., & Lumpe, A. T. (1993). A content analysis of the presentation of the mole concept in chemistry textbooks. *Journal of Research in Science Teaching, 30*(4), 321–337.

Staver, J. R., & Lumpe, A. T. (1995). Two investigations of student understanding of the mole concept and its use in problem solving. *Journal of Research in Science Teaching, 32*(2), 177–193.

Stromdahl, H., Tulberg, A., & Lybeck, L. (1994). The quantitatively different conceptions of 1 mole. *International Journal of Science Education, 16*(1), 17–26.

Taber, K. S. (1998). The sharing-out of nuclear attraction: Or I can't think about physics in chemistry. *International Journal of Science Education, 20*(8), 1001–1014.

Toomey, R., DePierro, R., & Garafalo, F. (2001). Helping students to make inferences about the atomic realm by delaying the presentation of atomic structure. *Chemistry Education Research and Practice, 2*(3), 183–202.

Tsaparlis, G. (1984). The chemical bond as an atomic tug-of-war. *Journal of Chemical Education, 61*(8), 677.

Tsaparlis, G. (1989). *What a single molecule does not look like–Two analogies and their effect on learning*. Abstracts of papers of the American Chemical Society, 198, 176-CHED.

Tsaparlis, G. (1997a). Atomic and molecular structure in chemical education. *Journal of Chemical Education, 74*(8), 922–926.

Tsaparlis, G. (1997b). Atomic orbitals, molecular orbitals, and related concepts: Conceptual difficulties among chemistry students. *Research in Science Education, 27*(2), 271–287.

Tsaparlis, G. (2004). Atomic structure. In J. J. Lagowski (Ed.), *Chemistry: Foundations and applications* (Vol. 1, pp. 78–87). New York: MacMillan Reference-Thomson Gale.

Tsaparlis, G. (2009). Learning at the macro level: The role of practical work. In J. K. Gilbert & D. F. Treagust (Eds.), *Multiple representations in chemical education* (pp. 109–136). Dordrecht: Springer.

Tsaparlis, G., & Kampourakis, K. (2000). An integrated physical-science (physics and chemistry) introduction for lower-secondary level (grade 7). *Chemistry Education Research and Practice, 1*(2), 277–290.

Tsaparlis, G., Kolioulis, D., & Pappa, E. (2010). Lower-secondary introductory chemistry course: A novel approach based on science-education theories, with emphasis on the macroscopic approach, and the delayed meaningful teaching of the concepts of molecule and atom. *Chemistry Education Research and Practice, 11*(2), 107–117.

Tsaparlis, G., & Papaphotis, G. (2002). Quantum-chemical concepts: Are they suitable for secondary students? *Chemistry Education Research and Practice, 3*(2), 129–144.

Tulberg, A., Stromdahl, H., & Lybeck, L. (1994). Students' conceptions of 1 mole and educators' conceptions of how they teach the "mole". *International Journal of Science Education, 16*(2), 145–156.

von Glasersfeld, E. (1989). Cognition, construction of knowledge, and teaching. *Synthese, 80*(1), 121–140.

Chapter 4
Challenging Myths About Teaching and Learning Chemistry

Diane M. Bunce

Introduction

Teaching and learning are not always seen as two distinct processes. Often it is assumed that if something is taught, then the students will learn. If students don't learn, then the first response from instructors is often that students are not working, not working hard enough or not academically prepared for the current course. Sedlak (1987) referred to this problem as the teacher viewing his/her responsibility ending with the presentation of material and the student's responsibility beginning with learning what the teacher presented.

The teacher's opinion on *how* a subject should be taught is often accepted by most people because the teacher, as content specialist, is credited with knowing what should be taught and how it should be taught. This is especially true of university teachers who are doing research in the same field they are teaching. A similar assumption is often made with doctors. Here we expect that what the doctor says is accurate and true. We don't often question the doctor's opinion. When we are seriously ill, we might seek out a doctor who is a specialist and doing research on the disease we are suffering from. We believe this specialist will know more than our regular doctor because the specialist is involved in research on the problem and must therefore be an expert. This model of "professional as expert" is prevalent in both medicine and academia. The result of this thinking can be misinterpreted in terms of the teacher will know how to teach the material and if the students don't learn, it must be the students' fault. The main problem with this teacher–doctor analogy is that the doctor operates in a one-to-one relationship with the patient and uses both interactions with the patient and tests that supply the doctor with pertinent data to help analyze the problem and prescribe a treatment. In the classroom, the teacher is working with a one-to-thirty or one-to-two hundred or more relationship with the students and there are few if any two-way

D. M. Bunce (✉)
Chemistry Department, The Catholic University of America, Washington, DC 20064, USA
e-mail: bunce@cua.edu

I. Devetak and S. A. Glažar (eds.), *Learning with Understanding in the Chemistry Classroom*, DOI: 10.1007/978-94-007-4366-3_4,

interactions with the vast majority of the students. Therefore, the analysis by a teacher that a student is not learning may not be as accurate as a diagnosis made by the doctor. There is a need for more teacher–student interaction and open-ended assessments that would provide both additional and more targeted information on the problems that the student is experiencing. Open-ended assessments can involve students in examining their own learning and helping focus their attention on understanding, rather than only achieving correct answers. This approach has been recommended by teaching experts (Shepherd 2000).

Student learning is a complex process that may not be totally evident to the teacher or the student. The student should be encouraged by the teacher and taught how to monitor his/her own learning. This self awareness of the learning process (metacognition) should be promoted by the student's interaction with the teacher and teaching process. Metacognition is a process of the student engaging in "making sense" of new information, self-assessment, and self-reflection on what worked or didn't work in understanding new information (Bransford et al. 2000). The learning environment can be developed by the teacher to include opportunities for the student to engage in metacognitive activities. These opportunities include self-assessment (such as ConcepTest questions), reflection, and opportunities to explain the logic of the answer in addition to providing an answer.

In learning theories, emphasis has been placed on how the process of teaching can affect learning in the individual student. The Constructivist theory places the process of learning inside the mind of the individual and requires the learner's active participation in the learning process. The Information Processing movement has helped identify brain-based parameters of how information is entered into working memory, moved to long-term memory and interwoven in students' mental schemas. An often cited Information Processing theory of memory and learning is Baddeley's model of working memory (1986) which emphasizes the fact that the sooner information can be rehearsed will affect how well it is kept in working memory. The implications for teaching are that the teaching process should provide opportunities for students to rehearse material during the learning process. Once information enters the working memory, it must be integrated with and transferred to long-term memory if it is to be retained. Long-term memory represents our store of knowledge and is accessed when new information enters the working memory (Pellegrino et al. 2001). This Information Process model of learning has been broadened to include cognitive neuroscience models of learning. Cognitive neuroscience models include the idea of schemas as ways of organizing information in long-term memory. Schemas are constantly changing as new information is received and integrated with that already in long-term memory. The formation and restructuring of schemas enable individuals to develop a mental model that guides their future learning (Pellegrino et al. 2001). Teaching based on the mental model of schema formation and revision includes the learner in the learning process. As a result, teaching has to go beyond traditional lecturing and content assessment and include opportunities for students to rehearse, process, and self-assess their knowledge.

In order to integrate the information processing-neuroscience model of learning with lecturing, traditional lectures should be examined for opportunities for students to incorporate new knowledge in working memory with that already held in long-term memory schemas. Pedagogical techniques that help provide such opportunities include the use of interactive questioning in class; formal student reflection on assumptions and conclusions in lab reports; use of technological resources such as course websites that provide access to support materials online; group work using guided inquiry materials; assessment opportunities that emphasize the reasons for an answer and not just the answer; and the use of peer leaders in discussion groups. There is still a place for lecturing in this new view of teaching and learning but it becomes one of several processes available to students for learning, not the only process.

Teaching is a cultural ritual (Nuthall 2005) that was invented by humans to help humans learn. Myths about teaching and learning can be viewed as cultural artifacts of the traditional approaches. The myth, discussed earlier, that if the teacher presents the information, the student will learn unless he/she is not working or not working hard enough can now be viewed as too narrow an explanation. This ritual of what constitutes good teaching is not based upon research but rather is based upon our collective experience of teaching and learning. In this experience, we "know" what constitutes good teaching. It occurs when the teacher exhibits the "customary" teaching behavior which up to recent times in science has been primarily lecturing. This is the teaching format that most of us experienced in our academic careers. In this model of teaching, successful learning is often viewed as students being able to repeat on tests and quizzes what the teacher has said during lecture. It is the customary way of doing things and thus constitutes our teaching and learning culture. This culture is promoted by books that offer advice to new teachers (McKeachie 2002; Bligh 2000) however, these "good" teaching practices can be part of the culture promoting the status quo. The presence of an entrenched teaching culture is difficult to confront in teacher training programs because most pre service teachers have primarily experienced lecturing as the way science is taught. Lecture is what pre-service or new teachers experience most often. It is familiar and thus easy to emulate. The same is true for science graduate students who do not typically take education courses but go on to become our new college professors. They experience the culture of teaching through the use of lecture in both their own university courses and in many college departments where they are hired and thus they strive to emulate it as new professors.

Can research challenge the current culture of teaching and learning? The answer depends on what questions or myths the research chooses to challenge and how effective it is in testing them. Research that asks the easy questions and uses standard research methodologies and previously established tools may promote rather than challenge the current teaching culture. In order to test whether the teaching/learning culture is valid and the myths that exist are real, the research should be theory-based and tailored specifically to the question or myth being explored. This means that research methodologies may be more unorthodox and tools used may need to be designed, validated, and tested for reliability for use in each specific research project. This approach should result in the research being more difficult to configure but more successful in challenging the current culture of teaching and learning.

Specific Myths About Teaching and Learning

How Long Can Students Pay Attention in Lecture?

A myth in relation to lecture that served to explain why more learning does not occur during lecture is that students pay attention for less than the length of a standard 50- or 55-min lecture. Typical attention spans of 10–20 min (Sousa 2006) to 30 min (Bligh 2000) are proposed in teaching books. To address this problem, McKeachie (2002) suggests, as does Sousa, that passive lectures be broken into shorter segments interspersed with teacher–student interaction. Research on the topic has been scarce. Johnstone and Percival (1976) published one of the few studies that measured student attention in lecture directly through observations of individual students in a class. Their conclusion was that the amount of time that students pay attention during lectures is cyclic and the length of each cycle depends on where it occurs in the lecture.

A recent experiment (Bunce et al. 2010) showed a significant difference in the attention students reported as the lecture proceeded. This research utilized an atypical research methodology that took advantage of the personal response device (clicker) as both an interactive learning device and a research instrument. Students in three different general chemistry classes (engineering, nursing and nonscience majors) participated in the study. Students used clickers during lecture to individually record when they were aware that their attention had lapsed. Students were instructed to press "1" if the lapse was for a minute or less, "2" for 2–3 min lapses, and "3" for a lapse of 5 min or more. This research design utilizing clickers as both a research tool as well as a teaching tool allowed for the daily collection of data in the three different courses for 4 weeks with minimal disruption to teachers or students. While Johnstone and Percival's (1976) data consisted of researchers recording attention by observing student facial expressions in lecture, the clicker data relied on the participants recording their own attention lapses.

The data show that after a short initial period of settling down at the beginning of the lecture, students pay attention for about 4.5–5.0 min and then again in 2–3 min cycles interspersed with lapses in their attention. Most of the self-reported attention lapses in this research were of short duration (usually 1 min or less). Most interestingly, the experiment included a comparison of lecture segments vs. segments when interactive questioning via personal response systems (clickers) or chemical demonstrations of comparable length were used. The data show that there were significantly fewer lapses in attention reported by students both during interactive questioning via clickers and chemical demonstrations compared to those reported during lecture. There was no significant difference in the lapses of attention reported by students between the two nonlecture segments (clicker and chemical demonstration). Both clicker and chemical demonstrations appear to engage student attention equally well. An added benefit of using either clickers or chemical demonstrations is that there is a decrease in attention lapses during a comparable length lecture segment *following* the clicker or demonstration segment compared to the number of attention lapses reported in the lecture segment *before* the clicker or demonstration.

This research demonstrates that student attention cycles between attention and inattention in ever shortening cycles as a lecture progresses. It also shows that using interactive segments such as clicker questions or chemical demonstrations can have both an immediate and a delayed benefit on student attention. This information can be used by teachers to structure their lectures into smaller segments interspersed with interactive segments such as clicker questions and/or demonstrations to maximize student's attention.

Is the Use of Clicker Questions More Effective than Frequent Online Quizzes?

Myths can occur about the teaching and learning of newer pedagogical approaches to learning as well to traditional approaches. One myth concerning the use of personal response devices (clickers) in lecture is that the use of clickers will automatically result in increased student achievement. Clicker use is becoming widespread in secondary and college classrooms and is estimated to be used by approximately 8 million users (Interactive Clickers 2006). The main use of clickers is as a means of implementing ConcepTests (Mazur 1997) in lecture. ConcepTests are conceptual questions that can be asked during a lecture. Students respond by choosing an answer. The clicker device software analyzes the student data in real time and constructs a graph providing teacher and students with a visual representation of how many students have chosen each answer to questions asked using clickers. With this information the teacher can decide whether the material just presented has been learned by the students and then the teacher can proceed with the next concept or recognize that the concept is not well understood by students and review or re-teach the original concept. It seems logical that such an interactive pedagogy would result in both increased student learning and more student-centered and responsive teaching. However, most research on this topic has not dealt directly with the effect of clickers on student learning. Most studies (MacArthur and Jones 2008) measure student attitude or engagement when using clickers. The effect of using clickers as a way of delivering ConcepTests and their effect on student learning has not been directly studied.

Bunce et al. (2006) investigated the relative impact of ConcepTests and clickers versus the impact of online daily quizzes on student achievement in general chemistry. Both the in-class clicker questions and the online daily quiz questions were keyed to questions on regularly scheduled hour exams. Student scores on corresponding questions on the end of the year standardized exam were also examined. The results showed that students scored significantly higher on regularly scheduled hour exams if they had a corresponding online quiz question on that topic compared to a corresponding ConcepTest clicker question. There was no significant difference between the effect of ConcepTest clicker questions and the online quiz questions on student achievement on the standardized final exam.

The explanation for this may have more to do with the implementation of the two pedagogies than the actual pedagogy itself. In this experiment, the ConcepTest clicker questions were used in class but students did not have access to either the question or answer after class. By contrast, the online quizzes with the correct answers were available on demand to students throughout the semester. A questionnaire helped document student study behavior in preparation for the regularly scheduled tests in terms of their review of online quiz questions. Students reported relying on reviewing the online quizzes in preparation for the regularly scheduled tests but did not review the ConcepTest clicker questions because they did not have access to them. Although students did significantly better on questions that had an online quiz counterpart compared to questions that had a ConcepTest clicker question on the hourly teacher-written tests, there was no significant difference between the two on the standardized final exam.

The seemingly small difference in implementation between having semester-long access to either the ConcepTest questions and the online quiz questions might have been expected to have inconsequential effects. The reality is that if both a teaching/learning tool and its implementation are in line with how the learner uses them, then there can be significant effects on student learning. Using just the tool may not be enough to affect change. The implementation of that tool in line with how students use it can be equally important. In this case, it was hypothesized that the ability to review and reflect on the content of the online quiz questions was responsible for the increase in student achievement on corresponding questions on the regularly scheduled tests. The inability of students to review and reflect on ConcepTest clicker questions is thought to have interfered with gains in student achievement on corresponding questions on the regularly scheduled tests. It is not known if students reviewed the online quiz question in preparation for the standardized final exam but it is conceivable that due to time constraints in preparing for a final exam, this was not a well used resource. In answer to the original question about whether the use of clickers increases student achievement, the result of this research is that the way the tool is implemented must be congruent with how students use it. If not, no tool, no matter how theoretically sound is likely to result in significant change in student achievement.

Can Students Successfully Answer Essay Questions in Chemistry?

In chemistry, like other sciences, success is often measured by the number of problems a student can solve. In chemistry, these questions are meant to be applications of chemistry concepts. Often chemistry tests include a high percentage of multiple choice questions that test the recall of concepts or facts necessary to solve open ended application problems which may also appear on the test. Some educators have questioned whether these numerical application problems adequately measure student conceptual chemistry understanding (Moore 1997). Meaningful learning as defined by Ausubel (Novak and Gowin 1984) might be better tested through essay-type questions that ask students to construct a logical argument to address a conceptually-based application question. The use of essay-type questions has several drawbacks in large introductory college chemistry courses. These drawbacks include the following: students do not typically do well on such essay-type questions; the grading of such questions is labor intensive; and the teaching assistants that typically help grade chemistry tests in large classes may not be able to reliably judge how well students are addressing the question asked. The latter two problems can be addressed through course management and in-service teaching assistant training programs, but the first problem, that of students typically not doing well on essay questions, is worthy of further exploration.

From a cognitive point of view, there are two aspects to the use of essay-type questions on assessments that should be addressed, namely, the number and level of difficulty of the chemistry concepts being tested and the way the question is phrased. Both of these variables affect the cognitive demand the question places on the learner. According to Cognitive Load Theory (Sweller 1994), the concept being assessed is an intrinsic variable and the way the assessment is designed or phrased is an extrinsic variable. Both intrinsic and extrinsic variables must be taken into consideration in the assessment process. Intrinsic variables such as chemistry concepts include all the previous knowledge that the current concept is based on. A question that seemingly addresses a single chemical concept may require the learner to understand three to five previously learned chemistry concepts that support the understanding of the current concept. The number of prerequisite chemistry concepts involved directly affects the level of cognitive demand of the question being asked. The cognitive demand of the extrinsic variable is affected by *how* the question is asked. If the information needed to solve the problem is presented directly in the stem of the problem, then the cognitive demand is lower. If the student is expected to provide an answer using assumptions that are not explicitly included in the problem as stated or must be derived from information that is given in the problem, the cognitive demand is increased.

The net result of Cognitive Load Theory on essay questions is that the cognitive demand of a chemistry question can quickly exceed the ability of many students to successfully solve it. But the transition of students from novice to expert understanding can be hampered if students are not taught how to address essay-type question logic. The use of essay questions helps students organize the information in their long-term memory (Wandersee et al. 1994) and thus increases the likelihood that the information will be more accessible to the students in other situations. Russell's research (2004) has shown that students need multiple exposures to both critiquing and writing essay questions in order to develop the skill of producing persuasive and logical arguments. Before students can adequately address essay-type questions, it is important to determine if they can recognize a complete and cogent (logical) answer to such questions.

Bunce and VandenPlas (2006) explored this question with undergraduate students enrolled in a nonscience majors' course. Students were shown sample essay questions and possible answers online 24 h prior to taking an exam on the same material. Students were then asked to analyze the answers provided to these essay questions for completeness and cogency using a Likert scale. In some cases, students were also asked to either plan or construct answers to other chemistry essay questions. The data show that in most cases students were able to correctly identify the correct essay question answer but their overall ability to rate the completeness and cogency of these answers on a Likert scale was moderate (2.7 and 2.9 out of 5, respectively). When one of the questions students were asked to examine was then included on a test within 24 h, there was no significant correlation between the students' ability to *recognize* a complete and cogent answer previous to the test and their ability to *create* a complete and cogent answer on the test.

These results suggest that being able to recognize correct answers to essay questions is not enough. Students need more intensive training in constructing and analyzing answers to essay questions for correctness, completeness and cogency. Essay questions of differing cognitive demand should be included in this training. Students' ability to answer essay questions may have more to do with students' understanding of what is expected of them in terms of completeness and cogency than it does in terms of understanding of the chemistry concepts. These variables should be addressed separately in the teaching experience. It may not be that students cannot answer essay questions. It may be more a question that they don't realize what is expected of them in order to answer such questions to a satisfactory level determined by their teachers. Being able to correctly and adequately address essay questions is further complicated by the different cognitive demand that essay questions can make on a student. All of these variables should be considered when including essay questions as part of a student assessment. The issue is more complicated than may be initially apparent.

How Long do Students Retain Their Chemical Knowledge After a Testing Situation?

Another myth that exists among teachers is that students forget what they have learned soon after completing a test. This phenomenon is seen at both the secondary and undergraduate levels. Wheeler et al. (2003) report that undergraduates experience an extinction of knowledge within 48 h of a testing situation. Anderson's et al. research (2004) modifies this assertion by reporting that students will regain some of this knowledge if multiple tests are given following the initial test. Regaining lost knowledge is seen as a result of students using multiple testing occasions to help them develop a stronger conceptual schema. Stronger schemas result in increased long-term retention according to Anderson et al. Hockley (1992) makes a distinction between forgetting discrete pieces of information vs. forgetting information that is associative. According to Hockley (1992), associative information is forgotten less quickly than discrete information. Anderson et al. (2004) explains this in terms of associative information having more retrieval paths from long-term memory than discrete information.

Curricula and teaching methods can affect whether chemistry is taught as a list of discrete facts or as an associated body of knowledge. Curricular that is organized using a spiral approach present a limited amount of information and then revisit and build upon it as new topics are introduced. This can be done either explicitly by including outlines of topics presented and revisited in the table of contents or implicitly in the organization of material from one chapter to the next. Teaching too, can either emphasize connections between concepts or simply present each new concept as it occurs in the syllabus. Students will address the learning of concepts in a manner that is consistent with the way the concepts are presented by the teacher and/or textbook (linear or spiral, discrete, or associative). In keeping with the theory of memory that includes the formation and utilization of schema, how the concepts are taught may affect how long students remember the information following a testing occasion.

Bunce et al. (2009) looked at the decay of student learning in three courses including both undergraduate- and secondary-level general chemistry courses (undergraduate nursing and nonscience major general chemistry and a secondary level honors general chemistry course). The curricula and teaching methods used in these courses included both explicit and implicit spiral curricula and teaching methods. Follow-up tests were used at different time intervals following a regularly scheduled testing occasion. The follow-up tests included two questions that appeared on the scheduled exam. Students did not have access to either answer keys or discussion of the test questions on the exam until all follow up tests were completed. Students cycled through the short-, medium-, and long-term delayed follow-up tests for the three scheduled tests used in this study. This meant that each student who participated answered two questions that appeared on one of three scheduled exams either after a short, medium, or long delay. The regularly scheduled exams were completed as paper and pencil tests. The delayed follow-up tests were completed either online or as paper and pencil tests as decided by the teacher of the course.

The results show that students (undergraduate nursing general chemistry and secondary level honors general chemistry) who experienced an explicit spiral curriculum where the chemistry was presented in an overall associative framework, showed no significant decay in their knowledge from the scheduled testing occasion through the long-term delay (10–17 days). Students, who experienced an implicit spiral curriculum where the associative framework of the chemistry concepts was not explicit (undergraduate nonscience major), experienced a significant decay in knowledge between the scheduled testing occasion and the first delayed follow-up test 2 days later. This result is consistent with the Wheeler et al. (2003) time frame of 48 h for the decay of knowledge. The results support the conclusion that students do forget knowledge after a scheduled test if that knowledge is not associated with new information that is being presented. The use of explicit spiral teaching methods and curricula where knowledge used in one testing occasion is seen as needed to understand new knowledge can help reduce the decay of student knowledge for at least two weeks (10–17 day) following a scheduled exam.

Conclusions

Myths about teaching and learning persist when they support the current culture of teaching and learning. Often this culture of teaching relies on the successful passing of knowledge from the expert (teacher) to the willing novice (diligent student) as the accepted way that teaching and learning take place. Any failures in this method are assumed to be the result of unwillingness or inability of the students to learn the content presented. In order to challenge this myth, the prevalent teaching and learning culture should be examined. If students are expected to learn what the teacher tells them then their inability to pay attention in lecture may result in failure. If the learning that they experience in lecture is not integrated at the level of understanding across topics versus memorization of individual topics, then knowledge may decay rapidly after students complete a scheduled exam. In addition, students might not be able to successfully answer essay questions on these topics if they have not been trained in what constitutes a correct, complete, and cogent answer. Finally, if the essay question is constructed with a high intrinsic (numerous concepts) and/or extrinsic (use of implicit assumptions) demand, it may exceed students' ability to successfully address it.

Introducing new pedagogies into our present teaching and learning culture will not necessarily result in higher student achievement if these pedagogies are not implemented with attention to the tenets of how the brain operates and learning occurs. For instance, the use of clickers as a way to provide multiple self-assessment opportunities for students may not be effective if students cannot review and reflect on both the questions asked and the answers chosen. The use of essay questions on assessments may not live up to the potential of helping students better organize their knowledge if the analysis of what constitutes a complete and coherent essay is not discussed and practiced with students outside of a testing situation and reinforced with an explicit spiral curriculum.

New pedagogies that are consistent with what we know about how the brain operates and learning occurs can be successful if implemented accordingly. The use of interactive ConcepTests as delivered by clickers or chemical demonstrations that break the flow of information in a lecture can be effective in increasing student attention both during their use and in subsequent lectures. The use of clickers and demonstrations in lecture can help increase student attention by providing students with nontesting opportunities to evaluate their understanding of a concept and reflect on their learning.

Myths are difficult to dispel if the research that challenges them is not grounded in theory. A theoretical framework can be used to help explain the results of research in terms of the basic premise of the teaching culture being challenged. Without this larger picture, each research result appears as a stand-alone artifact that at best provides little more than a passing acceptance on the part of the reader. By contrast, within a theoretical framework, individual research results can contribute to the mounting evidence needed to challenge our current teaching and learning culture. Designing experiments that challenge existing beliefs about teaching and learning must address the tough issues and be well designed in order to provide convincing results. To ask the tough questions, the research should facilitate the collection of pertinent data. Such research will necessitate the development and use of new and targeted valid/reliable instruments. It is not adequate to rely on instruments that were designed for another purpose without proving that they can provide the necessary data to address the questions asked. Data from such experiments should be analyzed using the power of sophisticated statistics that adequately control for threats of drawing conclusions that are not valid. These statistics must be implemented with attention paid to whether they are the appropriate statistic for the question asked and if the necessary assumptions of the data pool are met before the statistic is applied. Without these safeguards, research could be produced that is not valid and reproducible. As a result of misusing statistics, more myths could be produced and the teaching/learning process could continue to be misunderstood. A consequence of poorly conceived and/or analyzed research is that students will continue to be categorized as lazy or noncaring when in reality they might not be able to effectively learn and demonstrate that learning due to inappropriate teaching and assessment methods. The purpose of research should be to challenge myths, not produce new ones.

References

Anderson, J. R., Bothell, D., Bryne, M. D., Douglass, S., & Qin, Y. (2004). An integrated theory of the mind. *Psychology Review, 111*(4), 1036–1060.

Baddeley, A. D. (1986). *Working memory*. Oxford: Oxford University Press.

Bligh, D. A. (2000). *What's the use of lectures?*. San Francisco: Jossey-Bass.

Bransford, J. D., Ann, L. B., & Cocking, R. R. (Eds.). (2000). Learning: From speculation to science. In *How people learn: brain, mind, experience, and school*. Washington, DC: National Academy Press.

Bunce, D. M., & VandenPlas, J. R. (2006). Student recognition and construction of quality chemistry essay responses. *Chemistry Education Research and Practice, 7*(3), 160–169.

Bunce, D. M., Flens, E. A., & Neiles, K. Y. (2010). How long can students pay attention in class? *Journal of Chemical Education, 3*(7), 160–169.

Bunce, D. M., VandenPlas, J. R., & Havanki, K. L. (2006). Comparing the effectiveness on student achievement of a student response system versus online WebCT quizzes. *Journal of Chemical Education, 83*(3), 488–493.

Bunce, D. M., VandenPlas, J. R., & Soulis, C. (2009). Decay of student learning in chemistry over time. *Journal of Chemical Education, 88*(9), 1231–1237.

Hockley, W. E. (1992). Item versus associative information: Further comparisons of forgetting rates. *Journal of Experimental Psychology. Learning, Memory, and Cognition, 18*(6), 1321–1330.

Interactive clickers can increase student responses. (2006). *Curriculum Review, 45*(4).

Johnstone, A. H., & Percival, F. (1976). Attention breaks in lecture. *Education in Chemistry, 13*(2), 49–50.

MacArthur, J. R., & Jones, L. L. (2008). A review of literature reports of clickers applicable to college chemistry classrooms. *Chemical Education Research and Practice, 9*(3), 187–195.

Mazur, E. (1997). *Peer instruction: A user's manual.* Upper Saddle River, NJ: Prentice Hall.

McKeachie, W. J. (2002). *McKeachie's teaching tips* (11th ed.). New York: Houghton Mifflin.

Moore, J. W. (1997). Assessment, achievement, and understanding. *Journal of Chemical Education, 74*(5), 477.

Novak, J. D., & Gowin, D. B. (1984). *Learning how to learn.* New York: Cambridge University Press.

Nuthall, G. (2005). The cultural myths and realities of classroom teaching and learning: A personal journey. *Teachers College Record, 107*(5), 895–934.

Pellegrino, J. W., Chudowsky, N., & Glaser, R. (Eds.). (2001). Advances in the sciences of thinking and learning. In *Knowing what students know: The science and design of educational assessment.* Washington, DC: National Academy Press.

Russell, A. A. (2004). Calibrated peer review: A writing and critical-thinking instructional tool. In S. Cunningham & Y. S. George (Eds.), *Invention and impact: Building excellence in undergraduate science, technology, engineering, and mathematics (STEM) education.* Washington, DC: American Association for the Advancement of Science.

Sedlak, M. (1987). Tomorrow's teachers: The essential arguments of the Holmes Group Report. *Teacher College Record, 88*, 314–326.

Shepard, L. A. (2000). The role of assessment in a learning culture. *Educational Researcher, 29*(7), 4–14.

Sousa, D. A. (2006). *How the brain learns* (Vol. 3). Thousand Oaks: Corwin Press.

Sweller, J. (1994). Cognitive load theory, learning difficulty, and instructional design. *Learning and Instruction, 4*(4), 295–312.

Wandersee, J. H., Mintzes, J., & Novak, J. D. (1994). Research on alternative conceptions in science. In D. Gabel (Ed.), *Handbook of research on science teaching and learning* (pp. 177–210). New York: Macmillan.

Wheeler, M. A., Ewers, M., & Buonanno, J. F. (2003). Different rates of forgetting following study versus test trials. *Memory, 11*(6), 571–580.

Section I
Teaching and Learning Chemistry

Students' Characteristics on Chemistry Learning

Part II of Section I, "Students' Characteristics on Chemistry Learning", of this book comprises two chapters dealing with students' characteristics that can significantly influence chemistry teaching and learning. Student's attributes such as motivation and interest for learning chemistry, different mental abilities (i.e. intelligence, visualisation abilities, working memory capacity, formal reasoning ability...), social skills and others, should be considered when teachers organise their school lessons, authors design the teaching material, policy makers prepare national curriculums and teacher educators conduct pre- and in-service teacher education programmes.

Chapter 5 addresses the students' characteristics from the working memory point of view in the chemistry learning process. It emphasises that the working memory capacity is a key factor in enabling understanding to take place. Working memory is that part of the brain functions where understanding, thinking and problem solving occur. While the capacity of working memory cannot be changed, it is possible to use it more efficiently. One way is to introduce pre-learning and another to re-design the learning process so that the working memories of learners are less likely to overload. Recent evidence also suggests that limited working memory capacity is a major factor in the development of positive attitudes. Limited working memory capacity can often make understanding impossible, leaving the student no alternative other than memorisation of chemical concepts. This can cause attitudes to change and to become negative. All this can be correlated to today's chemistry education problems that are thoroughly analysed also in this book in all the chapters. This chapter illustrates an important aspect that all chemical education researchers and teachers ought to be aware of.

The theme of the previous chapter is somehow continued in Chap. 6 entitled "Educational models and differences between groups of 16 year-old students' in gender, motivation and achievements in chemistry" by Devetak and Glažar who described other aspects of students' characteristics using statistical analysis of numerous variables regarding the triple nature of chemical concepts presentations. Students can be differentiated into groups during classroom activities also according to their abilities, academic achievements and interests. On the basis of these classifications students can be engaged in specific learning tasks. Different models can be formed according to students' academic achievements, motivation for a specific subject, gender and also other variables that can influence classroom dynamics. Three predictive educational models developed by discriminant analysis are presented. The assessment of the differences between male and female students, low and high intrinsic motivation for learning chemistry and between the groups of students regarding their chemistry academic achievements was performed on a set of 42 independent variables. The first model of predictive variables shows that general academic achievement, motivation for chemistry and achievement in physics discriminate significantly between high and low achievers in school chemistry. The second one shows that variables describing intrinsic motivation for chemistry learning on different levels of concepts' representations discriminate significantly between students with high and low motivation for chemistry learning. In the third predictive model, the strongest discrimination between male and female students is shown in the intrinsic motivation for physics and motivation for foreign language learning. This rather complicated statistical analysis can show that teachers should be aware that girls need more attention in chemistry learning than boys. Teachers should extrinsically motivate using macroscopic level of chemical concepts and put those concepts into the context for those students who are low achievers in physics and have low intrinsic motivation for learning chemistry. After that they ought to use the triple nature of chemical concepts representations to stimulate also the low motivated students to develop mental models that show the understanding of chemistry concepts on the phenomenological, interpretational and symbolic levels.

Chapter 5
The Learning of Chemistry: The Key Role of Working Memory

Norman Reid

Introduction

While some very general work is taught at primary school stages, chemistry is usually only taught as a discipline from about age 12. It is an interesting study to look at the textbooks which were used up to about 1960 in most countries. Chemistry teaching was built around '*preparations and properties*' where endless lists of compounds were discussed, with the methods to make them outlined and their properties described. Success in chemistry meant that the school students of that day had to memorise and recall accurately.

The early 1960s saw the beginnings of a revolution in school chemistry. One of the earliest countries was Scotland where a new syllabus was published in 1962 (Curriculum Papers 512). The chemistry content was greatly updated but the major change was in the emphasis. There was an overt attempt to encourage the development of *understanding* and the examinations reflected this. Thus, the school students were encouraged to carry out many experiments in class, under careful direction, which aimed to enable students to understand why matter behaved in the way it did. This approach proved to be successful in that chemistry was (and still is) a very popular subject at school level in Scotland (Scottish Qualifications Authority, http://www.sqa.org.uk/sqa/57518.4241.html).

With such new curricula, problems started to become apparent, in varying degrees in different countries. Chemistry was now seen as difficult by many school students. In most countries, we began to see a trend where the numbers opting to study chemistry started to fall (for England: Osborne et al. 1998, 2003; Ramsden 1998; Jenkins and Nelson 2005), despite the fact that qualifications in chemistry at school level were valuable for future studies or for careers. Did the problem lie in the chemistry being taught or in the way students were learning the chemistry? It turned out to be both and this chapter seeks to uncover what much careful research

N. Reid (✉)
University of Glasgow, St Andrews Building, Glasgow, G3 6NH, Scotland, UK
e-mail: dr_n@btinternet.com

I. Devetak and S. A. Glažar (eds.), *Learning with Understanding in the Chemistry Classroom*, DOI: 10.1007/978-94-007-4366-3_5,

has shown to be the sources of the problems. Recent research has now provided many of the answers and these will be discussed as well.

Chemistry is Difficult

In a very early study, Johnstone et al. (1971) asked students of chemistry at two Scottish universities to look back at their school experiences. From this, they were able to identify the key areas where students had found problems. Four broad areas were apparent (Table 5.1).

Most of the new chemistry curricula had introduced the topic of the the amount of substance (Mole). Some aspects of the topic seemed intrinsically difficult but was it possible for us to teach it differently to gain greater student's confidence and success? Thermochemistry and thermodynamics were also introduced. These can offer very powerful insights into the nature and direction of chemical reactions but the students found the area difficult. Electrochemistry and redox often proved very confusing for students while the introduction of organic chemistry left students with a bewildering array of carbon, hydrogen and oxygen atoms.

Over many years, Johnstone et al. started to explore these broad areas to see if they could identify the nature of the problem and how to make life easier for the students. Although this resulted in a series of research papers (e.g. Duncan and Johnstone 1973; Johnstone and Kellett 1974a; Garforth et al. 1976a, b; Johnstone et al. 1977a, b), his work led to the publication of two textbooks which attempted to present chemistry on the basis of what his research had uncovered. Later, a paper summarised many of the research findings (Johnstone 2000).

Both textbooks were quite radical in nature and made a considerable impact at the time. One was a general text for middle school (age 14–16) chemistry students (Johnstone et al. 1980) and this ran to numerous reprintings. In this text, only the bare minimum of atomic and molecular structure was introduced before the organic chemistry was studied. The other areas followed. It was, perhaps, the first textbook deliberately to base its approach on the known research evidence related to how students learn. The other textbook was a brilliant monograph on thermodynamics (Johnstone and Webb 1977). Sadly, when both texts went out of print,

Table 5.1 Areas of difficulty

Curriculum Area	Example
Equations and the amount of substance (Mole)	Equations and the the amount of substance (Mole): volumetric and gravimetric work, Avogadro and the amount of substance (Mole)
Computational topics	Thermochemistry and thermodynamics
Electrochemistry and redox	E° and ion electron ideas and equations
Organic topics	Esters, proteins, amines and carbonyls, aromaticity

subsequent textbooks tended all to revert to traditional approaches and students problems continued.

In the 1970s, Kellett was looking at the difficulties that students had when studying organic chemistry at school level. Although she originally considered that it might be a perception problem (students could not *see* organic structures) (Johnstone and Kellett 1974a), she started to appreciate that the problem lay in the *amount of information* that a student had to take in and handle *at the same time* (Johnstone and Kellett 1974b). This turned out to be the key breakthrough.

Why is Chemistry Difficult?

While most studies have explored the areas *where* chemistry is difficult and have come up with all kinds of ingenious attempts to make life easier for our students, the real answer must lie in seeking to find out *why* the difficulties occur. Suggestions that chemistry is difficult because it is highly abstract or very conceptual may be true, but they do not take us forward in any way that can lead to making it more accessible. As teachers, we cannot change the nature of chemistry. However, we can change the way it is taught. The real question is *how* should we change it so that we retain the true nature and rigour of chemistry but, at the same tine, make it more accessible to young learners. The brilliant insight of Kellett led to a quite ingenious experiment being set up. However, before this is outlined, we need to look at some findings which had arisen from psychology research.

Medical research had shown that human memory had more than one component. Miller (1956) developed ways to measure the capacity of what he called short-term memory and found, amazingly, that the capacity of this part of the brain was very small. It is now known that it grows with age but the final capacity is fixed genetically. Miller found that the average adult (aged 16 or over) could hold seven pieces of information at the same time and that almost all adults have a capacity lying between 5 and 9. He described the pieces of information as '*chunks*'. The size of each chunk could vary but the key thing as that the individual person saw each chunk as *one* piece of information.

While Miller talked in terms of '*short-term memory*', much subsequent work uses the phrase, '*working memory*' (Baddeley 2000). This recognises that this part of the brain not only holds information temporarily but it is also the location where thinking, understanding and problem solving take place. It is a 'holding-thinking' space. It is finite in capacity and, if there is too much to hold, then little space is left for thinking and understanding. Was this the key to the reason why chemistry is so often perceived as difficult?

While the average working memory capacity of a 16-year old is SEVEN, 14-year olds have an average capacity of SIX and 12-year olds an average capacity of FIVE and so on. This offers an immediate explanation why certain topics cannot be introduced too early in a school student's career. As the working memory grows with age, increasing numbers of ideas can be handled at the same time, opening up more complex areas of knowledge to the learner.

Working memory limitations also offer an immediate explanation why mathematics is so often a problem when applied to chemistry or physics. The working memory simply cannot cope with the chemistry concepts *at the same time* as using the ideas from mathematics. When the mathematics is more or less automated (and requires minimal working memory space), then success is possible. However, this takes time and considerable practice.

With that background in mind, let us now return to the experiment. In this work, a large number of first-year university examination questions were examined in detail to see how many ideas had to be held at the same time for a student to have some prospect of success. Student's success in these questions was explored. The success rate was plotted against what the research called the 'information load' (the number of ideas to be held at the same time). The researchers expected to obtain some kind of linear relationship (see Fig. 5.1).

What, in fact, they obtained was closer to Fig. 5.2. The relationship was not linear but was more like a titration curve.

The researchers went further. They measured the working memory capacities of the students and divided the students into three groups: those with above average capacities (mainly 8), those with average (which is 7) and those with below average (mainly 6). They then plotted the three curves for each sub group. Figure 5.3 shows the kind of outcomes they obtained. The curves are drawn more

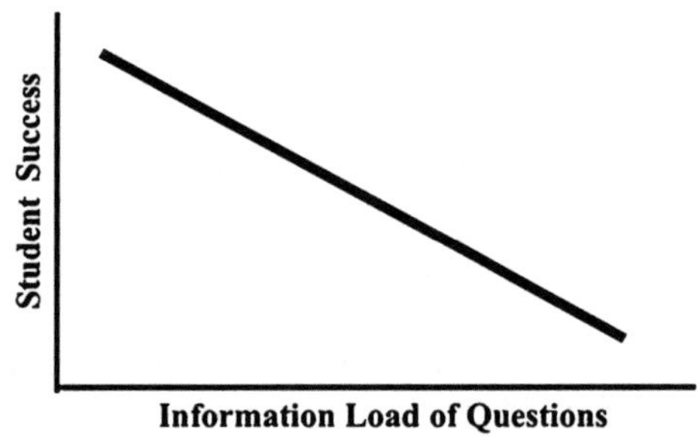

Fig. 5.1 Predicted relationship

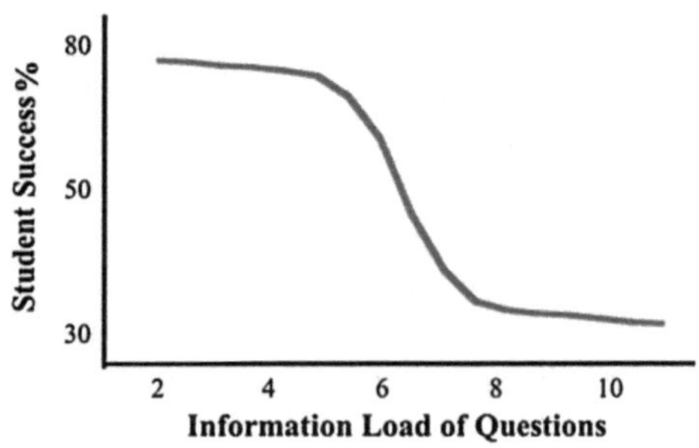

Fig. 5.2 Experimental relationship between load and success

Fig. 5.3 Experimental relationship between load and success, with working memory capacity

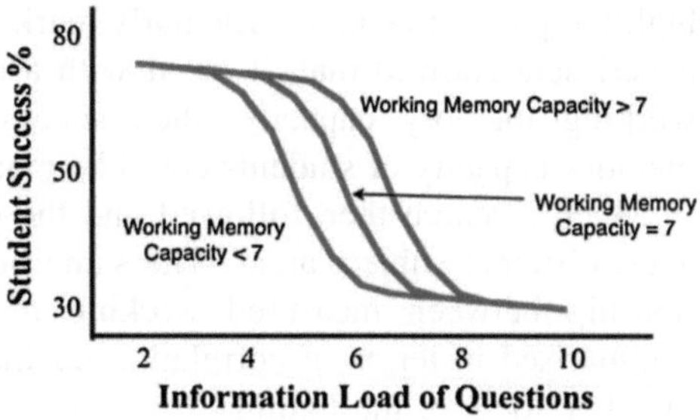

precisely here than the original experiment, simply for clarity but the pattern is unaltered.

What this experiment showed was that it was the *actual capacity of working memory which was controlling success* in the test items. It revealed much more as well. The performance of the students with a measured working memory of 7 started to drop dramatically when the information load of the question reached 6. It has been found that we cannot work right up to the actual limits. We seem to need some room for manoeuvre in using our working memories. For those who had less than 7 (mainly 6), performance dropped at 5 while those with more than 7 (mainly 8) showed a performance drop at 7. The whole experiment is described in two papers (Johnstone and Elbanna 1986, 1989).

Enabling working memory to work more efficiently turned out to be an important idea. Miller (1956) had used the word '*chunk*'. A chunk was what the individual person saw as one unit of information. Was it possible that some students were able to group items of chemistry information together so that they saw them as one chunk of information? This only took up one space in the working memory, leaving enough space to handle the other chunks.

The possibility is that we teachers might be able to teach our students to chunk information, thus reducing load on limited working memory capacity. It was found that teaching what might be called '*chunking skills*' is not easy in that we all tend to do it in so many different ways. However, specific chunking strategies can be introduced and we shall return to this later. In passing, developing chunking skills is almost certainly the explanation of what has been called cognitive acceleration (Shayer and Adey 2002). It is interesting that the authors admit they do not know why cognitive acceleration works. However, the observation that cognitive acceleration only benefits about half of the pupils suggests strongly that their materials are giving opportunities for the school students to develop chunking skills. A careful scrutiny of the teaching materials they produced is certainly consistent with this explanation. Some students respond to the approaches offered while others do not, simply because there are so many diverse ways to chunk.

Let us return to the Johnstone and Elbanna experiment. When the papers were first published, there was some surprise that such a simple idea explained many learning difficulties in chemistry. There was also surprise that the outcomes were so predictively precise. Nonetheless, the results of the experiment seem to be highly reproducible and particularly marked for mathematics (Christou 2001). The experiment showed that, if faced with a task where the load was more than the working memory capacity, then success was highly unlikely. The working memory capacity of students could be measured reliably and easily (Reid 2009a).

Much research then followed and the experiment was repeated with different ages, different subject areas. The same outcomes were obtained. Later, the relationship between measured working memory capacity and performance was summarised in terms of correlation coefficients. Reid (2009a, p. 134) has drawn together some of the findings for school courses (see Table 5.2).

Table 5.1 needs some interpretation. First, it shows that working memory capacity is related to performance in many subjects at many ages (the Johnstone and Elbanna study shows it is actually cause-and-effect). Second, two tests were used to measure the working memory capacity: the digit span backwards test (DSBT) and the figural intersection test (FIT). Although very different, the same effect is observed. The digit span backwards test involves the recall of a series of number in reverse order while the figural intersection test requires students to find the area of common overlap between increasing numbers of geometrical shapes.

Third, the correlation coefficients obtained vary considerably. Indeed, in one study in mathematics, the researcher designed a test in mathematics for 12-year olds in such a way that no question placed a load on the working memory above the minimum in the sample (190). She obtained a correlation value very close to zero although the test was not easy and the pupils did not perform that well. However, it tested mathematical ability and *not* the capacity of their working memories (Reid 2002).

This leads to an important principle that '*performance will only correlate if one or both of two conditions are fulfilled:* (a) The learning process is such that those with higher working memory capacities have an advantage; (b) The assessment questions place demands on the working memory such that those with higher working memory capacities have an advantage' (Reid 2009b, p. 246).

Working memory capacity can have a profound effect on the marks obtained in an examination as Danili discovered. She found that students with below average working memory capacities performed, on average, 16 % less well when compared to those with above working memory capacities. The effect of working memory capacity can, therefore, be very considerable. The student with a less than average capacity faces a very large disadvantage in the examinations which we typically set today. Indeed, the study of Ali and Reid (2012) showed the highest correlation found so far, the capacity of working memory accounting for nearly half of the marks gained. The test used in the schools in this study was clearly measuring working memory capacity to a quite unacceptable extent.

Working memory capacity is not neatly linked to what we might call ability. It is simply the capacity of part of the human brain and varies slightly from person to person. The problem lies in our teaching and our assessment. The way we teach chemistry and, even more importantly, the way we test chemistry is giving a considerable advantage to those students who happen to have higher working memory capacities.

Table 5.2 Some correlations of working memory with performance

Age	Country	Sample	Subject	Test used	Pearson correlation	Probability	Source
13-15	India	454	Science	DSBT	0.34	$p < 0.001$	Pidikiti 2005
13	Kuwait	641	Science	FIT	0.23	$p < 0.001$	Hindal 2007
15	Greece	105	Chemistry	FIT	0.34	$p < 0.001$	Danili 2004
13	Taiwan	151	Physics	FIT	0.30	$p < 0.001$	Chen 2005
13	Taiwan	141	Biology	FIT	0.25	$p < 0.001$	Chu and Reid 2012
13	Taiwan	141	Genetics	FIT	0.62	$p < 0.001$	Chu and Reid 2012
16-17	The Emirates	809	Physics	DSBT	0.11	$p < 0.01$	Al-Ahmadi 2008
16-17	The Emirates	349	Physics	DSBT	0.32	$p < 0.001$	Al-Ahmadi 2008
16	Greece	90	Mathematics	DSBT	0.40	$p < 0.001$	Christou 2001
11	Pakistan	700	Mathematics	FIT	0.69	$p < 0.001$	Ali and Reid 2012

An Interim Summary

After 25 years of using the concept of working memory, Baddeley (2002) published a paper with the enigmatic title: '*Is Working Memory still Working?*' After reviewing the evidence, Baddeley argues that working memory is still working. The concept is certainly valid. However, Reid (2009b, p. 250) remarks that, '*for some pupils, it is perhaps having a struggle*'.

Working memory is a psychological and physical space in the brain where incoming information is held temporarily, into which information may be drawn from long-term memory, and where information can be manipulated. In educational terms, it is where the learner thinks, understands, makes sense of information, solves problems. Information can be transferred from the working memory and stored in long-term memory, leaving the working memory space free for further tasks. Because working memory has finite capacity, it is a controlling step for all learning, when learning is seen as understanding. This controlling nature is critical: it is a kind of 'bottle neck' for learning. Indeed, the work of Kirschner et al. (2006) demonstrates the imperative of taking working memory limitations into account in considering any new approach to learning.

Chemistry, by its very nature is conceptual and often abstract. It is world of atoms, molecules, electrons and protons and energy. It is full of representations like equations, along with numerous abstract ideas like the amount of substance (Mole), free energy, reaction rates and electron spin, delocalisation and quantum numbers. Almost by definition, if a concept is to be understood, many ideas must be held *at the same time* by the learner. The only place where such ideas can be held is the working memory and capacity is highly limited. St Clair-Thomson and

Botton (2009) offer a useful overview of the structure of working memory in the context of science education.

The new curriculum developments introduced in the 1960s and which have continued on until today have rightly emphasised understanding. However, in the way the material is taught and the way we assess it in typical examinations, we have inadvertently given a massive disadvantage to those school students who have less than average working memory capacities as well as placing a considerable potential cognitive overload on all learners. This explains why chemistry is perceived as difficult by so many. The real question is what we teachers can do about it.

Before looking at that, we need to expand further the understandings of how the brain works when understanding is the goal. This leads us into the world of processing information.

Learning as Information Processing

It is strange quirk of history that the research on how the brain works ran in parallel with the development of the modern computer. The language of the latter found its way easily into the former. The human being can be seen as a highly complex processor of information. Through our senses, every waking moment, we take in enormous amounts of information. Piaget has shown that the natural way the learner tries to work is to try to make sense of the information coming in. Of course, we all sometimes make mistakes, interpreting things wrongly and sometime drawing wrong conclusions. However, the natural process is that, as we learn: we correct and expand our understandings continually (Atkinson 1983).

The way the brain was working in taking in and interpreting this endless flow of information has been studied for over 40 years. Pascual-Leone (1970) focussed on the load of information, laying the foundation for later understandings while the work of Atkinson and Shiffrin (1971) on information processing is considered a key foundation in cognitive research.

In the context of chemistry education, Johnstone (1997) developed a model which will be used here (Fig. 5.4). It has to be stressed that the various models only differ in small details, the essential ideas being well established from a long series of research studies. Indeed, the literature is replete with information processing models used as the basis for some very elegant research. The power of the model lies in its ability to predict in relation to learning. In the sciences and in mathematics, the model has offered a powerful way to understand and predict what is happening during the process of understanding as well as success in assessment (see Reid 2009b).

We are bombarded with information all the time. We select what we take in and the basis of selection is controlled by what we already know as well as our attitudes. What is taken into the brain goes into the working memory. Here, we attempt to make sense of the incoming information, drawing in information held in

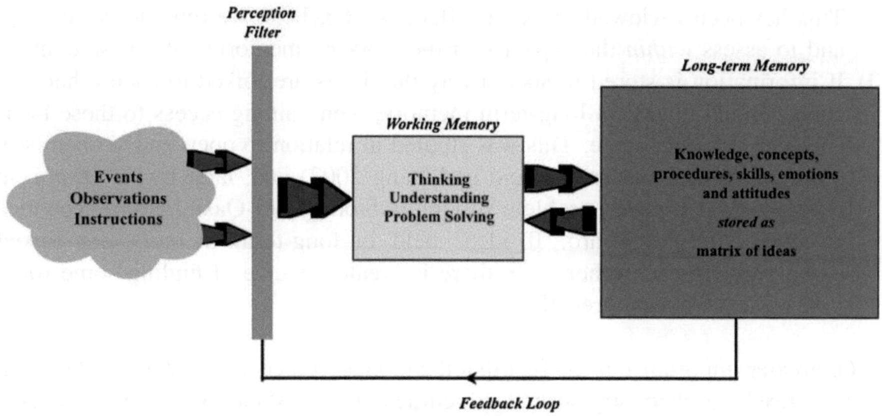

Fig. 5.4 An information processing model (derived from: Johnstone 1997)

long-term memory as required. We may store incoming information in long-term memory. If we understand what we are learning and can link it to what we already know, then the new information is added to a network of inter-related ideas in long-term memory, enriching all the ideas.

The model absorbs most of the key findings from other research. For example, the feedback loop captures the insights of Ausubel et al. (1978) while the fact that each individual seeks to understand incoming information in the working memory and linking it in their own way to what is already held in long-term memory underpins constructivism. It is the natural process of the learner to try to make sense of new information. Each constructs meaning in their own way.

The idea of teaching in a constructivist way is very misleading. Constructivism is related to the natural process of understanding in all learners. It is nothing really to do with teaching. Each learner will *inevitably* construct their own meaning. Indeed, writing from a cognitive load perspective, in a brilliant analysis, Kirschner et al. (2006, p. 78) raise very serious doubts about the whole enterprise of 'constructivism' as a predictive tool: '*The cognitive description of learning is accurate but the instructional consequences suggested by constructivism do not necessarily follow*'. The fundamental problem is that such approaches have not taken into account the rate determining control of working memory on all understanding and problem solving.

The power of the model lies in its ability to predict. For example:

(1) If the perception filter works well, then less unnecessary information enters the working memory and information overload is less likely. This leads to better learning and better test results. Many studies have looked at this and demonstrated that the prediction is supported but, perhaps, the work of Danili is particularly useful in chemistry (Danili and Reid 2004);
(2) If the working memory has to handle more than its capacity allows, learning more or less ceases and the assessment task provides more or less impossible.

This has been reviewed by Reid (2008). As teachers, we must learn to teach and to assess *within* the capacity of the working memories of our students.

(3) If information is stored in such a way that ideas are linked to each other in a sophisticated matrix in long-term memory, then gaining access to these ideas is easier at a later stage. This was studied in relation to open-ended chemistry problem solving by Yang (Reid and Yang 2002) and, later by Al-Qasmi, in relation to open-ended problem solving in biology (Al-Qasmi 2006). The idea is really straightforward. If ideas held in long-term memory are linked extensively to each other, then there is greater chance of finding some route into what we need to recall.

Of greater importance is the fact that the model predicts how we as teachers can re-think teaching chemistry so that difficulties in understanding are reduced. This will be discussed in a moment. Before that, we turn to a very useful insight from Johnstone when he looked, in the context of his information processing model, at the learning of chemistry in general terms (Johnstone 1999).

He noted that, in chemistry, there are three kinds of activities:

(1) There is what he called *macro chemistry*: here we see colour, detect smells, observe reactions and describe materials.
(2) There is what he called the *micro* level: this is the world of atoms, molecules, bonds, electrons and so on. None of this is directly accessible to the senses.
(3) Finally, there is the *symbolic*: as chemists, we represent the world of chemistry by means of formulae, equations, diagrams and mathematical representations.

The experienced chemist can move happily at all three levels. The learner simply cannot do this as the amount of information involved simply overwhelms limited working memory capacity. Johnstone (1999) pictured this in terms of triangle (Fig. 5.5).

Since the publication of the triangle for chemistry, the idea has carried forward into biology where Chu and Reid (2012) developed a biology tetrahedron while Ali and Reid (2012) has suggested a tetrahedron for mathematics learning (see Reid 2009b). However, in all three subjects, the same principle is being applied.

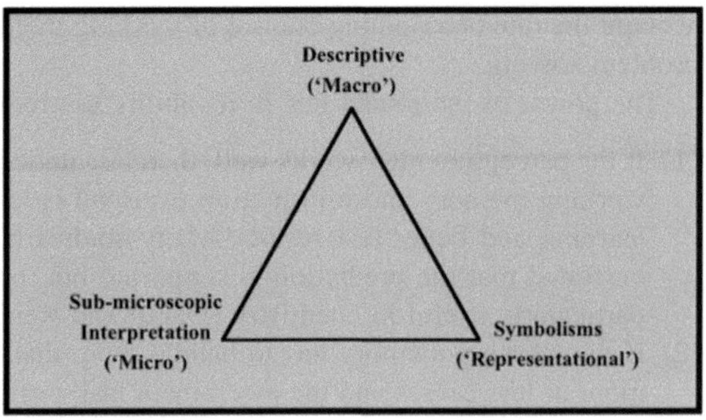

Fig. 5.5 Chemistry learning triangle

The novice learner cannot cope with the cognitive load in trying to work at all the corners *at the same time*.

Johnstone (1999) argues that early courses in chemistry must concentrate on the macro, the descriptive. When the school students are more familiar with the way materials behave in descriptive terms, then the micro level can be gently introduced as a way of explaining why the familiar macro-chemistry takes place in the way it does. The symbolic must be introduced carefully, always ensuring that it is perceived as a way of simplifying what the student already knows. An example of a totally descriptive approach but involving rigorous chemistry for 13-year old students is described by Reid (1999). In this course,

> The pupils started to look at their world (the air, water, the sea, rocks and minerals, the atmosphere) with a simple agenda: what elements could be found and what was mankind doing with what was there? Many fundamental chemical ideas just arose naturally, e.g. the concept of bonding, reactivity, physical properties of matter, energy and bonds, states of matter. The course was descriptive, based on the world around, applications orientated and it avoided quantitative aspects.

This approach was found to highly effective with the school students and stands in complete contrast to the typical kind of approach: atoms are introduced, with atomic structures, arrangements of electrons, to be followed by bonding and the whole panoply of molecular theory. This approach places excessive demands on limited working memory capacity as well as being perceived as largely irrelevant by most school students at early stages. Indeed, considering that the vast majority of younger school students will never become chemists or, even, scientists of any description, it is an irrelevant approach.

Improving Understanding: Pre-learning

We now return to the model of information processing again. We can now use the model to explore how we might make learning more accessible and allow the school students to understand more of what they are doing. Firstly, we look at the perception filter. The question is how to make it function more efficiently.

Four studies in first-year university chemistry offer major insights. Two relate to labwork and two to learning from formal presentations. We shall look first at labwork where the potential for information overload is very considerable: new chemicals, unfamiliar equipment, written and verbal instructions and chemistry understandings all to be handled *at the same time*. Johnstone and Wham (1982) found that working memory overload was a major problem. Later the idea of pre-learning was used to reduce the problem.

In pre-learning, knowledge already learned is revised and brought to the surface. This offers the learner the key landmarks for the new material which is to be presented. Therefore, in the laboratory, the learner has the key ideas brought to the surface and, thus, the selection of what is important is much more efficient.

Table 5.3 The effects of pre-learning

Year	Pre-learning	Upper group average	Lower group average	Difference between groups	Statistical significance
1993–1994	Yes	50.9	48.8	2.1	Not significant
1994–1995	Yes	49.2	49.0	0.2	Not significant
1995–1996	No	46.9	38.7	8.2	Significant
1996–1997	No	48.2	42.0	6.2	Significant
1997–1998	No	46.7	41.3	5.4	Significant
1998–1999	Yes	49.8	47.7	2.1	Not significant

Pre-learning has a very large effect in making laboratory learning much more effective. One experiment in physics found a rise of 11 % as a result of the pre-laboratory experiences (Johnstone et al. 1998). An earlier study was conducted in chemistry, with even larger numbers and, yet again, the power of pre-learning to improve understanding was very marked (Johnstone et al. 1994). This is exactly in line with what the information processing model predicts. The work on pre-labs was later followed up by Reid and Shah (2010) while Carnduff and Raid (2003) collated examples of pre-lab exercises for university chemistry and offered guidance on how such pre-lab exercises could be developed.

In another experiment, a large university first-year chemistry class was followed for six successive years. They were given pre-learning experiences in the first 2 years, these being discontinued in the next three and then, finally, pre-learning was re-introduced in a paper form known as 'chemorganisers' in the final year. The original pre-lectures took the form of a series of short activities based on previous knowledge and this was undertaken *before* each lecture course. When these were discontinued, the extra time was given over to the lectures. The aim of pre-learning was to bring to the surface previous ideas so that these ideas then enabled the selection filter to work more efficiently. The new material then was more easily understood as the working memory was less likely to overload. The full experiment is described in Sirhan and Reid (2001).

This experiment is unusual in that those who were *least* well qualified (The Lower Group) gained the most (Table 5.3). Those who were least well qualified did not have a set of clear landmarks in their long-term memories. The revision of key ideas led to these landmarks becoming clearer. The perception filter then could select more efficiently in the light of these landmarks and working memory was less overloaded.

In the final study, Hassan et al. (2004) looked at the underlying key ideas in an introductory university organic chemistry course. These ideas would be established from school courses. They were able to relate very precisely which key ideas were well established from school and show how this affected university performance in quite specific ways.

Such pre-learning is evident in many school classrooms where, by skilful use of questions and recapitulation, the class is reminded of previous learning which, in turn is then able to inform the perception filter. Working memory overload is less likely and subsequent understanding is enhanced. This offers a simple explanation of a practice which good teachers have used for generations.

Improving Understanding: Reducing the Load

The information processing model predicts that, if teaching is re-cast in order to reduce the overload on working memory directly, then better understanding will take place. This is difficult to test in that it is not easy to control all the variables. However, three experiments are described in the literature and, together, these make compelling support for the hypothesis. In all three, there was no change in subject matter covered, no change in time taken, no change in teachers employed, no training of the teachers involved. The changes were in how the material was presented, mainly in the form of written text in that, in the countries involved, this reflected the normal way teaching took place.

The explicit aim of these studies was to re-design the teaching approach so that pressure on the working memory was reduced. In places, the teaching order was changed while speed and sequencing of the presentation of ideas was modified. Complex areas were broken into smaller parts and ideas were developed and expanded step by step. Graphics were used where these were likely to reduce the information overload. All this was achieved by thinking through each difficult theme and then making sure that it was presented in a step-by-step way, thus reducing the load on working memory. It is worth remembering that working memory causes a problem when too much has to be thought about *at the same time*. By careful sequencing of ideas, by reminder and illustration, by a stepwise approach, the working memory is not faced with too much at the same time. It was predicted that learning will increase (Reid 2008).

Danili re-designed a large section of chemistry teaching at middle secondary school level (aged 15–16) in Greece, specifically to reduce working memory overload problems (Danili and Reid 2004). About 100 pupils followed the approach in the traditional way while a similar sized and matched group used materials which had been modified to reduce working memory load. There were no changes to content, time allocations or teachers. The experimental group improved performance by 22 % while the control group improved performance by only 13 % in pre and post tests, the difference being statistically highly significant.

In a much larger experiment, this time in the Emirates, with a total sample of 800, pupils in 2-year groups towards the end of their studies in school chemistry experienced being taught by new materials covering major sections of the school syllabus. The new materials aimed to minimise working memory overload, to use relevant applications, to encourage understanding, not memorising, and to link new material to previously taught material in a meaningful way. All of this was based on the information processing model considered here.

Four areas of the curriculum were covered: in Year 10 (age 16 and 17), two major topics were included: The Periodic Table; Chemical Equations. 400 students were divided into two groups, each group completing only one of the two topics using the new approach and completing the other using a traditional approach. In year 11, the same system was used but the topics were: Organic Chemistry; Acids and Alkalis (Table 5.4).

Table 5.4 Improved performance (Hussein and Reid 2009)

Year	Topics	Groups	Average mark	Differences
10	Periodic table	Experimental group	79.2	18.2
		Control group	61.0	
	Chemical equations	Experimental group	80.2	9.2
		Control group	71.0	
11	Organic chemistry	Experimental group	71.0	14.0
		Control group	57.0	
	Acids and alkalis	Experimental group	75.0	10.7
		Control group	64.3	

Each group was measured after completing each topic and the examination outcomes are shown in Table 5.4.

The remarkable thing is that, with large samples and large areas of the curriculum, the average performance rose for the four areas by so much, completely transforming the examination performance of these pupils. In this experiment, there was no contact at all with the many teachers involved. The teachers only had to give the new materials to the students and allowed the students to follow these instead of the normal textbook. Attitudes were also measured and it was found that the student attitudes towards their studies in chemistry had improved quite dramatically. Indeed, the statistical analyses showed that the improvement was one of the most marked ever observed. As with the examination performance, attitudes had been *transformed* simply by applying the ideas predicted from the information processing model (Hussein and Reid 2009).

The third experiment took place in Taiwan where the entire syllabus in genetics was re-cast specifically to reduce the load on working memory (Chu and Reid 2012). Here, with large samples of school students aged about 13, there was observed the same marked improvement in performance and also considerable changes in attitudes.

In each of three experiments described here, the researchers deliberately try to re-cast the teaching approaches so that there was less demand on the limited capacity of working memory. They used a variety of approaches. In all three, there were very large improvements in examination and test performance compared to control groups. In two of the experiments, attitudes in relation to their studies were observed to improved quite remarkably. Attitude changes will be considered again later but let us first look further at the vexed problem of the amount of substance (Mole).

The Problem of the Amount of Substance

The literature is replete with references, descriptions and carefully conducted research all of which demonstrate that the amount of substance (Mole) concept is one which causes school and university students considerable difficulty. Nearly 40 years ago, Johnstone et al. (1971) noted the amount of substance (Mole) as one of the difficult areas in school chemistry, following this up with further exploration of the difficulties a year later.

It appears that many learners can cope with the idea of the amount of substance (Mole) as it relates to gram formula masses and some can manage to apply it to simple reaction equations. Many can make sense of the amount of substance (Mole) as it relates to gas volumes under given physical conditions and some can also apply that to simple equations. However, in all these situations, there is clear evidence that coping does not necessarily imply understanding. As soon as the amount of substance (Mole) is brought into solution, problems increase even further.

Introducing the Concept of Power

The major problems arise when substances are dissolved in water and concentrations and volumes are involved. It is a classic case of information overload. Is there a way round this? A suggested way has been described (Reid 1982) but this monograph is now not easily obtained.

It was suggested that we need to invent what is called: '*neutralising power*' (when thinking of acids and bases). The ability of an acid to neutralise a base can be thought of as depending on three factors:

the volume used
its molarity
its 'power'.

For simple acidimetry, *power* is defined as the *number of hydrogen ions produced or absorbed by one molecule of the acid or base*. Thus, hydrochloric acid and sodium hydroxide both have a power of ONE while sulfuric acid and calcium hydroxide have a power of TWO. [The molecule was defined in terms of the written formula.] This leads to the relationship:

$$\underset{\text{(acid)}}{V_1 \times M_1 \times P_1} = \underset{\text{(base)}}{V_2 \times M_2 \times P_2}$$

The usefulness of such a relationship in obtaining 'correct' answers is obvious. The argument against using such a relationship is that it could be seen as removing the necessity to *understand* the chemistry of the reaction. However, this is *not* true in that the reaction has to be understood *before* using the concept of 'power'.

Consider the following problem

> Calculate the molarity of potassium hydroxide if 25 ml is exactly neutralised by 10 ml of 0.1 M sulfuric acid.

The student needs to know the formula of the acid and base and, hence, deduce the 'power' of the acid as 2 and the 'power' of the alkali as 1. The rest is easy. If an acid like ethanoic (acetic) is used, the student has to *understand* the reaction and that only one hydrogen is involved in the formation of water. The method has the great advantages in that it gathers all the variables into one easily remembered relationship which can be applied in a straightforward fashion. This generates confidence for the first-time learner, so important for future success.

However, there is another even greater advantage. The same relationship can easily be extended to redox reactions as well. Power is now defined in terms of the *electrons lost or gained per molecule or ion of reactant.*

Consider the following analysis.

If 20 ml of 0.02 M potassium permanganate is acidified and treated with excess potassium iodide solution, iodine is released. When this iodine is titrated against 0.1 M sodium thiosulfate solution, using starch indicator, what volume of thiosulfate would be used?

Ion electron equations have to be developed and balanced:

$$MnO_{4^-} + 8H^+ + 5e^- \rightarrow Mn^{2+} + 4H_2O$$
$$2I^- \rightarrow I_2 + 2e^-$$
$$2S_2O_3^{2^-} \rightarrow S_4O_6^{2^-} + 2e^-$$

From these, the relevant values for 'power' are easily observed under these conditions:

Power of permanganate ion = 5

Power of iodine molecule = 2

Power of thiosulfate ion = 1 (note: electrons per molecule or ion)

The whole calculation can be done easily:

$$\underset{\text{(permanganate)}}{V \times M \times P} = \underset{\text{(iodine)}}{V \times M \times P} = \underset{\text{(thiosulfate)}}{V \times M \times P}$$

The student can see easily that all that is need is the permanganate-thiosulfate relationship, the iodine not being relevant. The calculation reduces to:

$$\underset{\text{(permanganate)}}{V \times M \times P} = \underset{\text{(thiosulfate)}}{V \times M \times P}$$
$$20 \times 0.02 \times 5 = V \times 0.1 \times 1$$

giving the volume of thiosulfate as 20 ml.

The relationship can be extend to complexometric titrations where power is defined in terms of ligand attachment points while another useful application is to give the relationship to technicians. For example, in finding the volume to be added to 900 ml of 0.585 M sulphuric acid to obtain exactly 0.500 M acid:

$$\underset{\text{original acid}}{V \times M \times P} = \underset{\text{desired acid}}{V \times M \times P}$$
$$900 \times 0.585 \times 2 = V \times 0.500 \times 2$$

This gives the required volume as 1053 ml and the 900 ml must be diluted to 1053 ml.

In passing, the product, VMP, is meaningful and, in the units used above, is the *number of millimoles of the reactant*. Thus, the method could be extended to calculations where masses are also involved.

Reaction of Learners

The approach has been tried out with twenty 15-year old students, twenty 16-year-old students and a small group of the 17-year-old school students. Not only were they all highly successful in calculations (including redox) but they seemed confidently to *understand* what they were doing.

In almost every application of the amount of substance (Mole) concept, the working memory space for the novice learner quickly becomes overloaded, causing almost no learning to occur: understanding is virtually zero; confidence plummets. The formula (VMP) is a remarkable 'chunking' device, reducing the working memory space demand immediately. 'Chunking' is the ability to group several variables, facts or ideas together into a meaningful unit so that working memory space is not overloaded and was first described by Miller (1956) and applied extensively by Johnstone (1997). The whole area has recently been reviewed extensively by Reid (2008, 2009a, b).

In the VMP relationship, six variables are brought together in a way that is easy to remember and easy to apply. Nonetheless, *the relationship cannot be applied blindly: there MUST be understanding of the chemistry*. In this way, the use of the method is consistent with the psychology of the learner but is also a means of encouraging sound understanding. As an added bonus, it leads to quick success, with the concomitant rise in confidence.

Working Memory and Attitudes

There are many reports of poor attitudes towards chemistry among school students (Schibeci 1984; Ramsden 1998; Reid 2006). Is it possible to explain this drop in positive attitudes in such a way that it gives clear direction to a better way forward? There are several research studies which do offer key insights.

In a major study in physics, Skryabina looked in detail at the way attitudes towards physics changed with time, from age 10 to age 20. One of the problems with most attitude studies is the use of inappropriate methodologies (see Reid 2006 for an analysis of this issue). Skryabina (2000) used some very imaginative approaches and built up a picture in fascinating detail. The work was set in Scotland and much has been reported (Reid and Skryabina 2002a). Both chemistry

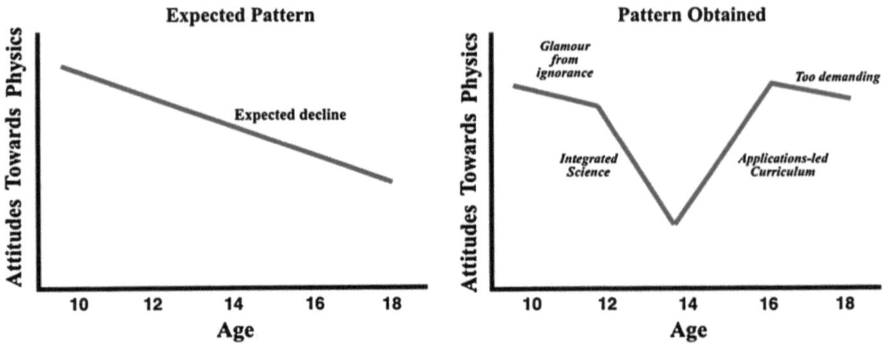

Fig. 5.6 Attitudes to physics in Scotland (derived from Skryabina 2000)

and physics are highly popular subjects at school and university levels on Scotland but Skryabina expected some decline in positive attitudes with age.

The early positive attitudes with primary school children rapidly deteriorated in the first 2 years of secondary school, this being attributed to the use of an integrated teaching approach. For those who continued on, their attitudes rose steadily during the next 2 years (aged 14–16), and the curriculum structure was found to be the reason. During the final 2 years, attitudes did decline but very very slightly, the reasons being that the course was excessively difficult.

It has to be stressed that Fig. 5.6 offers a very simplified picture of what Skryabina (2000) found. The graphs cannot be treated quantitatively. However, they do show the general trend of what was found and a possible interpretation of the findings.

There were clear messages from this observed pattern. The foolishness of asking a teacher who was not committed to, and qualified in, physics to teach physics (the way integrated science usually works) is obvious. This is consistent with other evidence which shows that integrated teaching is a highly ineffective way to teach the sciences and usually causes marked attitude deterioration. An interesting summary of many of the issues can be found in Venville et al. (2002).

The course structure in the years 14–16 was an applications-led curriculum. This can be described in the following way: the biology, chemistry or physics to be taught and its teaching order are determined by the learners—their needs, what is perceived by them to be related to their context and lifestyle (Fig. 5.7).

This type of curriculum structure has been discussed (Reid 1999, 2000) but few have followed up the ideas. These papers give examples and outline the principles in detail as well as offering some evidence to support this approach. In the case of the physics course here, the themes covered included topics like Telecommunications, Health Physics, Transport, Leisure, Space Physics. Major areas of life, of direct relevance and importance to the school students, were the themes. The physics was unpacked so that the students could make sense of these aspects of life. The end goal in terms of the physics covered was little different from

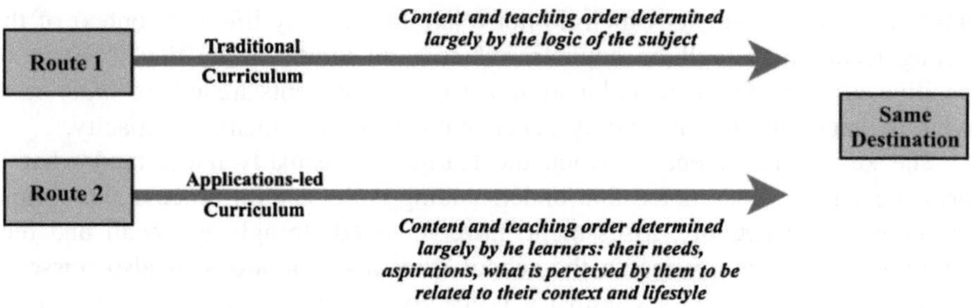

Fig. 5.7 Approaches to the curriculum

traditional courses. The way the material was covered was very different: the goal was the same; the route to get there was completely different.

The implications for chemistry are very considerable. Chemistry lends itself to such an approach and the paper by Reid (1999) describes an approach which did offer some highly positive outcomes. However, in some of the previous work (especially Hussein and Reid 2009), there were hints that working memory might also be involved in this. This was explored further by Jung in South Korea and her work is now described.

Jung and Reid (2009) have shown that it is possible to see attitude development in terms of three factors shown in Fig. 5.8.

The role of the teacher is critical and it has been shown that positive attitudes are encouraged by teachers who are competent in their subject and are supportive of the learner (Skryabina 2000). However, it is worth considering further the nature of the communication and the way the brain processes information. This can be thought of in terms of the way the chemistry is presented and whether the subject matter can be understood. The power of the applications-led curriculum has been discussed already.

Of great importance, the school student needs to see that the chemistry they study offers interpretations and understandings of the world of the student. Abstract ideas, unrelated to the world of the student, may well seem irrelevant

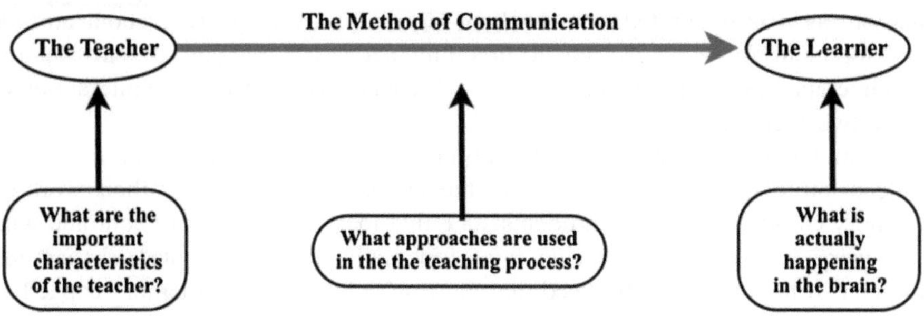

Fig. 5.8 Attitude development in education (from Jung and Reid 2009)

while the use of contexts which do not relate to the daily life and context of the young leaner may well be counterproductive in terms of the development of positive attitudes. Fundamental to all this is that the students are able to *understand* what is taught and this is heavily dependent on working memory capacity.

The key is the extent to which the learner can actually understand what is presented. It has been noted that understanding is the natural process. Too often, teaching and, especially, assessment, have focussed strongly on recall and recognition. This does not match the school student's aspirations. It also misses a wonderful opportunity for chemistry is ideally placed to enable the students to understand how their world actually works, in terms which are of direct importance and relevance to them. Working memory capacity is critical in enabling understanding to occur. Therefore, it might be expected that working memory capacity might well relate to the development of positive attitudes.

Festinger (1957) has offered brilliant insights into how attitudes can develop and change. He related attitude development tightly to the idea of dissonance. The idea is very common in life. When faced with information which is inconsistent with what we understand, then dissonance is set up. Festinger demonstrated that the possibility of attitude change or development is controlled by what he called '*total dissonance*' and this involves taking into account what is consonant as well as what is dissonant. Dissonance seems to be a natural process throughout life.

The important observation is that dissonance involves thinking: weighing up ideas, considering options and making judgements. This takes place in the working memory. This implies that the working memory is a critical factor in attitude development. More precisely, dissonance occurs in the working memory as former knowledge, feeling or experience are drawn from long-term memory to interact with new knowledge, feeling or experience. The role of the working memory is critical for it is here that all thinking, understanding and problem solving take place. If the working memory is overloaded, then dissonance is impossible. If learning is reduced to rote learning or is the passive reception of information, then dissonance is highly unlikely.

In the process of learning, information is processed cognitively by the learner and the information processing model of Johnstone (1993) offers valuable insights into the processes involved. However, as information is processed, held attitudes may affect information selection and the way it is handled. Equally, new information, as it is integrated into the long-term memory may bring about attitude development. These two aspects occur simultaneously in real educational situation and interact with each other continuously. We shall now consider the interaction of these two factors.

Attitudes have a powerful and continuous influence on the learning process. Indeed, attitudes may influence what the learner allows to enter their working memory (Reid 2008). Many students state that they do not want to continue with chemistry because they perceive it as too mathematical, too abstract and too difficult. It might be hypothesised that those with low working memory capacities tend to demonstrate lower understanding and, in order to pass their chemistry examinations, they may well be forced to resort to rote learning. It might then be hypothesised that high dependence on rote learning leads to little intellectual satisfaction, thus encouraging the development of less positive attitudes towards chemistry and aspects of the learning experience.

Jung explored the relationship between aspects of attitudes and measured working memory capacity working with 714 South Korean school students aged between 12 and 15 who were following an integrated science course. Jung and Reid (2009) asked,

(a) Are there significant relationships between students' beliefs and attitudes and their working memory capacity?
(b) Is there any relationship between students' ideas about various aspects of learning science and working memory capacity?

When she related measured working memory capacity to the response patterns on a five point scale to questions of interest, enjoyment and perceived importance, she obtained low but very significant ($p < 0.001$) correlations using Kendall's Tau-b correlation (Table 5.5).

What Table 5.6 tell us is that the students who said they enjoyed studying science found it interesting and thought it was important tended also to be those with *higher* working memory capacities. The correlation values, although low, are statistically highly significant.

She then went on to ask them if they were interested in science (forcing them to respond 'yes' or 'no') and relating their responses to their measured working memory capacity. She looked separately at two age groups and divided each age group into three subgroups: those with above average working memory capacity [high], those with average working memory capacity [mid] and those with below average working memory capacity [low]. The results are shown in Table 5.6.

The data in Table 5.6 brings a clear message. There is a strong pattern that those with the *higher* working memory capacity tend to be the groups where there

Table 5.5 Attitudes and working memory (Jung and Reid 2009)

Sample = 714, Aged 12–15, South Korea	Kendall's Tau-b
I am enjoying studying science	0.17
Science is interesting	0.13
Sciences is an important subject for my life	0.16

Table 5.6 Working memory capacity and interest in science

Are you interested in science?	Age 12–13			Age 14–15		
	Working memory capacity			Working memory capacity		
	High ($N = 100$) (%)	Mid ($N = 166$) (%)	Low ($N = 98$) (%)	High ($N = 95$) (%)	Mid ($N = 172$) (%)	Low ($N = 83$) (%)
YES	66	55	39	48	36	22
NO	33	42	57	53	64	78

Table 5.7 Preferred ways of learning and working memory capacity

	Age 12–13			Age 14–15		
	Working memory capacity			Working memory capacity		
	High ($N = 100$) (%)	Mid ($N = 166$) (%)	Low ($N = 98$) (%)	High ($N = 95$) (%)	Mid ($N = 172$) (%)	Low ($N = 83$) (%)
I have tried to understand science	71	54	50	70	61	37
I have tried to memorise science	24	34	39	21	35	45

are far more with an interest in science. The pattern is highly significant statistically. Is it possible that there is cause-and-effect relationship? The hypothesis is that loss of interest is caused by possessing a lower than average working memory capacity.

To gain further insights into this, she asked the students how they preferred to learned in their science studies. She offered them only two alternatives and related their responses to their measure working memory capacity (Table 5.7).

The results here suggest what is happening. Those with higher working memory capacities are tending to try to understand much more. Those with lower working memory capacities may well be unable to understand and are having to resort to memorisation. This process is not the natural way to learn and attitudes towards the subject tend to deteriorate.

This leads to a simple hypothesis (Fig. 5.9).

The findings from Jung and Reid (2009) suggest very strongly that working memory capacity is important in the development of positive attitudes towards chemistry. Because those with lower than average working memory capacities find to difficult to understand, they resort to memorisation to pass examinations and then lose interest in chemistry as this is not their natural way of learning.

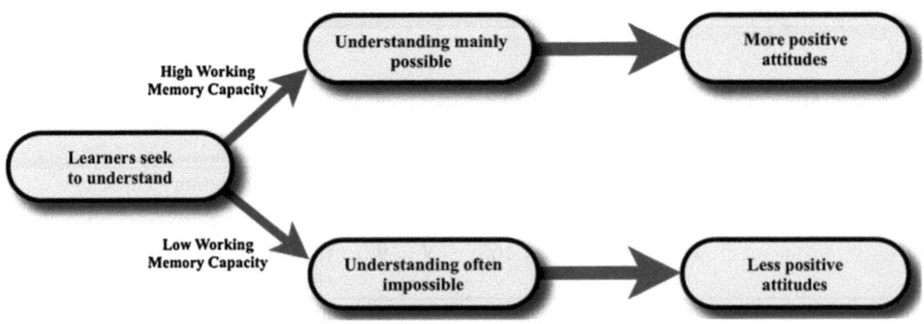

Fig. 5.9 A working hypothesis

Conclusions

This chapter has noted that understanding chemistry can be difficult because, being highly conceptual, understanding makes high demands on limited working memory space. Indeed, the study of chemistry is often not popular because students often have to resort to memorisation to pass examinations because they are *unable* to understand due to limited working memory capacity. Understanding always takes place in working memory and, if there is overload, understanding is impossible. Positive attitudes towards the study of chemistry depend on being able to understand and also being able to perceive that what is being taught is of relevance and value.

The aim of all chemistry teaching is to give young people at school stages an insight into the place of chemistry in the development of modern day society as well as offering an elegant insight into the way the world is constructed and changes take place. If we do not take the limiting capacity of working memory into account, then understanding will prove elusive and positive attitudes will rapidly deteriorate. Some of the clear evidence has been presented here and, on the basis of clear research evidence, ways have been suggested by which the exciting world of chemistry can be made more accessible for our students.

References

Ali, A. A., & Reid, N. (2012). Understanding mathematics some key factors. *European Journal of Educational Research, 1*(3), 283–299.

Al-Qasmi, S. (2006). *Problem solving in biology at university level*, PhD Thesis. Glasgow: University of Glasgow.

Atkinson, C. (1983). *Making sense of Piaget*. Boston, MA, USA: Routledge.

Atkinson, R., & Shiffrin, R. (1971). The control of short-term memory. *Scientific American, 225*, 82–90.

Ausubel, D. P., Novak, J. D., & Hanesian, H. (1978). *Educational psychology: A cognitive view*. New York: Holt Reinhart and Winston.

Baddeley, A. D. (2000). *Working memory*. Oxford: Oxford University Press.

Baddeley, A. D. (2002). Is working memory still working. *European Psychologist, 7*(2), 85–97.

Carnduff, J., & Reid, N. (2003). *Enhancing undergraduate chemistry laboratories*. London: The Royal Society of Chemistry.

Christou, K. (2001). *Difficulties in solving algebra story problems with secondary pupils*, MSc thesis. Glasgow: University of Glasgow.

Chu, Y.-C., & Reid, N. (2012). Genetics at school level: Addressing the difficulties. *Research in Science & Technological Education, 31*(1), 1–25.

Curriculum Papers. 512 (1962). *Alternative chemistry for ordinary and higher grade*. Edinburgh: Scottish Examination Board.

Danili, E., & Reid, N. (2004). Some strategies to improve performance in school chemistry, based on two cognitive factors. *Research in Science and Technological Education, 22*(2), 203–226.

Duncan, I. M., & Johnstone, A. H. (1973). The mole concept in chemistry. *Education in Chemistry, 10*(6), 213.

Festinger, L. (1957). *A theory of cognitive dissonance*. Stanford, CA: Stanford University Press.

Garforth, F. M., Johnstone, A. H., & Lazonby, J. (1976a). Ionic equations and examinations at 16+. *Education in Chemistry, 13*(2), 41.

Garforth, F. M., Johnstone, A. H., & Lazonby, J. (1976b). Ionic Equations - Difficulties in Understanding and Use. *Education in Chemistry, 13*(3), 72.

Hassan, A. K., Hill, R. A., & Reid, N. (2004). Ideas underpinning success in an introductory course in organic chemistry. *University Chemistry Education, 8*(2), 40–51.

Hussein, F., & Reid, N. (2009). Working memory and difficulties in school chemistry. *Research in Science and Technological Education, 27*(2), 161–186.

Jenkins, E. W., & Nelson, N. W. (2005). Important but not for me: Students attitudes towards secondary school science in England. *Research in Science and Technological Education, 23*(1), 41–57.

Johnstone, A. H. (1993). The development of chemistry teaching - a changing response to a changing demand. *Journal of Chemical Education, 70*(9), 701–705.

Johnstone, A. H. (1997). Chemistry teaching, science or alchemy? *Journal of Chemical Education, 7*(3), 262–268.

Johnstone, A. H. (1999). The nature of chemistry. *Education in Chemistry, 36*(2), 45–48.

Johnstone, A. H. (2000). Teaching of chemistry - logical or psychological? *Chemistry Education: Research and Practice in Europe, 1*(1), 9–15.

Johnstone, A. H., & El-Banna, H. (1986). Capacities, demands and processes: A predictive model for science education. *Education in Chemistry, 23*(3), 80–84.

Johnstone, A. H., & El-Banna, H. (1989). Understanding learning difficulties - a predictive research model. *Studies in Higher Education, 14*(2), 159–168.

Johnstone, A. H., & Kellett, N. C. (1974a). Condensation and hydrolysis - an optical problem? *Education in Chemistry, 11*(4), 111.

Johnstone, A. H., & Kellett, N. C. (1974b). Learning difficulties in school science - towards a working hypothesis. *European Journal of Science Education, 2*(2), 175–181.

Johnstone, A. H., & Wham, A. J. B. (1982). Demands of practical work. *Education in Chemistry, 19*(3), 71–73.

Johnstone, A. H., & Webb, G. (1977). *Energy chaos and chemical change*. London: Heinemann.

Johnstone, A. H., MacDonald, J. J., & Webb, G. (1977a). Misconceptions in school thermodynamics. *Physics Education, 12*(4), 248.

Johnstone, A. H., MacDonald, J. J., & Webb, G. (1977b). Chemical equilibrium and its conceptual difficulties. *Education in Chemistry, 14*(6), 169.

Johnstone, A. H., Morrison, T. I., & Sharp, D. W. A. (1971). Topic difficulties in chemistry. *Education in Chemistry, 6*(8), 212–213.

Johnstone, A. H., Morrison, T. I., & Reid, N. (1980). *Chemistry about us*. London: Heinemann.

Johnstone, A. H., Sleet, R. J., & Vianna, J. F. (1994). An information processing model of learning: Its application to an undergraduate laboratory course in chemistry. *Studies in Higher Education, 19*(1), 77–88.

Johnstone, A. H., Watt, A., & Zaman, T. U. (1998). The students' attitude and cognition change to a physics laboratory. *Physics Education, 33*(1), 22–29.

Jung, E. S., & Reid, N. (2009). Working memory and attitudes. *Research in Science and Technological Education, 27*(2), 205–224.

Kirschner, P. A., Sweller, J., & Clark, R. E. (2006). Why minimal guidance during instruction does not work: An analysis of the failure of constructivist, discovery, problem-based, experiential, and inquiry-based teaching. *Educational Psychologist, 41*(2), 74.

Miller, G. D. (1956). The magical number seven plus or minus two: Some limits on our capacity for processing information. *Psychological Review, 63*(2), 81–97.

Osborne, J., Driver, R., & Simon, S. (1998). Attitudes to science: Issues and concerns. *School Science Review, 79*(288), 27–33.

Osborne, J., Simon, S., & Collins, S. (2003). Attitudes towards science: A review of the literature and its implications. *International Journal of Science Education, 25*(9), 1049–1079.

Pascual-Leone, J. (1970). A mathematical model for the transition rule in Piaget's developmental stages. *Acta Psychologica, 32*(1), 301–345.

Ramsden, J. (1998). Mission impossible?: Can anything be done about attitudes to science? *International Journal of Science Education, 20*(2), 125–137.

Reid, N. (Ed.). (1982). *Chemistry for the Certificate of Sixth Year Studies*, Memorandum 4, Scottish Curriculum Development Service, 1–106.

Reid, N. (1999). Towards an application-led curriculum. *Staff and Educational Development International, 3*(1), 71–84.

Reid, N. (2000). The presentation of chemistry: Logically driven or applications led? *Chemistry Education: Research and Practice, 1*(3), 381–392.

Reid, N. (2006). Thoughts on attitude measurement. *Research in Science and Technological Education, 24*(1), 3–27.

Reid, N. (2008). A scientific approach to the teaching of chemistry (The Royal Society of Chemistry Nyholm Lecture, 2006–2007). *Chemistry Education Research and Practice, 9*(1), 51–59.

Reid, N. (2009a). The concept of working memory. *Research in Science and Technological Education, 27*(2), 131–138.

Reid, N. (2009b). Working memory and science education. *Research in Science and Technological Education, 27*(2), 245–250.

Reid, N., & Skryabina, E. (2002). Attitudes towards physics. *Research in Science and Technological Education, 20*(1), 67–81.

Reid, N., & Yang, M. J. (2002). Open-ended problem solving in school chemistry: A preliminary investigation. *International Journal Science Education, 24*(12), 1313–1332.

Reid, N., & Shah, I. (2010). Laboratory work for trainee chemistry teachers. *Journal of Science Education, 11*(1), 8–12.

Reid, P. A. (2002). *Problem solving by primary school children with particular reference to dyslexics*, M.Sc. thesis. Glasgow: University of Glasgow.

Schebeci, R. A. (1984). Attitudes to science: An update. *Studies in Science Education, 11*(1), 26–59.

Scottish Qualifications Authority (undated). Business intelligence services. Retrieved November 28, 2009 from http://www.sqa.org.uk/sqa/57518.4241.html. Accessed Nov 2013

Shayer, M., & Adey, P. (Eds.). (2002). *Learning intelligence: Cognitive acceleration across the curriculum from 5 to 15 years*. Buckingham: Open University Press.

Sirhan, G., & Reid, N. (2001). Preparing the mind of the learner - part 2. *University Chemistry Education, 5*(1), 52–58.

Skryabina, E., (2000). *Students' attitudes to learning physics at school and university levels in Scotland*, Ph.D. thesis. Glasgow: University of Glasgow.

StClair-Thomson, H., & Botton, C. (2009). Working memory and science education: Exploring the compatibility of theoretical approaches. *Research in Science and Technological Education, 27*(2), 139–150.

Venville, G. J., Wallace, J., Rennie, L. J., & Malone, J. (2002). Curriculum integration. *Studies in Science Education, 37*(1), 43–83.

Chapter 6
Educational Models and Differences between Groups of 16-year-old Students in Gender, Motivation, and Achievements in Chemistry

Iztok Devetak and Saša A. Glažar

Introduction

Learning chemistry is strongly connected with building knowledge through understanding and concepts linking in students' long-term memory by interpreting multimodal representations of chemical phenomena (Ainsworth 1999; Russell and McGuigan 2001). Students who recognized relationships between different representations demonstrated better conceptual understanding than students who lacked this knowledge (Prain and Waldrip 2006). Students should be also able to translate one representation into another and co-ordinate their use in representing scientific knowledge (Ainsworth 1999). Russell and McGuigan (2001) argue that learners need opportunities to generate various representations of a concept, and to recode these representations in different modes, as they refine and make more explicit their understanding (e.g., teacher should use a physical model, virtual model, and submicrorepresentation of a specific molecule). diSessa (2004) also points out that the quality of the representation ought to be evaluated according to its purpose. Waldrip et al. (2006) argue that, in order to maximize the effectiveness of designed representational environments, it is necessary to take into account the diversity of learner's background knowledge, expectations, preferences, and interpretive skills.

Representations of the chemical concepts could be defined on three levels (i.e., macro, submicro, and symbolic level). Adequately merged, these representations can help students to develop a conceptual understanding of chemical phenomena. The Interdependence of Three Levels of Science (ITLS) concepts model shows these connections between different representations and the role of visualization methods used in the process of mental model construction of chemical phenomena that students ought to develop. (e.g., using computer to view the experiment and animation to illustrate the particle animation in the reaction mixture and product

I. Devetak (✉) · S. A. Glažar
Faculty of Education, University of Ljubljana, Kardeljeva pl. 16, 1000 Ljubljana, Slovenia
e-mail: Iztok.devetak@pef.uni-lj.si

I. Devetak and S. A. Glažar (eds.), *Learning with Understanding in the Chemistry Classroom*, DOI: 10.1007/978-94-007-4366-3_6,

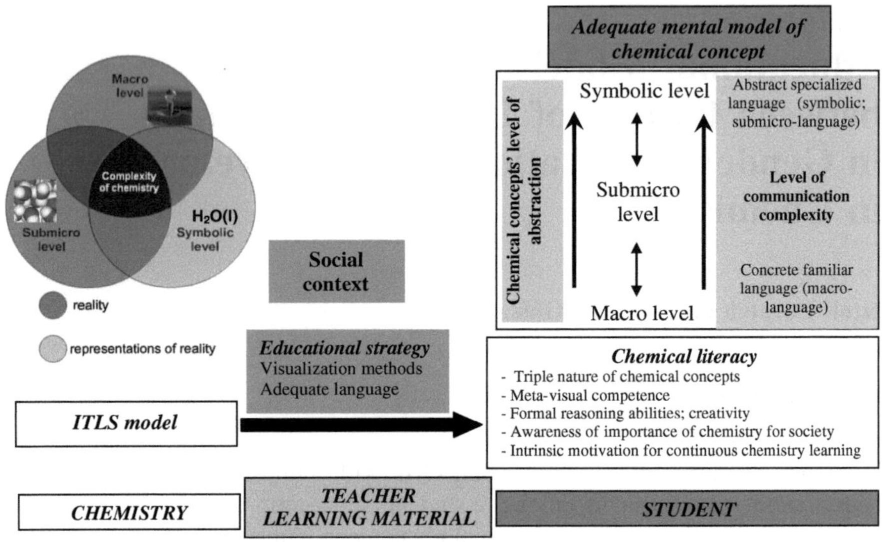

Fig. 6.1 Teaching and learning chemistry model (adapted from Juriševič et al. 2008; Gilbert and Treagust 2009)

formation). The ITLS model draws on different educational theories, such as Paivio's dual coding theory, Mayer's SOI model of meaningful learning, Johnstone's model of information processing, cognitive theory of multimedia learning, and Mayer's theory of effective illustrations (for more details, see Devetak et al. 2009b) (Fig. 6.1).

To illustrate chemical concepts on the level of particles, submicrorepresentations (SMRs) can be used and can be presented as static or dynamic modes of representations. Research shows (Bunce and Gabel 2002; Tien et al. 2007; Kelly and Jones 2008; de Berg 2012; Becker et al. 2013) that those students who were exposed to SMRs during the educational process more adequately understand the nature of the particle interactions compared to those who learned the same concepts only by reading textbooks. Studies in the last two decades (Williamson and Abraham 1995; Johnson 1998; Chittleborough et al. 2002; Solsona et al. 2003; Papageorgioua and Johnson 2005; Stains and Talanquer. 2007; Tien et al. 2007; Kelly and Jones 2008; Devetak et al. 2009; Davidowitz et al. 2010; Devetak and Glažar 2010a; Kern et al. 2010; de Berg 2012; Ramnarain and Joseph 2012; Becker et al. 2013) also show that students have many difficulties in understanding the submicro and symbolic levels of chemical concepts, and that previous knowledge of a specific topic has an influence on integrating new science concepts into students' mental structure. It is also important to emphasize that a lot of different factors influence students' achievement on different pictorial test questions (Halakova and Prokša 2007; Sanger and Phelps 2007; Stains and Talanquer 2008) and that the students' knowledge evaluation part of the educational process needs further research. Research so far also shows that teachers use mostly the symbolic level of chemical concepts to teach chemistry (Williamson and Abraham 1995; Chittleborough et al. 2002). It is important to introduce different presentations to illustrate abstract science concepts to students at the beginning of science education—age 10 or 11 (Longden et al. 1991)—thus also the application of SMRs (Papageorgioua and Johnson 2005).

Thiele and Treagust (1994) report that students who cannot visualize the chemical phenomena and/or do not have properly developed formal reasoning abilities cannot properly understand chemical concepts. They argue that those concepts are hard for those students to understand, and are unattractive and pointless for them. This means that the learning content has not been understood in depth and it is difficult to build on (e.g., students should be aware of the importance to understand basic concepts related to chemical reaction if they are able to understand ozone concentration depletion problems in the stratosphere). According to some research results (Wu and Shah 2004), the significant correlation between spatial ability and chemistry problem-solving skills is based on general reasoning abilities or intelligence rather than on visuospatial thinking. Statistically significant correlations were proven between formal reasoning abilities and students' chemical knowledge (CK) especially on submicro level (Haidar and Abraham 1991; Williamson and Abraham 1995). Wu and Shah (2004) also reported no statistically significant correlations between students' achievements on the test with static SMRs and spatial abilities. They anticipated that the knowledge achievement depends more on students' prior knowledge and the general cognitive factor than on visualization abilities.

A negative relationship toward chemistry does not enable proper concept change and/or modification of students' mental models of chemical phenomena. Students often do not have a proper knowledge base to upgrade their knowledge of increasingly more abstract chemical concepts when they progress on the educational vertical (Treagust et al. 1998). This situation again does not lead students toward building CK with deeper understanding (e.g., basic atom structure understanding is the basis for understanding chemical bonds). According to Ryan and Deci (2000), intrinsic motivation is an individual's inherent inclination from which stems his/her tendency to learn about particular areas of life regardless of the presence of external enticements. This construction encourages humans to '… assimilate, control, generate spontaneous interests and to research, which makes it essential for the individual's social and cognitive development while on the other hand it represents the fundamental source of personal satisfaction and life energy.' (p. 70). Highly intrinsically motivated students are more successful in learning new concepts and show better understanding of the learning matter (Stipek 1998). Rennie (1990), drawing upon the research on science learning, also concluded that higher results in science are related to learners' active engagement in learning tasks, to their positive attitudes toward the subject and to a highly positive self-concept in science, which all imply the learner's intrinsic motivation to learn. This is especially important, since many researchers (Anderman and Young 1994; Zusho et al. 2003) report that the decrease in intrinsic motivation with years of schooling is particularly noticeable in mathematics and science and reaches its peak in the period of early adolescence. Keig and Rubba (1993) pointed out that motivation can be a potential source of variance on students' chemistry knowledge achievements. These claims were confirmed by Tuan et al. (2005) and Devetak and Glažar (2010a) as they reported that from 7 to 16 % of variance on the science knowledge test could be explained by students' motivation, and that 9.4 % of the achievement test score variance can be accounted for by students' level of intrinsic motivation for learning chemistry, respectively. But on the other hand Nieswandt (2007) reported no statistically significant effect of students' affective variables (situational interest, attitudes toward chemistry and students chemistry-specific self-concept) on their understanding of Grade 9 (age 15 and 16) chemistry concepts.

It is also important to emphasize that there are statistically significant differences in achievements in chemistry knowledge tests between 16-year-old males and females (Devetak and Glažar 2010a). Research (Anderman and Young 1994; Meece and Jones 1996) also shows that gender differences in motivation for science learning are connected with achievements on the standardized test of science knowledge. It was also established that girls show lower interest in science, that science is boring for them, especially because they just have to learn everything by heart. Simpson and Oliver (1990) argued that girls possess lower levels of self-confidence in demonstrating their science knowledge. On the other hand, Meece and Jones (1996) did not confirm these results; they established that there is no difference between girls and boys regarding the interest in learning science and they also pointed out that gender influence on motivation and its effect on the manifestation of science knowledge are more complex processes than other researchers try to show.

Problem of the Research, Research Question, and Research Hypothesis

The basic purpose of the research was to design predictive models that can explain the differences between students' gender, level of motivation for learning chemistry, and school chemistry achievements. After determining which variables influence students' classification into different groups, specific and more targeted educational strategies for chemistry teaching and learning can be suggested.

According to the purpose of this study, the main research question was: Which independent variables discriminate between two groups of students regarding specific criteria (e.g., gender, motivation for learning chemistry, and school chemistry achievement)?

From the research questions, three research hypotheses were set up:

H1: It can be expected that numerical variables measuring motivation for learning science subjects and languages and school achievements discriminate between 16-year-old male and female students

H2: It can be hypothesized that that numerical variables measuring motivation for learning chemistry on different levels of chemical concepts' presentation discriminate between 16-year-old students who are more and less intrinsically motivated for chemistry learning in general

H3: It can be estimated that numerical variables measuring motivation for learning chemistry in general, school achievements at biology and physics and formal reasoning abilities discriminate between more and less successful 16-year-old students at school chemistry.

Method

Participants

A total of 386 secondary school students (60.6 % females; 39.4 % males) participated in the study. On average, the students were 16.3 years old (M = 195.4 months; SD = 5.7 months). All students attended second year of the general type of secondary school (grammar school). The chemistry curriculum of the grammar school is common to all students. The students were attending the fourth year of chemical education in the period when testing was performed (2 years in higher elementary school—age 13 and 14, and 2 years in secondary school—age 15 and 16). The sample represented a predominantly urban population with mixed socioeconomic status. Parents' basic educational background was diverse (3.1 % with completed primary school; 45.1 % with completed secondary school; 43.0 % university and 7.3 % with other formal education). Among the parents only 11.6 % had finished some kind of science or technology education.

Instruments

Students' abilities to read and draw the SMRs (chemical knowledge at three levels of concepts' presentation—macro, submicro, and symbolic) were measured using the diagnostic instrument for determining CK. The instrument comprised 19 items. Eight items required reading and 11 items required drawing SMRs in solving the chemistry problems considering the ITLS model. The CK included four different topics: pure substances and mixtures (4 items), chemical reactions (6 items), water solutions (4 items), and electrolyte chemistry (5 items)—see Appendix 1. The CK showed satisfactory measuring characteristics (i.e., internal consistency reliability—Cronbach's alpha was 0.80; discriminate indexes for every item between 0.21 and 0.80 were all statistically significant). Kurtosis and skewness coefficients show normally distributed data. Students had 60 min to solve the CK.

Four different tests and a questionnaire (Test of Logical Thinking (TOLT), Rotations (RO), Patterns (PA), and Intrinsic Motivation for Learning Science questionnaire (IMLS)) were administered to the students to measure their mental abilities and motivation for chemistry learning.

The level of students' formal reasoning abilities was obtained by the TOLT (Tobin and Capie 1981). The TOLT is a 10-item group paper–pencil test. The authors of the test reported a strong correlation ($r = 0.82$; $p < 0.0001$) between performance on tasks during Piagetian clinical interviews that are considered a traditionally preferable method in measuring individuals' formal reasoning abilities and the results on the TOLT. The TOLT has high internal consistency reliability (Cronbach's alpha was 0.85). The test consists of two items designed to measure each of the five modes of reasoning (i.e., controlling variables, proportional, correlational, probabilistic, and combinatorial reasoning). The test scores from 0 to 1 points (concrete reasoners), 2 to 3 points (transitional reasoners), and 4 to 10 points (formal reasoners) were used as a basis for classifying the students. Students had 38 min to solve the test.

The students' visualization abilities were measured with two tests: PA and RO (Pogačnik 1998, 2000), where the PA measures students' speed of perception and the RO measures students' spatial relations abilities. Both tests were developed based on the Cattell–Horn theory of mental abilities. The PA is a 36 item group paper–pencil test. It requires individuals to find and mark exactly the same pattern among the four similar patterns on the right side of the paper to the one on the left part of the paper as quickly as possible. The PA has high internal consistency (Cronbach's alpha was 0.86). Correlations between some other instruments for determining individuals' perception abilities *(BTI-Or; BTI-Pr, Beta 6* and *4)* determine that the instruments' validity is higher and statistically significant. Students had 4.5 min to solve the test. The RO is a 90 item group paper–pencil test. The RO requires individuals to find and encircle those patterns on the right side of the paper that are only rotated in comparison with the left pattern. Individuals have to cross those patterns that are not only rotated in the plane but represent a different pattern. Cronbach's alpha for the RO was 0.94. Correlations between some other instruments for determining individuals' perception abilities *(BTI-Pr, Beta 4)* were also high and statistically significant. Students had 6 min to solve the test.

The last independent variable, the intrinsic motivation for learning chemistry, was measured by the IMLS questionnaire. There are many questionnaires to measure students' attitudes or interests in science and/or chemistry (e.g., Moore and Foy 1997; Tuan et al. 2005; Coll et al. 2002; Nieswandt 2007). All these instruments show a rather general structure of students' attitudes toward science, but they lack the dimension with reference to the ITLS model and separately for different science school subjects. These questionnaires do not show sufficient specific characteristics regarding the research questions asked in this study and would need extensive revision for adapting the instrument to secondary level. For these reasons, a new instrument for measuring intrinsic motivation, a 125-item IMLS questionnaire, was developed. The response to each item is on a five-point

Likert-type scale ranging from 1, as strongly disagree, to 5 as strongly agree. The internal consistency (Cronbach α) of IMLS was 0.78. Students had 20 min to complete the questionnaire. Three sample items of each component of intrinsic motivation from the IMLS questionnaire are included in Appendix 2.

Research Design

The research was a nonexperimental, cross-sectional, and descriptive study (Bryman 2004).

Students had received no special teaching about using SMRs in the chemistry classroom. The chemical concepts comprised in the CK were not instructed using SMRs by the teachers who taught the students participating in the study.

Chemical knowledge and IMLS were designed specifically for this study. The CK was administered to two university chemistry and chemical education teachers. Their responses provided scientifically correct answers and content validation for the instrument. The IMLS was distributed to two experts in science education and one in educational psychology. Their evaluation of the instrument confirmed that the IMLS can measure students' intrinsic motivation for learning, and their analysis provided validation for the questionnaire. The Slovene translation of the TOLT was used for the study.

After all the instruments had been developed or chosen in relation to the purpose of the study, a pilot study was conducted with 77 students. The CK, TOLT, and IMLS were used in the pilot study. Taking into account the statistical analysis of the results from the pilot study, the CK and IMLS were modified.

All instruments were applied on the research sample at the end of the school year 2005/06. The testing took students about 135 min on two separate days. Students solved the IMLS and CK in the first week, and in the second one they solved the TOLT, RO, and PA. The last testing was conducted by a trained psychologist. All instruments were applied in a group and under normal examination conditions.

In this chapter three different discriminant analyses were used to form predictive models according to 42 predictor variables. Discriminant analysis can be used when you wish to explore the predictive ability of a set of independent variables on one categorical dependent measure. This means that you want to show which variables best predict group membership (Pallant 2005). The canonical discriminant analysis was performed to determine which of the 42 predictor variables discriminate between students participating in the study (Table 6.1).

Students were classified into two groups based on genderand level of motivation for learning chemistry, and four groups according to their school chemistry achievements. The average values of selected variables were used for students' classification. A stepwise procedure was used in discriminant analysis based on Wilks' lambda (λ). The statistical significance of covariance matrices was determined by Box's M test. In the following step, the structure matrix was calculated

Table 6.1 Predictive variables and instruments

Variable abbreviation	Variable description	Instrument
Ach.	General school achievement	IMLS
Ach.Bio.	School achievement (biology)	IMLS
Ach.Chem.	School achievement (chemistry)	IMLS
Ach.Phy.	School achievement (physics)	IMLS
Ach. FL1	School achievement (foreign language 1—usually English)	IMLS
Ach.FL2	School achievement (foreign language 2—other than English)	IMLS
Ach.Math.	School achievement (physics)	IMLS
MVg	Motivation for learning	IMLS
MVchem	Motivation for learning chemistry	IMLS
MVbio	Motivation for learning	IMLS
MVphy	Motivation for learning	IMLS
MVfl	Motivation for learning	IMLS
MVmath	Motivation for learning	IMLS
MVmac	Motivation for learning symbolic macrolevel of chemistry	IMLS
MVsub	Motivation for learning submicrolevel of chemistry	IMLS
MVsym	Motivation for learning symbolic level of chemistry	IMLS
TOLT	Formal reasoning abilities	TOLT
PA	Visualization abilities; speed of perception abilities	PA
RO	Visualization abilities; spatial relations abilities	RO
TNZs	Chemical knowledge	CK
TNZsα	Chemical knowledge—reading SMRs	CK
TNZsβ	Chemical knowledge—drawing SMRs	CK
TNZsub	Chemical knowledge—items at submicrolevel	CK
TNZsubα	Chemical knowledge—items at submicrolevel—reading SMRs	CK
TNZsubβ	Chemical knowledge—items at submicrolevel—drawing SMRs	CK
TNZmac-sub	Chemical knowledge– items connecting macro- and submicrolevel	CK
TNZmac-subα	Chemical knowledge—items connecting macro- and submicrolevel—reading SMRs	CK
TNZmac-subβ	Chemical knowledge—items connecting macro- and submicrolevel—drawing SMRs	CK
TNZsub-sym	Chemical knowledge—items connecting submicro- and symbolic level	CK
TNZsub-symα	Chemical knowledge—items connecting submicro- and symbolic level—reading SMRs	CK
TNZsub-symβ	Chemical knowledge—items connecting submicro- and symbolic level—drawing SMRs	CK
TNZmac-sub-sym	Chemical knowledge—items connecting macro- submicro- and symbolic level	CK
TNZmac-sub-symα	Chemical knowledge—items connecting macro- submicro- and symbolic level—reading SMRs	CK
TNZmac-sub-symβ	Chemical knowledge—items connecting macro- submicro- and symbolic level—drawing SMRs	CK
TNZund	Chemical knowledge—items testing understanding	CK
TNZuse	Chemical knowledge—items testing using	CK

(continued)

Table 6.1 (continued)

Variable abbreviation	Variable description	Instrument
TNZana	Chemical knowledge—items testing analysis	CK
TNZpsm	Chemical knowledge—items testing pure substances and mixtures	CK
TNZcr	Chemical knowledge—items testing chemical reactions	CK
TNZsol	Chemical knowledge—items testing solutions	CK
TNZads	Chemical knowledge—items testing acid, bases, and salts	CK
TNZcomb	Chemical knowledge—items testing combination of different knowledge	CK

to show the correlations of each variable with discriminant function. According to the criteria (Box's M test, covariant matrix, the significance of discriminant function, centroids and the percentage of correct classification of students into groups) which determine the relevance of the discriminant function, the discriminant functions were selected and are presented below. The larger the eigenvalue, the more the variance in the dependent variable is explained by that function and/or, the better is the discrimination among groups. In the next step, the percent of variance explained by each function (the squared canonical correlation is the percent of variation in the dependent variable discriminated by the independent variables) was considered. Canonical correlation represents the relation between the discriminant scores and the levels of the dependent variable. A high correlation indicates a function that can discriminate well (Green et al. 2000).

Results

Differences Between Male and Female Students

In the first discriminant analysis, we determined significant differences in achievements between male and female students, regarding the 42 manifest variables. Box's M test is 21.91 ($F = 1.44$; $p = 0.119$).

There is only one discriminant function because there were two groups of students (male and female) analyzed (Table 6.2). Wilks' lambda is high and the Chi square test of significance of the function ($\chi^2 = 113.64$; $p = 0.000$) assessing whether there are significant differences among male and female students across the predictor variables, is significant. The discriminant function has an eigenvalue of 0.356 and the canonical correlation of 0.512. This shows moderate correlation between both groups of students and the determined discriminant function, and that the 26 % discriminant function variance can be accounted for by the differences between males and females.

Five variables were determined in five steps (Table 6.3).

Table 6.2 Statistical parameters for the discriminant function

Function	Eigenvalue	Canonical correlation	Variance explained (%)	Wilk's lambda	χ^2	p
1	0.356	0.512	26.2	0.738	113.64	0.000

Table 6.3 Structure matrix of correlations between discriminating variables and discriminant function

Discriminating variable	r
Motivation for physics	−0.617
Motivation for foreign language	0.510
School achievement (foreign language 2)	0.325
School achievement (chemistry)	−0.230
Motivation for biology	0.141

It can be summarized from Table 6.3 that discriminating variables, such as motivation for physics, foreign language, and biology, school achievements in foreign language and chemistry significantly discriminate between male and female students in the second year of secondary school. The level of correlations of the first two variables (motivation for physics and motivation for foreign language) show significant meaning of the variable to discriminate between the two groups of students. Boys are more intrinsically motivated for learning physics (negative correlations shows that girls are unlikely intrinsically motivated for learning physics and obtain good grades in chemistry), while girls show higher motivation for learning foreign language. According to the moderate level of correlation, it is possible to determine also the variables indicating foreign language (benefit to female students) and chemistry school (benefit to male students) achievements, and also motivation for biology although the correlation is low, but it still significantly contributes to the discriminant function. It can be summarized that boys are strongly motivated by the physical–chemical part of school science, while girls show higher interest toward the more humanistic, or language-oriented section of the education.

Table 6.4 summarizes group centroids on the discriminant function. It is important for the centroids to be well apart to show whether the discriminant function is clearly discriminating. The closer the mean values, the more errors of classification can be expected. The values addressing the group centroids of this discriminant function are for male and female students and show that the discriminant function discriminates well between boys and girls who participated in this study.

The results presented in Table 6.5 can be used to estimate how well the classification functions derived from all cases could be predicted in a new sample. The average classification results of the discriminant function indicate that 74.2 % of students were correctly classified into the groups according to their gender. This means that the actual classification of students does not correspond with the predictive one in 25.8 % on the basis of the discriminate function.

Table 6.4 Centroids (mean values) for canonical discriminant function according to the selected groups of students

Group	Centroids
Girls	0.477
Boys	−0.741

Table 6.5 Structure of classification of students into the selected groups according to gender

Actual group	Predictive group			
	Girls		Boys	
	f	%	f	%
Girls	174	73.1	64	26.9
Boys	37	24.2	116	75.8

Differences Between More and Less Motivated Students for Learning Chemistry

In the second discriminant analysis significant, differences in the 42 manifest variables between more and less motivated students for learning chemistry were determined. Students were divided into two groups according to IMLS questionnaire scores. Box's M test is 19.98 ($F = 1.31$; $p = 0.184$).

There is only one discriminant function, since we analyzed two groups of students (more and less motivated ones for chemistry learning) (Table 6.6). Wilks' lambda is high and the Chi square test of significance of the function ($\chi^2 = 236.54$; $p = 0.000$), assessing whether there are significant differences among students' level of intrinsic motivation for learning chemistry across the predictor variables, is significant. The discriminant function has a high eigenvalue of 0.884 and the canonical correlation of 0.685 is strong. This shows strong correlation between both groups of students and the determined discriminant function. These parameters show that almost 47 % of discriminant function variance can be accounted for by the differences between more or less motivated students for learning chemistry in general.

In five steps five variables were determined (see Table 6.7).

As can be seen from Table 6.7, the discriminating variables, such as motivation for different levels of chemical concept presentations, discriminate most between more and less motivated 16-year-old students for learning chemistry in general. The results show that correlations between discriminating variables, such as intrinsic motivation for learning chemistry at the symbolic level and the discriminant function are the strongest ($r = 0.861$). It can be confirmed that those students who are more intrinsically motivated for learning chemistry at different levels of chemical concepts are also more intrinsically motivated for learning

Table 6.6 Statistical parameters for the discriminant function

Function	Eigenvalue	Canonical correlation	Variance explained (%)	Wilk's lambda	χ^2	p
1	0.884	0.685	46.9	0.531	236.54	0.000

Table 6.7 Structure matrix of correlations between discriminating variables and the discriminant function

Discriminating variable	r
Motivation for symbolic level of chemical concepts	0.861
Motivation for submicro level of chemical concepts	0.842
Motivation for macro level of chemical concepts	0.729
Student achievements on items evaluating pure substances and mixtures understanding	0.062
Formal reasoning ability	−0.023

Table 6.8 Centroids (mean values) for the canonical discriminant function according to the selected groups of students

Group	Centroids
Less motivated	−0.913
More motivated	0.963

chemistry in general. From these results, we can conclude that secondary school students perceive chemistry as a science of symbols. It can be also estimated from the discriminating variables which correlate strongly with this canonical discriminant function, that chemistry courses are, according to students' opinion, dedicated mostly to the symbolic level of chemical concept presentations (Table 6.8).

Group centroids on the discriminant function regarding students' motivation for learning chemistry in general, which are well apart, show that the discriminant function is clearly discriminating between more and less motivated students for learning chemistry.

The results presented in Table 6.9 can be used to estimate how well the classification functions derived from all cases could be predicted in a new sample. The average classification results of the discriminant function indicate that 83.9 % of students were correctly classified into the groups according to their motivation for learning chemistry. This means that the actual classification of students does not correspond with the predictive one in 16.1 % on the basis of the discriminate function.

Table 6.9 Structure of classification of students into the selected groups according to students' motivation for learning chemistry

Actual group	Predictive group			
	Less motivated		More motivated	
	f	%	f	%
Less motivated	169	83.7	33	16.3
More motivated	30	15.9	159	84.1

Differences Between More or Less Successful Students in School Chemistry

In the third discriminant analysis, we determined the significant differences in the 42 manifest variables between students regarding their school chemistry achievements. Students were divided into four groups according to school chemistry grades (pass, good, very good, and excellent), and three canonic discriminant functions were obtained. Box's M test is 46.087 ($F = 2.52$; $p = 0.000$) and it is statistically significant; however, other criteria indicated that the discriminant function could be further interpreted.

There are three discriminant functions, since we analyzed four groups of students classified according to their school chemistry grade (pass, good, very good, excellent) (Table 6.10). Wilks' lambda is low only for the first function, however the Chi square test of significance is significant only for this function ($\chi^2 = 286.651$; $p = 0.000$) assessing whether there are significant differences among students' grade in school chemistry across the predictor variables. According to these data, only the first canonical function is suitable for interpretation. The discriminant function has a high eigenvalue of 1.143 and the canonical correlation of 0.73 is strong. This shows a strong correlation between both groups of students and the determined discriminant function, meaning that the four groups of students regarding their school chemistry achievements are well differentiated according to the group of predictive variables. These results show that 53 % of the discriminant function variance can be accounted for by the differences between students whose success in school chemistry is different. Three variables were determined in three steps (see Table 6.11).

From the results in Table 6.11, we can conclude that discriminating variables such as general school achievement, achievement in physics, and motivation for chemistry learning, discriminate most between students regarding their school chemistry achievements in the first grade of secondary school. Among the extracted variables, the highest discriminating values are shown between students with different school chemistry success variables such as general school achievement, and school physics achievement. According to the moderate level of correlations ($r = 0.4$), the variable of motivation for chemistry learning, which contributes a relevant part of the first discriminant function, can be also taken into account.

Table 6.10 Statistical parameters for the discriminant functions

Function	Eigenvalue	Canonical correlation	Variance explained (%)	Wilk's lambda	χ^2	p
1	1.143	0.730	53.3	0.461	286.65	0.000
2	0.009	0.095	0.90	0.989	4.271	0.371
3	0.002	0.049	0.24	0.998	0.880	0.348

Table 6.11 Structure matrix of correlations between discriminating variables and first discriminant function

Discriminating variable	r
General school achievement	0.873
Motivation for chemistry learning	0.395
Achievement in physics	0.650

Table 6.12 Centroids (mean values) for the canonical discriminant function according to the selected groups of students (their school chemistry achievements)

Group	Centroids
Sufficient	−1.800
Good	−0.700
Very good	0.333
Excellent	1.523

Table 6.12 presents group centroids on the discriminant function determining students' achievements in school chemistry. The values addressing the group centroids of this discriminant function show a linear pattern of discrimination according to the increasing value of students' school chemistry achievements. Centroids between groups of students with pass and excellent achievements in school chemistry are well apart, and show that the first discriminant function clearly discriminates between those groups of students.

From the results presented in Table 6.13, we can see that the average classification results of the discriminant function indicate that 55 % of students were correctly classified into the groups according to their school chemistry

Table 6.13 Structure of classification of students into selected groups according to students' school chemistry achievements

Actual group	Predictive group							
	2		3		4		5	
	f	%	f	%	f	%	f	%
2	44	84.6	7	13.5	1	1.9	0	0
3	33	28.7	36	31.3	41	35.7	5	4.3
4	13	9.5	14	10.2	72	52.6	38	27.7
5	1	1.2	4	4.8	18	21.4	61	72.6

achievements in the first grade of secondary school. This means that the actual classification of students does not correspond with the predicted one in 45.1 % on the basis of the first discriminate function. This classification shows that it is important to take into account the fact that students' achievements in school chemistry cannot be considered as a good predictive variable to classify students into groups regarding their chemistry knowledge tested by the instrument (CK) which we used in this study. This discriminate analysis also shows that the students school chemistry achievements' variance has a different source than the variance of the test results obtained in this research. This can be also concluded according to the parameters of all discriminant functions, and it is possible to estimate that school chemistry achievements do not have great discriminate value.

Discussion and Implications for Education

The purpose of this study was to identify differences between groups of 16-year-old students classified into different groups regarding their gender, motivation for learning chemistry, and school chemistry achievements. Three hypotheses were set up and according to the results of the study all the hypotheses can be confirmed.

The first hypothesis stated: "*It can be estimated that numerical variables measuring motivation for learning science subjects and language and school achievements discriminate between 16 year-old male and female students.*" This hypothesis can be confirmed because, the discriminate analysis shows that the most discriminating value between male and female students is shown by variables such as: motivation for physics learning (MVphy) that shows in the favor of boys, and the variable of learning foreign language (MVfl) that is in favor of girls. Variables such as school chemistry achievements (Ach.Chem) (in favor of boys) and foreign language school achievements (Ach.FL1) (in favor of girls) show a lower level of discrimination value between boys and girls. It can be concluded that the differences between boys and girls are more obvious in motivation for physics and less for chemistry.

There are no differences between boys and girls in motivation for biology. Similar results were obtained by other researchers (Steinkamp 1984; Dweck 1986 cited by Meece and Jones 1996). They concluded that girls show lower interest in science than boys, because girls think that it is important to learn science concepts only by rote, which is intellectually not challenging. The research also shows that girls do not demonstrate a high level of self-confidence in solving science problems and they think that mathematics and science in general are more in the domain of male students.

The second hypothesis says that *"it can be hypothesised that that numerical variables measuring motivation for learning chemistry on different levels of chemical concepts presentation discriminate between 16 year-old students who are more and less intrinsically motivated for chemistry learning in general"*. This hypothesis has been confirmed, because students are moderately intrinsically motivated for chemistry $(\overline{X} = 37.25;\ \max = 70)$, and the variables that differentiate between those students are: (1) motivation for symbolic (MVsym), (2) submicroscopic (MVsub), and (3) macroscopic (MVmac) level of chemistry concepts. From the results of discriminative analysis, we can infer that students with higher intrinsic motivation for learning chemistry are better motivated for the symbolic level of chemistry concept, which is in fact the reflection of school chemistry practice (e.g., chemisty teaching is higly verbal, students usually do not have opportunities to experiment, participate in the activities in the social context, do not learn by inquiry…). The results of the interviews with teachers, reported in the study by Devetak et al. (2009), show that elementary and secondary school teachers mainly use symbols of elements, chemical formulae, and equations in presenting chemistry concepts in teaching and learning environment that does not stimulate activities for constructing students' knowledge based on their previous understanding of specific concepts. The large groups (up to 35 students in a class), frontal, and mostly verbal, approach to teaching, make chemistry influence on students understanding of chemistry. Such educational strategies are negatively reflected on students' interest in chemistry, whereby students perceive chemistry as a science of symbols, formulae, and equations, without any contextual importance for their professional development or everyday live, and so, they do not consider these as an important factor in building their own knowledge base. Similar findings have been reported by Lee (1999), Mulford and Robinson (2002).

The third hypothesis which estimates that *"numerical variables measuring motivation for learning chemistry in general, school achievements at biology and physics and formal reasoning abilities discriminate between more and less successful 16-year-old students at school chemistry"* has also been confirmed, because the third discriminative analysis divided students according to their achievements in chemistry (grades from sufficient to excellent). The following variables were excluded: (1) general learning achievement (Ach.), (2) general motivation for learning chemistry (MVchem), and (3) grades achieved in physics (Ach.Phy). Those students who achieved better results in chemistry also had a better general learning outcome and were better at physics, and were also more motivated for learning chemistry. This analysis showed that the factors which influence the results in chemistry are the general learning outcomes and achievements in physics. Motivation for learning chemistry has lesser discriminative power; however, it does indicate that students with higher intrinsic motivation for learning chemistry will generally achieve better results in chemistry, which also confirms the results referring to correlations between student achievements in chemistry and student intrinsic motivation (Devetak and Glažar 2010a).

It is important to emphasize that chemistry knowledge variables measured by CK does not influence on the predictive models, discriminating between students gender, motivation for chemistry learning, and school achievement in chemistry, selected to be analyzed in this chapter.

According to the results from our study, some implications for teaching can be suggested. Between the two genders, intrinsic motivation for learning physics and foreign language have the strongest discriminative power; boys have stronger motivation for learning physics and girls are more motivated for foreign languages. Considering a moderate correlation level between the variable and discriminative function, it is possible to exclude the achievements in foreign language and chemistry, however, there is a significantly smaller correlation between discriminative function and intrinsic motivation for biology.

Considering that the highest discriminatory power between students with higher or lower intrinsic motivation for learning chemistry is particularly evident in the variable motivation for the symbolic level, it is possible to infer that students correlate chemistry mainly with the symbolic component of chemistry concepts. Considering this, teachers should make efforts to develop students' abilities for solving problems that integrate all the three components of the STRP model, which requires deeper comprehension of chemistry concepts. In developing strategies for solving such problems it would be also necessary to emphasize the submicrolevel of chemistry concepts at higher cognitive levels (e.g., using tasks comprising SMRs where students have to integrate knowledge of different concepts and have to use analysis, synthesis, and evaluation of data to solve them). Teachers should primarily encourage reading and understanding of SMRs, and later upgrade this knowledge with students' active drawing to present chemical concepts and phenomena at a particulate level. Teachers should extrinsically motivate students using macroscopic level of chemical concepts and put those concepts into the context for those students who are low achievers and have low intrinsic motivation for learning chemistry. Teachers could also apply problem-solving strategies according to the specific forms of teaching, e.g., group or pair work (e.g., the *GALC* approach (Devetak and Glažar 2010b), projects and experiments, as well as field work. Building upon students' previous knowledge and experience a cognitive conflict would be triggered, which would also allow for rectifying any misconceptions formed before. By suitable application and active use of the STRP model it would be possible to create new and meaningful integration of all the three levels of chemistry concepts and, based on new experience, students would be able to form permanent and professionally correct correlations between different concept levels. New correlations in the students' mental model would allow them to implement different efficient strategies for solving chemistry problems, thus making chemistry more meaningful and linked with everyday life. This would also increase intrinsic motivation for learning sciences, particularly chemistry. Since girls are intrinsically less motivated for learning sciences, chemistry and physics, teachers should try to do more to motivate girls for learning these two subjects. Some researchers (Papageorgiou and Johnson 2005) believe that it is possible to implement the STRP model into teaching science already at an early stage, since at the age of 11 students are able to understand particulate matter. Therefore, teachers should make sure to use appropriate educational material to discourage the formation of a naïve mental model of the perception of substance. One method is using diagnostic questions to encourage students to engage in discussions in which they can be confronted with their own mental model of a particular science concept. If teachers use student responses to relevant questions during further steps of the teaching process this can support and maximize active learning. By being presented with carefully planned teaching contents, students would become able to upgrade their knowledge, generalize the knowledge, and apply it to other examples (highly developed scientific literacy), not only those presented in the class. Scientific literacy can only be developed if teachers use appropriate science language and encourage their students to express themselves in a professionally correct and accurate language when describing natural phenomena or science concepts.

It can be concluded that teacher should put more emphasize to development of such lessons supported by different learning materials that students can use in group work in an active way that the differences between students will be diminished as much as possible. This would lead to more in-depth understanding of chemical concepts in a students' meaningful context.

Appendix 1: Sample Items from the Diagnostic Instrument for Determining Chemical Knowledge (CK)

Appendix 1: Sample items from the diagnostic instrument for determining *Chemical Knowledge (CK).*

1. The scheme represents the reaction between substance A and B. Which equation correctly represents this reaction?

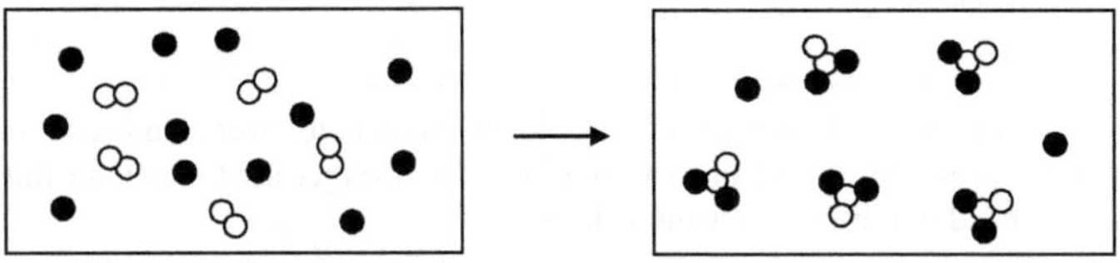

Mixture before the reaction   Mixture after the reaction

Legend: ● - Substance A; ⚮ - Substance B; - Product

A $A + 2B \rightarrow A_2B_2$
B $12A + 10B \rightarrow 6A_2B_2$
C $2A + 2B \rightarrow A_2B_2$
D $5A + 5B_2 \rightarrow 5A_2B_2 + 2A$
E $2A + B_2 \rightarrow A_2B_2$

Which substance was completely used during the reaction? ___________

Elaborate the answer: ____

2. Scheme A to C represents aqueous solutions of three different substances. Most of the water molecules were omitted for clarity.

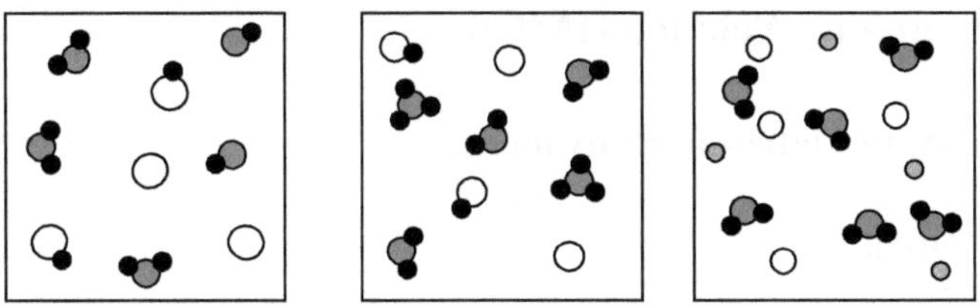

Legend:

- water molecule

● - hydrogen atom

Answer the following questions.

Which scheme represents an aqueous solution of acid? ___________
Which scheme represents an aqueous solution of base? ___________
Which scheme represents an aqueous solution of soluble salt? ___________

3. Draw the scheme of a chemical reaction product between two molecules of chlorine and two molecules of hydrogen in the box below.

Legend: ______________________________

Elaborate the answer: ______________________________

4. Scheme 1 represents the aqueous solution of an acid. Water molecules were omitted for clarity. Draw Scheme 2 representing the aqueous solution of a stronger acid, but with the same concentration. You need not draw water molecules.

Scheme 1 Scheme 2

Elaborate the answer: ______________________________

Appendix 2: Sample Items from the Questionnaire Intrinsic Motivation for Learning Science (IMLS)

1. Emotional component of intrinsic motivation:

I enjoy learning.

I am often bored during the:

… chemistry course.

… biology course.

… physics course.

… foreign language course.

… mathematics course.

I enjoy the chemistry course when:

… we observe chemical changes in experiments.

… we learn about particles (atoms, ions, and molecules).

… we learn and write chemical symbols, formulae, and equations.

2. Cognitive component of intrinsic motivation:

I often look for additional information about school science topics in books, magazines, on the Internet, CDs …

The media attract my attention when reporting on:

…chemistry topics.

...biology topics.
...physics topics.
...foreign language topics.
...mathematics topics.
I often think about:
...observation of chemical changes in experiments, *also out of school.*
... particles (atoms, ions, molecules), *also out of school.*
...learning and writing chemical symbols, formulae and equations, *also out of school.*

3. Challenge component of intrinsic motivation:
I persevere with learning.
New problems in:
... chemistry, *challenge me.*
... biology, *challenge me.*
... physics, *challenge me.*
... foreign language, *challenge me.*
... mathematics, *challenge me.*
If I do not understand something, connected with:
... observation of chemical changes in experiments, *I give up.*
... learning about particles (atoms, ions, and molecules), *I give up.*
... learning and writing chemical symbols, formulae, and equations, *I give up.*

References

Ainsworth, S. (1999). The functions of multiple representations. *Computers & Education, 33*(2–3), 131–152.

Anderman, E. M., & Young, A. J. (1994). Motivation and strategy use in science: Individual differences and classroom effects. *Journal of Research in Science Teaching, 31*(8), 811–831.

Becker, N., Rasmussen, C., Sweeney, G., Wawro, M., Towns, M., & Cole, R. (2013). Reasoning using particulate nature of matter: An example of a sociochemical norm in a university-level physical chemistry class. *Chemistry Education Research and Practice, 14*(1), 81–94.

Bryman, A. (2004). *Social research methods.* New York: Oxford University Press.

Bunce, D. M., & Gabel, D. (2002). Differential effects in the achievement of males and females of teaching the particulate nature of chemistry. *Journal of Research in Science Teaching, 39*(10), 911–972.

Chittleborough, G. D., Treagust, D. F., & Mocerino, M. (2002). *Constraints to the Development of First Year University Students' Mental Models of Chemical Phenomena. Teaching and Learning Teaching and Learning Forum 2002: Focusing on the Student.* Retrieved January 30, 2004, from http://www.ecu.edu.au/conferences/tlf/2002/pub/docs/Chittleborough.pdf

Coll, R. K., Dalgety, J., & Salter, D. (2002). The development of the chemistry attitudes and experiences questionnaire (CAEQ). Chemistry education; Research and practice in. *Europe, 3*(1), 19–32.

Davidowitz, B., Chittleborough, G., & Murray, E. (2010). Student-generated submicro diagrams: A useful tool for teaching and learning chemical equations and stoichiometry. *Chemistry Education Research and Practice, 11*(3), 154–164.

de Berg, K. (2012). A study of first-year chemistry students' understanding of solution concentration at the tertiary level. *Chemistry Education Research and Practice, 13*(1), 8–16.

Devetak, I., & Glažar, S. A. (2010a). The influence of 16-year-old students' gender, mental abilities, and motivation on their reading and drawing submicrorepresentations achievements. *International Journal of Science Education, 32*(12), 1561–1593.

Devetak, I., & Glažar, S. A. (2010b). Approach to developing the learning to learn strategy in chemistry. In M. Valenčič Zuljana & J. Vogrinc (Eds.), *Facilitating effective student learning through teacher research and innovation* (pp. 399–414). Ljubljana: Faculty of Education.

Devetak, I., Drofenik Lorber, E., Juriševič, M., & Glažar, S. A. (2009a). Comparing Slovenian year 8 and year 9 elementary school pupils' knowledge of electrolyte chemistry and their intrinsic motivation. *Chemistry Education Research and Practice, 10*(4), 281–290.

Devetak, I., Vogrinc, J., & Glažar, S. A. (2009b). Assessing 16-year-old students' understanding of aqueous solution at submicroscopic level. *Research in Science Education, 39*(2), 157–179.

diSessa, A. A. (2004). Metarepresentation: Native competence and targets for instruction. *Cognition and Instruction, 22*(3), 293–331.

Gilbert, J. K., & Treagust, D. F. (2009). Towards a coherent model for macro, submicro and symbolic Representations in chemical education. In J. K. Gilbert & D. F. Treagust (Eds.), *Multiple representations in chemical education* (pp. 333–350). Dortrecht: Springer.

Green, S. B., Salkind, N. J., & Akey, T. M. (2000). *Using SPSS for windows: Analyzing and understanding data*. Englewood Cliffs, NJ: Prentice Hall.

Haidar, A. H., & Abraham, M. R. (1991). A comparison of applied and theoretical knowledge of concepts based on the particulate nature of matter. *Journal of Research in Science Teaching, 28*(10), 919–938.

Halakova, Z., & Prokša, M. (2007). Two kinds of conceptual problems in chemistry teaching. *Journal of Chemical Education, 84*(1), 172–174.

Johnson, P. (1998). Children's understanding of changes of state involving the gas state, Part II: Evaporation and condensation below boiling point. *International Journal of Science Education, 20*(6), 695–709.

Juriševič, M., Devetak, I., Razdevšek-Pučko, C., & Glažar, S. A. (2008). Intrinsic motivation of pre-service primary school teachers for learning chemistry in relation to their academic achievement. *International Journal of Science Education, 30*(1), 87–107.

Keig, P. F., & Rubba, P. A. (1993). Translation of representations of structure of matter and its relationship to reasoning, gender, spatial reasoning, and specific prior knowledge. *Journal of Research in Science Teaching, 30*(8), 883–903.

Kelly, R. M., & Jones, L. L. (2008). Investigating students' ability to transfer ideas learned from molecular animations of the dissolution process. *Journal of Chemical Education, 85*(2), 303–309.

Kern, A. L., Wood, N. B., Roehrig, G. H., & Nyachwaya, J. (2010). A qualitative report of the ways high school chemistry students attempt to represent a chemical reaction at the atomic/molecular level. *Chemistry Education Research and Practice, 11*(3), 165–172.

Lee, K.-W. L. (1999). A comparison of university lecturers' and pre-service teachers' understanding of a chemical reaction at the particulate level. *Journal of Chemical Education, 76*(7), 1008–1012.

Longden, K., Black, P., & Solomon, J. (1991). Children's interpretation of dissolving. *International Journal of Science Education, 13*(1), 59–68.

Meece, J. L., & Jones, M. G. (1996). Gender differences in motivation and strategy use in science: Are girls rote learners? *Journal of Research in Science Teaching, 33*(4), 393–406.

Moore, R. W., & Foy, R. L. (1997). The scientific attitude inventory: A revision (SAIII). *Journal of Research in Science Teaching, 34*(4), 327–336.

Mulford, D. R., & Robinson, W. R. (2002). An inventory for alternative conceptions among first-semester general chemistry students. *Journal of Chemical Education, 79*(6), 739–744.

Nieswandt, M. (2007). Students affect and conceptual understanding in learning chemistry. *Journal of Research in Science Teaching, 44*(7), 908–937.

Pallant, J. (2005). *SPSS survival manual*. Berkshire: Open University Press.

Papageorgiou, G., & Johnson, P. (2005). Do particle ideas help or hinder pupils' understanding of phenomena? *International Journal of Science Education, 27*(11), 1299–1317.

Pogačnik, V. (1998). *Test hitrosti percepcije "Vzorci"*. Center za psihodiagnostična sredstva: Priročnik. Ljubljana.

Pogačnik, V. (2000). *Osebna izkaznica testa—Spacialni test Rotacije*. Ljubljana: Center za psihodiagnostična sredstva.

Prain, V., & Waldrip, B. (2006). An exploratory study of teachers' and students' use of multi-modal representations of concepts in primary science. *International Journal of Science Education, 28*(15), 1843–1866.

Ramnarain, U., & Joseph, A. (2012). Learning difficulties experienced by grade 12 South African students in the chemical representation of phenomena. *Chemistry Education Resesearch and Practice, 13*(4), 462–470.

Rennie, L. J. (1990). Student participation and motivational orientation: What do students do in science? In K. Tobin, J. Butler, & B. J. Fraser (Eds.), *Windows into science classrooms: Problems associated with higher-level cognitive learning* (pp. 164–198). London: The Falmer Press.

Russell, T., & McGuigan, L. (2001). *Promoting understanding through representational redescription: An illustration referring to young pupils' ideas about gravity*. Paper presented at the 3rd International Conference of the ESERA, Thessaloniki.

Ryan, R. M., & Deci, E. L. (2000). Intrinsic and extrinsic motivation: Classic definitions and new directions. *Contemporary Educational Psychology, 25*(1), 54–67.

Sanger, M. J., & Phelps, A. J. (2007). What are students thinking when they pick their answer? A content analysis of students' explanations of gas properties. *Journal of Chemical Education, 84*(5), 870–874.

Simpson, R. D., & Oliver, J. S. (1990). A summary of major influences on attitude toward and achievement in science among adolescent students. *Science Education, 74*(1), 1–18.

Solsona, N., Izquierdo, M., & De Jong, O. (2003). Exploring the development of students' conceptual profiles of chemical change. *International Journal of Science Education, 25*(1), 3–12.

Stains, M., & Talanquer, V. (2007). Classification of chemical substances using particulate representations of matter: An analysis of students' thinking. *International Journal of Science Education, 29*(5), 643–661.

Stains, M., & Talanquer, V. (2008). Classification of chemical reactions: Stages of expertise. *Journal of Research in Science Teaching, 45*(7), 771–793.

Steinkamp, M. (1984). Motivational style as a mediator of adult achievement in science. In M. Steinkamp & M. Maehr (Eds.), *Advances in motivation and achievement 2* (pp. 281–316). Greenwich, CT: JAI Press.

Stipek, D. (1998). *Motivation to learn: From theory to practice*. Boston: Allyn and Bacon.

Thiele, R. B., & Treagust, D. F. (1994). An interpretative explanation of high School chemistry teachers' analogical explanations. *Journal of Research in Science Teaching, 31*(3), 227–242.

Tien, L. T., Teichert, M. A., & Rickey, D. (2007). Effectiveness of a more laboratory module in prompting students to revise their molecular-level ideas about solutions. *Journal of Chemical Education, 84*(2), 175–180.

Tobin, K. G., & Capie, W. (1981). The development and validation of a group test of logical thinking. *Educational and Psychological Measurement, 41*(2), 413–423.

Treagust, D. F., Harrison, A. G., & Venville, G. J. (1998). Teaching science effectively with analogies: An approach for preservice and inservice teacher education. *Journal of Science Teacher Education, 9*(1), 85–101.

Tuan, H. L., Chin, C. C., & Shieh, S. H. (2005). The development of a questionnaire to measure students' motivation towards science learning. *International Journal of Science Education, 27*(6), 639–654.

Waldrip, B., Prain, V., & Carolan, J. (2006). Learning junior secondary science through multi-modal representations. *Electronic Journal of Science Education, 11*. Retrieved June 30, 2007, from http://ejse.southwestern.edu/volumes/v11n1/articles/art06_waldrip.pdf

Williamson, V. M., & Abraham, M. R. (1995). The effects of computer animation on the particulate mental models of college chemistry students. *Journal of Research in Science Teaching, 32*(5), 521–534.

Wu, H. K., & Shah, P. (2004). Exploring visuospatial thinking in chemistry learning. *Science Education, 88*(3), 465–492.

Zusho, A., Pintrich, P. R., & Coppola, B. (2003). Skill and will: The role of motivation and cognition in the learning of college chemistry. *International Journal of Science Education, 25*(9), 1081–1094.

Section II
Approaches in Chemistry Teaching for Learning with Understanding

Cooperative and Collaborative Learning

The first part of Section II focuses on cooperative and collaborative learning in the science classroom to promote students' leaning with understanding. As it has been emphasised for some time now learning occurs in a social context. Different aspects such as students' cultural, racial, ethnic and social backgrounds can influence collaborative and cooperative learning. In the late 1970s cooperative learning methods were integrated into science education to stimulate peer-to-peer teaching and learning hoping that these approaches would enhance students' academic achievements and stimulate interest for science learning and future careers in science and technology. The differences or similarities between cooperative and collaborative learning are explained by different authors. Both concepts are sometimes used for the same thing; small-group activities in the classroom where learning takes place, but some differences in organisation of the specific learning approach can be found. Collaborative learning can have fewer roles assigned, the teacher is not the centre of authority, group tasks are usually more open-ended, and complex, so collaborative learning is less structurally defined as cooperative learning.

In Chap. 7 Bodner, Metz and Lowrey Casey critically present 25 years of experience with interactive instruction in chemistry. The chapter begins by presenting the history of the traditional system of higher education in the USA that shaped the way science and mathematics courses were taught. The classic study of exemplary teaching at the high-school level that provides a way to understand the motives that lead a teacher to adopt a classroom environment that involves interactive instruction learning is presented. The chapter continues by reviewing some of the early work on cooperative learning that influenced the way in which the first implementations of this approach to university teaching in the 1980s were conducted. The authors also describe a study about the effects of interactive instruction on students' attitude and achievement in the introductory college-level classroom and describe subsequent efforts to bring interactive instruction into upper-level courses.

Cardellini in his Chap. 8 analyses problem solving through cooperative learning in the chemistry classroom at the university level. He presents cooperative learning as an instructional method that should incorporate five criteria, such as: positive interdependence, individual accountability, face-to-face interaction, development and appropriate use of interpersonal skills and periodic self-assessment of group functioning. A review of the literature on cooperative learning is given and the definition and structure of cooperative learning are presented. He describes in detail how he implements this teaching approach in his university-level chemistry classes, how to motivate and engage the students participating in the general chemistry course and how to teach chemistry to achieve the best results according to the students' abilities.

Chapter 9 entitled "The Learning Company Approach to Promote Active Chemistry Learning: Examples and Experiences from Lower Secondary Education in Germany" presented by Witteck, Beck, Most, Kienast and Eilks deals with the development and application of the learning company approach for lower secondary chemistry education in Germany. This approach is some form of cooperative learning. The authors try to interpret this approach as a methodological shift in order to create a different style of experimentation in the chemistry classroom. Students' following this learning approach should be motivated to perform self-regulated and self-organized experiments in a cooperative learning environment or situation. The authors developed three different lesson plans and evaluated them using a participatory action research. This chapter gives an overview of three separately tested lesson plans based on methods of separating matter, working out the different phenomena of chemical reactions, and introducing acid–base chemistry.

Chapter 7
Twenty-Five Years of Experience with Interactive Instruction in Chemistry

George M. Bodner, Patricia A. Metz and Kirsten Lowrey Casey

Introduction

There is no shortage of literature on cooperative learning. A recent search for information about cooperative learning using Google Scholar, for example, returned more than 1,500,000 references. Furthermore, excellent reviews of the application of innovative approaches to the teaching and learning of chemistry have recently been published (Eilks and Byers 2009; Byers and Eilks 2009; Eilks et al. 2009). The goal of this chapter is to trace a 25-year evolution in the practice of teaching chemistry that represents a basic shift in the way college- and university-level chemistry courses are taught. This chapter will also report on several studies of interactive approaches to instruction, explore some of the early literature on cooperative learning on which these studies were based, and examine answers to common questions the authors have encountered while advocating interactive instruction among chemistry faculty.

Throughout the twentieth century, the traditional model of instruction in science and mathematics courses was based on the assumption that knowledge can be transferred more or less intact from the mind of the teacher to the mind of the learner. Alternative modes of instruction introduced in recent years are based on a consensus among cognitive scientists and educators about the validity of a constructivist theory of knowledge that can be summarized as follows: *Knowledge is constructed in the mind of the learner* (Bodner 1986). This chapter examines some of the implications of the constructivist theory of knowledge, with particular emphasis on how a shift can be made from the traditional model of the instructor as "someone who teaches" to a constructivist perspective in which the instructor is "someone who tries to facilitate learning;" a shift from "teaching by imposition" in which an authority figure at the front of the classroom controls all aspects of the

G. M. Bodner (✉) · P. A. Metz · K. L. Casey
Department of Chemistry, Purdue University, West Lafayette, USA
e-mail: gmbodner@purdue.edu

I. Devetak and S. A. Glažar (eds.), *Learning with Understanding in the Chemistry Classroom*, DOI: 10.1007/978-94-007-4366-3_7,

teaching/learning process to an approach that involves "teaching by negotiation" between the instructor and his or her students.

Origin of the Present System of Instruction

The structure of most chemistry classrooms suggests that they were designed on the basis of the hidden assumption that little, if any, change has occurred in our general approach to education since Plato lectured at the first institution of higher learning, the Academy of Athens (*c*. 387 BC), or in the structure of higher education since the first universities were created in Bologna (*c*. 1088) and Oxford (*c*. 1167) almost a 1,000 years ago.

The early universities were free to govern themselves, but this freedom was achieved at the cost of having to raise their own finances. The teachers charged fees,[1] and therefore had to please their students to earn a livelihood. (Cambridge University, for example, was established in 1209 by a group of dissatisfied students who moved there from Oxford). These medieval universities focused on preparing men for careers in service to either Church or State. All students began with the same curriculum: grammar (Latin), logic, and rhetoric. They then went on to major in either law, medicine, or theology. Graduation depended on a single final examination, which many, if not most, students failed.

Whole-Class Noninteractive Modes of Instruction

The traditional approach to science and mathematics courses is based on a format introduced in medieval universities—a series of lectures in which scholars summarize the state of knowledge in their area of expertise. When this format was first implemented, there was no alternative. Books were rare; individual ownership of books even rarer. Thus, it is not surprising that the term *lecture* comes from a Latin stem meaning "to read."

Lectures are still the best way to introduce information when our role is the same as that of the 'masters' who taught at the early universities—when we bring together information from a number of sources to which the audience does not have access. Although this may be the case in certain graduate-level, special-topics classes, it is not the case in most undergraduate science and mathematics courses, where our students have access to excellent texts that provide them with more than enough information.

[1] These fees were placed in the pocket that can still be found in the bottom of the hoods worn with academic gowns at graduation ceremonies.

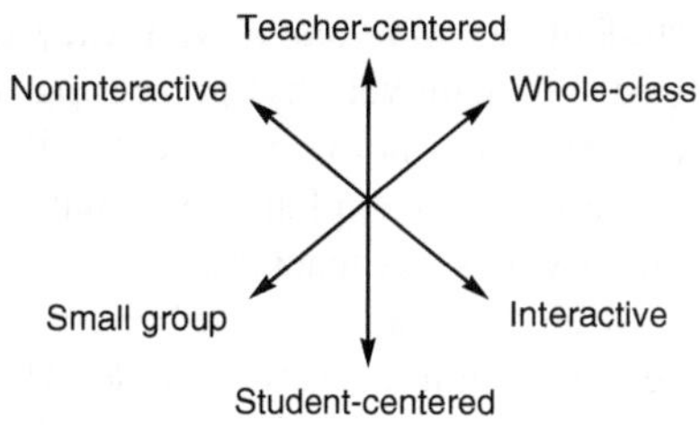

Fig. 7.1 A three-dimensional coordinate system that can be used to characterize the classroom environment

The traditional lecture approach to instruction can be characterized in terms of the three-dimensional coordinate system shown in Fig. 7.1.

Lectures represent a *whole-class*, *teacher-centered*, *noninteractive* mode of instruction. They address the whole class at the same time; all decisions about the topics to be covered, the order in which they are presented, and the amount of time devoted to each topic are controlled by the instructor; and interactions between the instructor and students are rare, if not nonexistent.

In her dissertation, Metz (1987) described the characteristics of her undergraduate chemistry courses as follows:

> All of the courses used a lecture format; a direct question that required a direct answer by the students was never asked by the instructor; there were no examples of the students being encouraged to ask questions; no demonstrations were performed; only written exams and quizzes were used to evaluate student progress; and students competed fiercely for grades.

In hindsight, she concluded that her chemistry studies were "… a matter of sheer academic survival and not a vehicle for intellectual growth and development."

Metz (1987) went on to describe her experiences in a course she took as a graduate student as follows:

> [This course was] the most frustrating of all…. My frustration as a teacher started to mount when a student asked some questions on a point he did not understand. The professor responded to the first, then said he could not answer other questions because he had material to cover and had limited time to do so…. Several days later when this same student raised his hand, he was ignored. No one ever again attempted to ask a question.

She then concluded her description of this course with a phrase that characterizes all too many whole-class, teacher-centered, noninteractive classrooms. "The next point of frustration was the lecture notes…. At times I felt the professor's notes became my notes without passing through either of our minds".

Whole-Class Interactive Modes of Instruction

As part of a study of exemplary practice in teaching, Tobin and coworkers described the classroom behavior of teachers they referred to as "Doug" and "Gary" (Tobin and Fraser 1987). Doug believed that: (1) it is important for students to *understand* what they learn and to *retain* this knowledge; (2) this can best be achieved by presenting the material in the syllabus, explaining this material, and helping students identify the logical structure of chemistry; and (3) the ability to explain new material to our students is a critical factor in our effectiveness as teachers.

Many who share these beliefs approach instruction from a whole-class, teacher-centered, noninteractive mode. They take the burden of learning onto themselves and dominate what happens in the classroom. When students fail, they conclude that they have failed as teachers, or that the students were either too lazy or did not have the cognitive skills to succeed.

An observer watching Doug's classes would find that he often used a "lecture" mode of instruction. (It is difficult to imagine a chemistry classroom in which this mode is not present, at least some of the time.) One of the keys to Doug's success, however, was his ability to also use an interactive mode of instruction. The focus of attention was still the whole class and the classroom environment was still predominantly teacher-centered, but now Doug was involved in a dialog with his students.

To help the students "understand" the material they were learning, Doug made extensive use of questions to develop the content of his lessons, to probe whether students understood the material, and to involve the students in the learning process. These teacher-initiated questions were asked in a nonthreatening atmosphere, in which the risk of an inappropriate response was minimized. Doug also encouraged his students to ask questions. (In some lessons, students asked more than 60 % of the questions, transforming the classroom environment along the coordinate system in Fig. 7.1 toward one that was more student-centered.) His students considered the most useful classes to be those in which the frequency of student-initiated questions was high. (Even students who did not ask questions valued the questions their peers asked.) Another important feature of the interactive mode of instruction in Doug's class was the tendency of the students to spontaneously take over the class and answer questions raised by other students.

Interactive Instruction in Large Classes

Some might argue that an interactive mode of instruction based on extensive use of questions was fine for Doug, who had only a limited number of students in each of his high-school classes. But what about those of us who teach hundreds of students at a time?

To study what happens when interactive techniques are used in large courses, Metz (1987) compared the performance of students in two large lecture sections of an introductory first-year chemistry course. The students in the study were all registered in the same course, but they were divided into two groups who attended class in different rooms. One section was a teacher-centered lecture, the other was a student-centered interactive class. The instructors who were selected to run these sections had the same amount of experience, but different natural teaching styles.

The course in which this study was carried out was the first semester of a general chemistry course primarily taken by students majoring in fields such as agriculture, foods and nutrition, the health sciences, and technology. The course was designed for students who need exposure to, but not necessarily a mastery of, chemistry. It consisted of two 50-min "lectures," a 3-h laboratory, and a 50-min recitation section per week.

Most students were exposed to a classical teacher-centered approach to instruction, where student–teacher interactions were kept to a minimum. Lectures in the teacher-centered classroom were presented using a chalkboard or an overhead projector. Demonstrations were done, but students were not actively involved. Few questions, except of a rhetorical nature, were asked and few questions were accepted from the students.

The instructor in the interactive class tried to maximize student involvement while minimizing his own. He started each class by briefly presenting the topic. He then provided the students with a whole-class or small-group activity and devoted his time to directing, clarifying, and providing feedback to students, as necessary.

Students in both sections were given the same exams, which were written by the individual who taught the lecture section. During the course of the semester, the students took four 50-min exams and a 2-h comprehensive final exam. A multiple-choice format was used for all of the exams, which is standard practice for this course. A typical mid-semester exam had 20 questions; the final exam contained 42 questions. The sample population for analysis of both student performance and student attitude consisted of the 384 students who regularly attended the two "lecture" classes offered each week.

A statistical analysis of the SAT math and verbal scores and the students' grade-point averages for both sections suggested that there were no significant differences between the two sections. When the total scores on each of the five exams were compared between the two sections, no statistically significant differences were found between the two modes of instruction. This is consistent with the equalization effect proposed by McLeish (1968), which suggests that the work students do in preparing for an exam brings the scores of groups exposed to different modes of instruction close to equality. We therefore concluded that interactive instruction was neither better nor worse than the traditional lecture when exam scores were used as a measure of effectiveness.

A fifteen-item attitudinal survey was used to assess students' feeling toward the method of instruction, the course, and chemistry in general. The survey was given one week before the end of the semester. It used a five-point Likert scale (ranging from strongly agree to strongly disagree). Once again, the sample population analyzed consisted only of those students who regularly attended the "lecture" classes. The word "lecture" was used as a generic term for method of instruction in the survey inasmuch as students in both sections referred to this portion of the course as the "lecture."

There was a significant difference in student attitude in the two sections on six items. In each case, the interactive class was viewed more favorably by the students. Students in the interactive class were more likely to indicate that they had little time to think about anything besides chemistry during class; to find it helpful when the lecturer asks questions that must be answered by the students before he or she continues; to feel challenged to think about the material being discussed during lecture; to feel comfortable about being called on to answer questions during class; and to want to take another chemistry course taught with the same format. They were also more likely to believe the course was taught at an appropriate level and that adequate material was presented in lecture to prepare them for exams, in spite of the fact that the very nature of interactive instruction means that considerably less material was covered during class in the interactive section.

Student behavior in the interactive class went through three stages. The initial stage, which lasted for three classes, was characterized by ambiguity and anxiety on the part of the students, who came into the course expecting to sit back and take notes. Student behavior in this stage included looking away from the instructor when questions were asked; waiting for others to answer questions; raising their hands instead of calling out answers; responding to questions only when they were sure of their answers; asking few questions themselves; and refraining from correcting errors made by the instructor or other students.

The second stage, which lasted until the first exam, involved a transition from teacher-dependence to student-independence. This stage was characterized by the following kinds of student behavior: asking questions; trying to answer questions even when not sure of the answer; answering questions without being called on; talking within a group; and talking between groups of students. During this stage, the students seemed to accept that the instructor was not going to tell them everything they needed to know for the exam and that it was up to them to take responsibility for their learning.

After the first exam, the students reached the final or working stage, in which they assumed an active role in class. They realized that other students in the class had similar problems with chemistry and that they could possibly help each other. It was during this stage that student–teacher and student–student interactions were at their highest. These interactions included correcting errors made by the instructor, other students, or themselves; checking with neighbors before asking the instructor for help; asking more *why* and *how* questions; moving over to work with other students without being told to do so; continuing to work on problems as a group even after a whole-class discussion was resumed; asking for more clarification of problems and explanations—even interrupting the instructor to do so; starting to work on problems before the directions were completed; answering questions from other students before the instructor could respond to the questions; and answering questions—even rhetorical ones—without waiting to be called on.

Teaching by Listening

For more than 20 years, the first author has been asked: How… do I create an interactive classroom environment in my course, which has so many students? I often give a concrete example from the early days of my efforts to transform a large-lecture section that enrolled more than 400 students into an interactive classroom, which was first described elsewhere (Bodner 1992).

The topic is the structure of ionic solids, which is often represented by the drawing in Fig. 7.2. This drawing is based on the fact that the monatomic negative ions in simple ionic solids are often larger than the corresponding positive ions by as much as a factor of two, or more. As a result, simple ionic solids often crystallize in a structure in which the negative ions pack to form a closest-packed (or closely packed) array described in the circles in Fig. 7.2.

The positive ions pack in holes between the planes of negative ions that form this array. There are two kinds of holes in a closest-packed structure, which are marked with "x's" and "o's" in the traditional figure used to capture this structure. Let's assume, for the moment, that the closest-packed planes of negative ions form a hexagonal closest-packed structure. This means that there is an identical plane of closest-packed negative ions both above and below the plane of the solid circles in Fig. 7.2, in the positions indicated by the circles formed with broken lines.

When I first taught this material, I would ask the students to focus on one of the holes marked with an "o" in Fig. 7.2, and then tell them the number of negative ions that touch—the positive ions that would occupy one of these holes.

One day, instead of telling the students the answer, I asked them: "If you were a positive ion in one of these holes, how many negative ions would you touch?" I then took a series of votes by asking students to raise their hands. What fraction of the students thought the correct answer was three? Four? Five? Six? The answer was obvious to me, and it may be obvious to you. I would therefore ask you to cover the next paragraph with the palm of your hand and write your answer in the margin of this chapter.

The results I obtained were fascinating. The majority of the students, who were science and engineering majors with a strong background in mathematics, thought

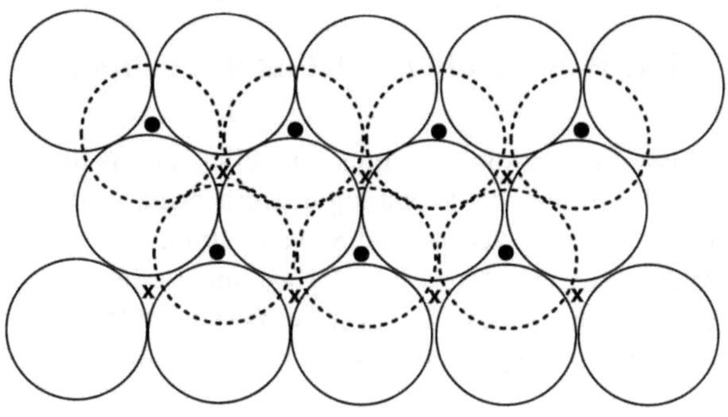

Fig. 7.2 Simple ionic solids consists of a closest-packed array of relatively large negative ions with relatively small positive ions in holes between the olanes of negative ions

the answer was five. This answer is absurd, if you assume the ions are spherical. It was so absurd that I couldn't figure out how *any* of the students had arrived at this number. After lecture, I realized what had happened. The students assumed that the positive ions packed in holes in the same plane as the negative ions. They therefore assumed that each of these positive ions touched the three negative ions that surrounded the hole within that plane and then touched one negative ion in the plane above and one negative ion in the plane below.

This is impossible, of course, because any spherical positive ion small enough to pack in the relatively small hole between the three negative ions in a given plane could not possibly touch the negative ions in the plane above or below. Although I had clearly stated that the positive ions packed in holes that lie *between the planes of the negative ions*, the students tried to incorporate the positive ions into the same plane as the negative ions.

If I had not become convinced that I needed to listen to what students say, I probably would have spent the remainder of my career telling each year's class the answer to a question they did not understand. Once I started listening to students, I recognized that I had structured my presentation of the material from my experience, not from theirs.

What do I do now? I provide the students with the same background information about the relative size of the negative and positive ions in simple ionic solids such as NaCl. I then ask them the same question, having them raise their hands to indicate whether they believe that there would 3, 4, 5, or 6 negative ions in contact with the positive ions in the holes marked with an "o". Each year, I get the same results; the vast majority of the students vote for an answer of "5". Without commenting on their answer, I show them a set of three images in which I construct a model based on styrofoam balls and a small macramé bead that clearly indicates how the positive ions pack in these tetrahedral holes, touching three negative ions below and one negative ion arranged toward the corners of a tetrahedron. I then ask them if they want to change their minds. The majority do.

The result is simple: Every time I listen to students, I learn new things about how to change the way I "teach" the material I am presenting. And each year I get indirect evidence that the students leave my course *understanding* rather than just *knowing* some of the content I "teach."

What is the Role of "Clicker Technology?"

Classroom response systems (or "clickers" as they are often called) have become popular in science, technology, engineering, and mathematics (STEM) classrooms in recent years (Judson and Sawada 2002; Fies and Marshall 2006; Caldwell 2007; MacArthur and Jones 2008; Kay and LeSage 2009). Arguments for the use of clickers are almost invariably based on the inadequacies of the traditional approach to large-lecture section instruction.

Carl Wieman, a Nobel prize winning physicist, for example, based on an article on the use of clickers on the assumption that "effective" physics instruction can be defined as an approach that "… changes the way students think about physics and physics problem solving and causes them to think more like experts—practicing physicists" (Wieman and Perkins 2005).

There is general agreement in the literature on personal response systems that "… students have a positive attitude towards the technology" (MacArthur and Jones 2008). However, there is little (if any) evidence to support the notion that clicker technology, by itself, leads to improved student learning (Martyn 2007). And there is little, if any, research to substantiate the common stated belief that "The value of the clicker is that it provides a way to quickly get an answer for which the student is accountable, and that answer is anonymous to the student's peers" (Wieman and Perkins 2005).

Our experience with interactive instruction over the course of 25 years is consistent with the following hypothesis offered by Wieman and Perkins (2005): "While the clickers provide some measure of what students are thinking, it is the specifics of the implementation—the change in the classroom dynamic, the questions posed, and how they are followed up—that determines the learning experience." While students might prefer the opportunity to provide anonymous responses to questions posed in class, there is no evidence to demonstrate that this anonymity has a beneficial effect on learning. Nor is there evidence to suggest that one needs to know the percentage of students choosing a particular answer to three decimal places in order to successfully use interactive instruction in the class. Before one requires students to purchase a handheld device that often costs on the order of $30, we would argue that one needs better evidence than is presently available that this is cost-effective, as opposed to asking students to simply raise their hands to indicate the answer they believe is correct.

Interactive Instruction in Organic Chemistry

For more than 10 years, we have been studying the implementation of an interactive approach to instruction in a large organic chemistry course taken by students majoring in pharmacy. The instructor is the author of an organic chemistry textbook that is now in its fifth edition (Loudon 2009). The intervention in this course began on the basis of a discussion with the instructor, who wanted an approach to the organic chemistry lecture that was consistent with the approach he took in one-on-one interactions with colleagues, with graduate students, and with undergraduates who came to his office.

We began the intervention by insisting that he no longer "lecture," in spite of the fact that he was recognized as one of the best "lecturers" in his department. The students were organized into groups of three who were encouraged to work together both in class and while studying for exams. One of the graduate students who worked on this project often cited a discussion of the nomenclature of alkanes as a characteristic example of the interactive classroom environment that was created (Lowrey 1996).

Lowrey (1996) noted that instructors traditionally approach alkane nomenclature by copying a list of nomenclature rules from the text onto the board, which are dutifully copied into the students' notebooks. Class time is often devoted to covering all of the rules listed in the book, so that equal time is spent on each rule without regard to their relative importance or level of difficulty. Examples are then shown to the students to illustrate how these rules are applied. It is not surprising that students often leave this class with the feeling that they need to memorize an overwhelming amount of information that involves purely linguistic rules that are unrelated to the underlying structure of organic compounds on the molecular level in order to "master" the subject of organic chemistry.

The students in the interactive class were told at the beginning of the semester that the instructor expected them to read, but not necessarily understand, the sections of the textbook he was going to cover before they came to lecture. They were also organized into groups of three, who were expected to sit next to each other and work together in class.

As might be expected, the discussion of alkane nomenclature occurred early in the first semester of the course. It was therefore the first example of a phenomenon that would come to characterize virtually every class the students attended, in which the instructor presented a problem to the students that he intended them to solve in their small groups.

Instead of enunciating a set of rules for naming organic compounds, the instructor began by drawing three structures on the blackboard. Two of the structures were different orientations of condensed structures of 2-methylpentane—one in which the molecule was written in a horizontal orientation with C_1 on the left and C_5 on the right, the other in which the molecule was drawn in a vertical orientation with C_1 at the top. The third structure was 2.2-dimethylbutane, written in a horizontal orientation with C_1 on the left. He then asked the groups to talk among themselves about whether these compounds were the same or whether they were different.

Because this was the first time the instructor presented a problem to the students to solve in their small groups, they were passive, at first. He therefore opened up the question to whole-class discussion and asked how many knew the answer. Approximately two-thirds of the students called out an answer. When he turned to one student, she noted that they were all structural isomers. The instructor asked the class to evaluate this answer, which lead to a discussion in which another student noted that two of these structures represented the same molecule. He asked this student to explain his answer, and then rephrased the student's wording and demonstrated the student's reasoning by counting off the carbons drawn on the board. He then returned to the first student and asked her: What lesson can be learned from this example? There was a lighthearted exchange, punctuated by laughter, in which the student eventually concluded that it is possible to draw the same molecule in different ways, the point she had missed earlier. The instructor reinforced this response by agreeing with her and repeating what she said.

He then drew the structures of a series of more complex alkanes and assigned different groups to work on each of these structures. The students turned their desks to work together and the volume of talking rapidly increased, while the instructor walked around the room, visiting various groups. When it appeared that most groups were finished, the instructor returned to the front of the class and regained their attention by asking how many of the groups had an answer. He then proceeded to work through the students' answers systematically, starting with the first structure. When students gave an answer, he would ask them to explain their answer. He then went over the rationale for the answers and validated the answers by telling the students when they were correct. Instead of presenting a series of rules for the students to memorize, the discussion of alkane structure and nomenclature focused on understanding the meaning of these linguistic rules at the molecular level. Having justified the need for a system of nomenclature, the students were then asked to spend time reading the appropriate section of the book where the rules of nomenclature were outlined in detail.

Throughout the semester, the instructor organized the material being presented around a series of problems he asked the students to solve (both in groups and as a whole class). It should be noted that these problems were not examples chosen to reinforce material that had already been discussed in class; the students were asked to solve the problems first and the instructor then led a discussion of the relevant material.

The instructor described this approach as follows: "Basically what we're doing is we're developing a concept by solving little problems" (Lowrey 1996).

Instead of making lists for students to memorize and reviewing material that can easily be found in the textbook, the instructor used problems to gauge the students' understanding of what they had read. He did not feel compelled to "cover" all of the material, but used class time to focus on ideas the students did not understand and to justify why the course material was worth learning.

Interactive Instruction in Physical Chemistry

Similar approaches have been taken to creating an interactive, cooperative learning environment in physical chemistry courses taken by both chemistry majors and by nonmajors. A detailed analysis of the results of this work has been reported elsewhere (Towns and Grant 1997; Towns 1998).

Small-Group Interactive Modes of Instruction

Although they may have involved the use of small groups, the techniques discussed so far tend to focus on the class as a whole. Tobin and Fraser's (1987) study of exemplary practices presented a case study of a chemistry teacher who used an alternative approach to instruction. Like Doug, Gary believed it was important for students to understand the material they were learning. But he believed this could only be achieved if they accepted responsibility for their own learning. He believed that his role was to: (1) provide students with opportunities to work independently or in small groups; (2) provide learning experiences for students and ample study time so that *they* could structure their own learning; (3) diagnose misconceptions that individual students constructed and helping them overcome these misconcepts; and (4) serve as a facilitator of student learning, rather than as an authority who was the source of facts, principles, and skills.

Gary therefore allocated a large proportion of his class time to individualized and small-group work. Like Doug, he constantly interacted with students, but his interactions were more likely to be one-on-one. It isn't surprising that his students felt very positive about the amount of individual assistance Gary provided.

Students in Gary's classes frequently organized themselves into small-groups, which were characterized by very high levels of student–student interaction. The students discussed ideas within these groups and asked each other questions about things they did not understand. They were quite willing to help one another and responded favorably to small-group activities when interviewed. Although the students frequently worked in groups, no obvious examples were observed of students copying from one another.

Cooperative, Competitive, and Individualistic Learning

The voluminous literature on cooperative learning generated by a Google search provides an important insight into the role that cooperative learning can play in science, technology, engineering, and mathematics (STEM) disciplines. For our purposes, however, it might be useful to focus on some of the early work in this area that has shaped the recent literature on cooperative learning.

The different modes of instruction discussed in this chapter can be analyzed in terms of a theory developed by Kurt Lewin (1935) and Morton Deutsch (1949). Lewin proposed a theoretical basis for understanding how individuals are motivated to work toward the accomplishment of their goals that was extended to interpersonal situations by Deutsch.

Deutsch distinguished between three ways in which the motivations of different individuals can be interrelated. A *cooperative* situation is one in which an individual can only achieve his or her goals if the other members of the group also attain their goals. In a *competitive* situation, individuals can only achieve their goals when others in the group cannot. In an *individualistic* situation, there is no correlation among the ability of members of a group to attain their goals.

This model provides a theoretical basis for understanding the effect of different modes of instruction. There is no doubt that learning requires work and that the extent to which students apply themselves to learning is a function of their motivation. When a student finds a particular subject interesting, the student often possesses an *intrinsic* motivation to learn. Although talented teachers can make almost any topic seem interesting, no subject is inherently interesting to all students. Schools have therefore developed ways of motivating students; the most common of these *extrinsic* motivators involve grades.

Grades are successful as extrinsic motivators only when they are given on a competitive basis. But, as Slavin (1984, 1988) noted, competition among students creates a situation in which:

- Students hope their classmates do poorly, so that they can do well.
- There is peer-group pressure not to do too well, thereby raising the curve.
- Students who excel are often looked on with disfavor by their peers.

At one time, individuals who believed that students must accept responsibility for learning, and that competitive situations can discourage students from doing their best, created approaches to teaching and learning based on individualized instruction in which students work at their own pace through a program of carefully sequenced activities. At first glance, this technique seems particularly suited for use in mathematics and the physical sciences, where each new skill builds on prior knowledge. Research, however, clearly suggests that this technique did not meet our expectations. There are many reasons for this (Slavin et al. 1984).

- Students find it boring because it forces them to interact with paper, not people.
- It isolates students from one another.
- Students tend to get bogged down as the tasks become familiar and therefore monotonous.
- There is no incentive to progress rapidly.
- The teacher is relegated to the role of nothing more than an answer-checker.

Vygotsky (1986) provided a theoretical framework for understanding how forcing students to work by themselves can actually limit the amount that can be learned. The best example of the problems with individualized instruction is the classic case study done by Erlwanger (1973) on "Benny," a young mathematics student who, on the basis of an extensive seatwork, constructed knowledge about the mathematics of fractions that some researchers would call a "less powerful" concept, but practicing teachers would label as "wrong."

This leaves us with a third goal structure to consider: cooperative learning, which is the basis for both Doug's whole-class interactive mode of instruction and Gary's use of small-group activities.

The Theory of Cooperative Learning

A theoretical model based on the work of Piaget has been proposed to explain why cooperative learning might improve student achievement (Damon 1984). This model suggests that the group discussions that occur during cooperative learning achieve the following:

- They expose inadequate or inappropriate reasoning, which can result in disequilibration that can lead to better understanding.
- They motivate individuals to abandon misconcepts and search for more powerful concepts.
- They provide a forum that encourages critical thinking.
- They lead to constructive controversy, which focuses students' thinking and increases the use of higher-order cognitive processes.
- They encourage students to vocalize ideas, which inevitably improves their performance.

According to this model, the most important element of cooperative learning is the fact that students work together in groups in which they are involved in discussions of the course content they are struggling to learn.

A separate model has been proposed from the perspective of the theory of motivation, which attributes the success of group learning to the structure of the goals of cooperative learning (Slavin 1984, 1987, 1988). This model proposes that cooperative learning activities, when properly carried out, create a situation in which the only way individual group members can attain their goals is if the group is successful. To meet their own goals, members of the group therefore help their classmates do whatever is necessary to succeed. According to this theory, cooperative learning encourages students to want their classmates to succeed, in sharp contrast to the situation when individuals compete for grades (Nicholls 1989).

Effect of Cooperative Learning

Cooperative learning has been shown to improve student achievement, enhance students' self-esteem, increase the use of higher-order cognitive skills, improve both cross-sex and cross-ethnic relationships, and reduce science and math anxiety.

A classic meta-analysis of 122 studies suggested that the average person in a cooperative learning environment performs at a level equal to the 78th percentile of students in competitive or individualistic environments (Johnson et al. 1981). When high-, medium-, and low-achieving students are compared, it is the low- and medium-achievers who seem to benefit most from cooperative learning (Johnson et al. 1985). It therefore is not surprising that cooperative learning has been found to improve students' self-esteem (Johnson and Johnson 1979). The performance of the high achievers is usually the same in both competitive and cooperative learning situations. When these students are interviewed, however, they report feeling more support and academic encouragement from their peers and the teacher when they work in groups.

Cooperative learning has been shown to enhance the use of higher-order reasoning strategies (Gabbert et al. 1986). Furthermore, the higher-order cognitive skills developed during cooperative learning activities are then transferred to individual learning situations. This can be understood by examining the research on cooperative approaches to teaching problem solving (Ross 1988). Both whole-class interactive and small-group approaches help students develop strategies for solving problems that their peers in competitive or individualistic environments only experience second-hand.

Some have argued that cooperative learning should be used—even if it had no impact on student's achievement (Slavin 1984)—because of the way it improves the relationships between males and females (Waring et al. 1985) and among different ethnic and racial groups (Sharan 1980; Johnson et al. 1984).

Mathematics and Science Anxiety

An equally compelling argument can be made for cooperative learning on the basis of the effect it can have on reducing students' anxiety about science and mathematics (Stodolsky 1985), by creating a relaxed, tension-free atmosphere in which a feeling of mutual trust prevails (Okebukola 1986).

Anxiety can be defined as a state of uneasiness or distress in the absence of a specific threat. It has physiological, cognitive, and psychological consequences. It can cause a person to sweat and feel nauseous; it occupies a large fraction of working memory, so the individual can no longer perform even the simplest tasks; and it feeds upon itself—anxiety produces failure, which in turn gives rise to feelings of anxiety.

It has been argued that the structure of students' early experiences with mathematics is what gives rise to their feelings of anxiety (Stodolsky 1985). Young children have positive attitudes toward mathematics—they frequently rank it on a par with reading as one of their best-liked subjects—but 17-year-olds rank it as their least-liked subject (Carpenter et al. 1981). This has been attributed to the fact that there appears to be only a single route to learning in math classrooms—the teacher explaining the material, which the students then practice. It isn't surprising that when they face mathematics in later life, students view it as a subject that comes from authority; that they reject the idea they can learn mathematics on their own; and that they believe mathematics is an area for which one either does or does not have talent—an idea that can be found in music, art, and sports, but not other traditional academic fields (Stodolsky 1985).

Implementing Cooperative Learning

If you choose to include an element of cooperative learning in your course, you might wish to consider the following answers to common questions, which summarize the results of a variety of research studies (Johnson and Johnson 1979; Sharan 1980; Slavin 1983; Johnson et al. 1984; Johnson et al. 1985; Waring et al. 1985).

How are groups constructed?

Groups should contain between three and five members. If the group is too small, one member can dominate the others. If it is too large, the group will ignore the contributions of one or more members. The group should be heterogeneous. It should include high-, medium-, and low-achievers; both males and females; and members of different ethnic groups, if possible. The more heterogeneous the group, the larger its resources for problem solving.

What are the essential elements of group work?

Cooperative learning assumes that the success of each individual depends on other members of the group. This can be achieved by sharing mutual goals, resources, and rewards. It can also be accomplished by giving each member of the group a specific task whose completion is essential to the group's success.

How do I get students to work in groups?

It might help to impose the following rules: Members of the group should share their information with each other; encourage other members of the group; bring out the ideas of other group members; argue their own point of view; be critical of ideas, not people; and make sure they understand the views of other members of their group.

How do I assess cooperative learning?

The group can produce a single product, which each member signs to indicate acceptance of the answers and understanding of why each answer was chosen, and the members of the group can all receive a grade based on the quality of this product. (For over 25 years, the first author has had the students in his general chemistry work on laboratory reports in groups of three who all receive the same grade on the laboratory report.) Alternatively, and some would argue preferentially, small-group activities can be used to enhance learning, and testing for grades can then be done on an individual basis. (At Purdue, group work is encouraged in lecture, in the laboratory, while students are working home-work assignments, and while they are writing laboratory reports. When the time comes to take exams, however, students work individually. Recently, we have studied what happens when groups of students are allowed to discuss an exam for up to 30 min and then split up to write their individual answers. When this was implemented in an organic chemistry course as a reward for learning how to work together in groups, we found a significant increase in the extent to which the individuals grading the exams believed that the students understood the questions they were answering.)

Some argue that the optimum use of small-group activities requires both cooperative learning and individual accountability. Authors who take this position frequently noted that this approach to instruction is known as *cooperative learning*, not *cooperative assessment*.

What is the relationship between groups?

It is possible to construct modes of instruction based on both *intergroup cooperation* and *intergroup competition*. The first can be fostered by asking the students to think about questions such as: "What do we need to do to make sure everyone in the class does a good job learning?" The second is much easier to achieve. Have members of the group reflect on the following question: "What can we do to make sure that our group is best?" Those who have played team sports know the value of intergroup competition. It is worth noting, however, that *intergroup cooperation* is less likely to accentuate the importance of status and ability within groups. It therefore enhances learning by students with less ability or by those from minority groups.

What do I have to do as the teacher to implement group work?

Your role is to notice when a group consistently leaves one member out and point out that they are losing valuable resources. More importantly, you will have to learn how to relinquish your role as the primary dispenser of knowledge and control. You must be willing to decentralize authority in the classroom.

How do I minimize the tendency of students to compete with each other for grades?

Students often bring a competitive ethos to our classes based on their prior experiences in school (Nicholls 1989). Traditional approaches to grading based on a bell-shaped curve reinforce this competition; students believe that they only way they can succeed is when others fail. The first author has had more than 50,000 students enroll in chemistry courses he has taught over a period of almost 40 years. Regardless of whether the course he teaches is general chemistry or an advanced-level course in organic or physical chemistry, he uses the same approach to grading. On the first day of the semester, the students are told that grades will be assigned by giving an "A" to any student who has a total number of points at the end of the semester that is equal to or greater than 90 % of the points earned by one of the top students in the course. A "B" is then given to students with between 80 and 90 % of this total, "C"'s are given to those with grades between 70 and 80 %, and so on. Over the years he has found that this approach often locates the natural breaks in the grade distribution. But, more importantly, he has found that this approach to grading minimizes students' beliefs that they can only succeed when others fail. It also makes it easier to standardize grading from semester to semester, or from year to year.

Conclusion

To some, "cooperative learning" means the small-group, interactive approach that characterized Gary's classroom or the aspect of small-group work incorporated into large lecture sections in Metz's study. In its broadest sense, however, it can include the teacher-facilitated "cooperative learning" that occurs during the dialog between students and teacher in Doug's class.

Cooperative learning should not be viewed as a threat to the instructor, or a way of replacing a dedicated teacher with a "teacher-proof curriculum." Incorporating cooperative learning into one's classroom can begin as the instructor shifts from the top of the coordinate system in Fig. 7.1 along one or more axes toward the bottom. For some, it is a shift away from whole-class toward small-group instruction, whereas for others it is a shift toward a student-centered classroom. For many, the simplest way of bringing cooperative learning into the classroom involves asking questions that both the instructor and the students know are not rhetorical; questions the instructor truly expects the students to answer. Many instructors, today, try to achieve this with mechanical devices, known as "clickers." But, as noted previously, there is no need to force students to purchase clickers and then bring them to class in order to create an active learning environment. In classes that are reasonable in size—less than 100 students—one can enter into a dialog with individual students who represent the work of the small group of students who sit in their immediate vicinity in class. For larger classes, it is possible to get students to "vote" by raising their hands.

What is most important in creating an atmosphere of cooperative learning is a shift toward an interactive environment, in which student–student and student–teacher interactions are maximized. Regardless of the axis (or axes) in Fig. 7.1 along which the shift takes place, the goal of cooperative learning is the same—to transform the instructor from "someone who teaches" to "someone who facilitates learning."

References

Bodner, G. M. (1986). Constructivism: A theory of knowledge. *Journal of Chemical Education, 63*(10), 873–878.

Bodner, G. M. (1992). Refocusing the general chemistry curriculum, why changing the curriculum may not be enough. *Journal of Chemical Education, 69*(3), 186–190.

Byers, B., & Eilks, I. (2009). The need for innovation in higher chemistry education—a pedagogical justification. In I. Eilks & B. Byers (Eds.), *Innovative methods for teaching and learning chemistry in higher education* (pp. 5–22). London: RSC Publishing.

Caldwell, J. E. (2007). Clickers in the large classroom: Current research and best-practice tips. *Life Sciences Education, 6*(1), 9–20.

Carpenter, T., Corbitt, M., Kepner, H., Lindquist, M., & Reys, R. (1981). Decimals: Results and implications from the second NAEP mathematics assessment. *Arithmetic Teacher, 28*(8), 34–37.

Damon, W. (1984). Peer education: The untapped potential. *Journal of Applied Developmental Psychology, 5*(4), 331–343.

Deutsch, M. (1949). An experimental study of the effects of cooperation and competition upon group process. *Human Relations, 2*, 199–232.

Eilks, I., & Byers, B. (2009). *Innovative methods for teaching and learning chemistry in higher education*. London: RSC Publishing.

Eilks, I., Markic, S., Bäumer, M., & Schanze, S. (2009). Cooperative learning in higher chemistry education. In I. Eilks & B. Byers (Eds.), *Innovative methods for teaching and learning chemistry in higher education* (pp. 103–122). London: RSC Publishing.

Erlwanger, S. (1973). Benny's conception of rules and answers in IPI Mathematics. *Journal of Children's Mathematical Behavior, 1*(2), 157–283.

Fies, C., & Marshall, J. (2006). Classroom response systems: A review of the literature. *Journal of Science Education and Technology, 15*(1), 101–109.

Gabbert, B., Johnson, D. W., & Johnson, R. T. (1986). Cooperative learning, group-to-individual transfer, process gain, and the acquisition of cognitive reasoning strategies. *Journal of Psychology, 120*(3), 265–278.

Johnson, R. T., & Johnson, D. W. (1979). Cooperative learning, powerful sciencing. *Science and Children, 17*(3), 26–27.

Johnson, D. W., Maruyama, G., Johnson, R., Nelson, D., & Skon, L. (1981). Effects of cooperative, competitive, and individualistic goal structures on achievement: A meta-analysis. *Psychological Bulletin, 89*(1), 47–62.

Johnson, D. W., Johnson, R. T., Tiffany, M., & Zaidman, B. (1984). Cross-ethnic relationships: The impact of intergroup cooperation and intergroup competition. *Journal of Educational Research, 78*, 75–79.

Johnson, D. W., Johnson, R. T., Roy, P., & Zaidman, B. (1985). Oral interaction in cooperative learning groups: Speaking, listening, and the nature of statements made by high-, medium-, and low-achieving students. *Journal of Psychology, 119*(4), 303–321.

Judson, E., & Sawada, D. (2002). Learning from past and present: Electronic response systems in college lecture halls. *Journal of Computers in Mathematics and Science Teaching, 21*(2), 167–181.

Kay, R. H., & LeSage, A. (2009). Examining the benefits and challenges of using audience response systems: A review of the literature. *Computers & Education, 53*(3), 819–827.

Lewin, K. (1935). *A dynamic theory of personality*. New York: McGraw-Hill.

Loudon, G. M. (2009). *Organic chemistry* (5th ed.). Greenwood Village: Roberts & Company.

Lowrey, K. A. (1996). *An alternative mode of instruction: A detailed analysis of three units in Medicinal Chemistry 204*. Unpublished M.S. thesis. Purdue University.

McLeish, J. (1968). *The lecture method*. Cambridge: Cambridge University.

MacArthur, J. R., & Jones, L. L. (2008). A review of literature reports of clickers applicable to college chemistry classrooms. *Chemistry Education Research and Practice, 9*(3), 187–195.

Martyn, M. (2007). Clickers in the classroom: An active learning approach. *Educause Quarterly, 30*(2), 71–74.

Metz, P. A. (1987). The effect of interactive instruction and lectures on the achievements and attitudes of chemistry students (Doctoral dissertation, Purdue University, 1987). *Dissertation Abstracts International, 49*, 474A.

Nicholls, J. G. (1989). *The competitive ethos and democratic education*. Cambridge: Harvard University Press.

Okebukola, P. A. (1986). Reducing anxiety in science classes: An experiment involving some models of class interaction. *Educational Research, 28*(2), 146–149.

Ross, J. A. (1988). Improving social-environmental studies problem solving through cooperative learning. *American Educational Research Journal, 25*(4), 573–591.

Sharan, S. (1980). Cooperative learning in small groups: Recent methods and effects on achievement, attitudes, and ethnic relations. *Review of Educational Research, 50*(2), 241–271.

Slavin, R. E. (1983). When does cooperative learning increase student achievement? *Psychological Bulletin, 94*, 429–445.

Slavin, R. E. (1984). Students motivating students to excel: Cooperative incentives, cooperative tasks, and student achievement. *Elementary School Journal, 85*(1), 53–63.

Slavin, R. E. (1987). Developmental and motivational perspectives on cooperative learning: A reconciliation. *Child Development, 58*(5), 1161–1167.

Slavin, R. E. (1988). Cooperative learning: Where behavioral and humanistic approaches to classroom motivation meet. *Elementary School Journal, 88*(1), 29–37.

Slavin, R. E., Leavey, M. B., & Madden, N. A. (1984). Combining cooperative learning and individualized instruction: Effects on student mathematics achievement, attitudes, and behaviors. *Elementary School Journal, 84*(4), 409–422.

Stodolsky, S. S. (1985). Telling math: Origins of math aversion and anxiety. *Educational Psychologist, 20*(3), 125–133.

Tobin, K., & Fraser, B. J. (1987). *Exemplary practice in science and mathematics education.* Perth: Western Australia Institute of Technology.

Towns, M. H., & Grant, E. R. (1997). I believe I will go out of this class actually knowing something: Cooperative learning activities in physical chemistry. *Journal of Research in Science Teaching, 34*(8), 819–835.

Towns, M. H. (1998). How do I get my students to work together? Getting cooperative learning started. *Journal of Chemical Education, 75*(1), 67–69.

Vygotsky, L. S. (1986). *Thought and Language. Revised and edited by Alex Kozulin.* Cambridge, MA: MIT Press.

Warring, D., Johnson, D. W., Maruyama, G., & Johnson, R. (1985). Impact of different types of cooperative learning on cross-ethnic and cross-sex relationships. *Journal of Educational Psychology, 77*(1), 53–59.

Wieman, C., & Perkins, K. (Nov 2005). Transforming physics education. *Physics Today, 58*(11), 36–41.

Chapter 8
Problem Solving Through Cooperative Learning in the Chemistry Classroom

Liberato Cardellini

> *The best answer to the question, "What is the most effective method of teaching?" is that it depends on the goal, the student, the content, and the teacher. But the next best answer is, "Students teaching other students".*
>
> Wilbert McKeachie

Introduction

Many chemistry instructors complain about their students' lack of interest in the subject and their low motivation to learn it. Students often enter my class without being able to solve simple stoichiometric problems, such as "*10.00 g of Na_2CO_3 react with 10.00 g of HCl. One of the reagents is completely consumed. Calculate the grams obtained of every product, explain the reasoning you used to do it, and outline a method to verify your results.*" Some remember a rote-learned algorithm and can solve the first part of the problem, but few can explain their logic or verify their results. They do not seem to believe that this activity deserves much effort, reflecting an attitude that arriving at the answer is more important than understanding the solution process.

Like learning itself, problem solving is an important and complex enterprise involving many cognitive processes, including knowledge retrieval from procedural and declarative memory, selection among alternative solution procedures, and validation or refutation of obtained solutions. As instructors of general chemistry at the university level, we routinely teach our students procedures for basic stoichiometric and reaction equilibrium calculations. We should also feel obliged to foster in them an appreciation for chemistry and learning that will motivate them to develop the capacity for independent high-level problem solving, and to teach them in a manner that promotes such development. The traditional

L. Cardellini (✉)
Università Politecnica delle Marche, Ancona, Italy
e-mail: l.cardellini@univpm.it

I. Devetak and S. A. Glažar (eds.), *Learning with Understanding in the Chemistry Classroom*, DOI: 10.1007/978-94-007-4366-3_8,

lecture-based instructional approach has been frequently shown by research to be deficient in achieving these goals.

What can be done to improve the interest of students and the standard of their learning? *Cooperative learning* is a well-tested and validated response to this question. I have had considerable success in my classes putting students participating in the general chemistry course to work on high-level problems in teams under conditions such that each member is held individually accountable for all the work done by his or her team. Cooperative learning has the potential of promoting the development of both cognitive and interpersonal skills, and it is one of the few instructional approaches that offer didactic advantages in large enrollment courses (Cooper 1995; Felder and Brent 2007). "When science students are given tasks that demand high levels of cognitive skills and/or personal characteristics such as perseverance and positive attitudes toward science, cooperative learning has the potential to contribute significantly to cognitive and affective development" (Lazarowitz et al. 1994).

Cooperative learning is probably the most exhaustively researched instructional method in all of education (Ledlow 2001). The widespread support for peer learning (Mazur 1997) has been stimulated by the success of cooperative learning. A robust and rapidly growing body of research, included some meta-analysis, confirms the effectiveness of cooperative learning in higher education (Slavin 1980; Johnson et al. 1981; Ziegler1981; Okebukola and Ogunniyi 1984; Lazarowitz et al. 1988; Felder 1996; Springer et al. 1997; Johnson et al. 1998, 2000; Prince 2004; Chiappetta and Koballa 2006; Felder and Brent 2007; Johnson and Johnson 2009).

Learning meaningfully requires the construction of new knowledge. The construction of new knowledge happens through the consideration of new ideas and the reasoned observation of events, interpreted and mediated through concepts that we already own. It can be seen as a dynamic process open to intellectual competition; a collection of progressive transitions between models having a different grade of explicative capacity, which encourage conceptual reorganization through cognitive disputes (Smith et al. 1981).

According to, cognitive development is a social process and the skill to reason increases through interaction with peers and experts (Vygotsky 1962). Working in groups also promotes the development of skills in critical reasoning. Students working cooperatively can engage in discussions with their peers in which they construct and extend conceptual understanding of what is being learned and develop shared mental models. "Cognitively it provides an opportunity for elaboration—putting material into one's own words" (McKeachie 1994).

What is Cooperative Learning?

Cooperative learning is an instructional approach to group work that involves students working in teams toward a common goal. Beyond developing cognitive skills, cooperative learning helps students develop important skills of teamwork, conflict management, and leadership, skills they need to be successful as professionals and in their personal lives. The most widely accepted cooperative learning model in higher education is probably that of David and Roger Johnson of the University of Minnesota. According to the Johnson and Johnson model (Johnson et al. 2006), a learning exercise can be classified as cooperative learning if the following elements are present: positive interdependence, individual accountability, face-to-face promotive interaction, appropriate use of collaborative skills, group processing.

Positive interdependence Team members are obliged to rely on one another to achieve the common goal. If any team members fail to do their part of work, everyone suffers consequences. Students take responsibility for their own learning and for the learning of their teammates. In problem solving, the instructor creates positive interdependence by giving students different roles and requiring group members to agree on the answer and on the strategies for solving each problem. In group problem solving, "communication will be greater where interdependence is highest" (Raven and Shaw 1970). It is considered by some to be the most important element for the success of cooperative learning (Gillies and Boyle 2009). Positive interdependence is successfully structured when group members perceive that they are linked with each other in a way that one cannot succeed unless everyone succeeds: group members have to know that they sink or swim together (Johnson et al. 1998).

Individual accountability All students in a group are held accountable for doing their share of the work and for the mastery of all the material to be learned. Individual accountability can be achieved by giving individual examinations covering the complete content of the assignment or project, and also using a variety of other techniques to be discussed.

Face-to-face promotive interaction Although some of the group work may be parceled out and done individually, some must be done interactively, with group members providing one another with feedback, challenging reasoning and conclusions, and perhaps most importantly, teaching and encouraging one another. One of the Ten Educational Commandments of Alex Johnstone is to give students the opportunity to teach because you don't really learn until you teach (Johnstone 1997). (See also, the McKeachie quotation that begins this chapter.)

Appropriate use of collaborative skills Students are encouraged and helped to develop and practice trust-building, leadership, decision-making, communication, and conflict management skills.

Group processing Team members set group goals, periodically assess what they are doing well as a team, and identify changes they will make to function more effectively in the future. Towns (1998) provides a series of statements that can facilitate group processing.

Criteria for Team Formation

Experts in cooperative learning distinguish between *informal cooperative learning* (often also referred to as *active learning*)—short exercises presented in class to non-fixed groups of two or more students—and *formal cooperative learning*—longer and more complex exercises presented to groups of students that work together through a significant part of the course (Johnson et al. 2006; Smith et al. 2005). Excellent learning outcomes result from both approaches (Prince 2004). Felder and Brent (2009b) discuss active learning structures and offer implementation suggestions; the remainder of this chapter concerns only formal cooperative learning.

In formal cooperative learning, students work in groups on problems, projects, laboratory reports, or on anything else the teacher deems suitable. The work may be done all or partially in class or outside. Techniques to meet the five defining criteria of cooperative learning can be found in the literature (Felder and Brent 1994; Nurrenbern 1995; Felder 1996; Slavin 1995; Millis and Cottell 1998; Johnson et al. 2006; Felder and Brent 2007).

Formal cooperative learning groups should be made of students with different levels of skills (Felder and Brent 2007). In well-functioning diverse groups, all the students benefit from such organization: weaker students have the benefit of being helped by their more gifted colleagues, and the stronger students (who generally are initially most resistant to working in groups) have the benefit of learning by teaching. As any professor knows, even when we understand an argument, the act of formulating explanations and thinking of examples and answering questions leads to an in-depth understanding that might not be reachable otherwise. Groups formed entirely of the best students tend to split the work and complete their parts separately instead of working as a real team, and as they do not have the need to explain to others, they do not achieve the in-depth learning that derives from teaching.

Another rule for group formation is to avoid isolating members of under-represented minorities at risk for dropping out (Oakley et al. 2004). Such students tend to take relatively passive roles within groups, either by their choice or because they are forced into such roles by their teammates. If, for example, women are a minority of the students in a chemistry curriculum, groups formed of all men, all women, equal numbers of each sex, or a majority of women are acceptable, but groups with only one woman should be avoided (Felder and Brent 2007).

Both of these rules of thumb—mixed skill levels and avoiding isolating members of at-risk minorities—are only achievable if the instructor forms the teams instead of leaving to students the task of organizing themselves. Research sustains this conclusion (Obaya 1999). When students form their own teams, friends tend to cluster together and better students seek each other. One way to form teams is to randomly form temporary training groups for the first 3 weeks of a course; give a written test during this period; and use the results to form the permanent teams. If the students object that they want to choose their own teammates, an effective response is that when they are in the working world they will not have that option and they might as well get used to assigned teams now.

In the literature there is no consensus on the optimal team size: it depends on the subject and the scope of the assignment. A team of two is obviously optimal in a physical or computer laboratory with two-person workstations. For assignments and projects, teams of three or four are generally considered optimal (Felder and Brent 2007): groups of two do not offer adequate diversity of ideas and approaches and they have no clear mechanism for conflict resolution, and in groups of five or more it is easy for one or more team members to be less than fully engaged.

Teams of three are considered ideal by several authors (Heller and Hollabaugh 1992; Robinson 1995; Laughlin et al. 2006), but not every class has a number of students exactly divisible by three, and so having teams of both three and four is ideal. If students often drop a course early in the semester, forming mostly teams of four decreases the chances that many teams will fall below critical mass.

In the first lesson of the course, after announcing the group work requirement and the advantages of working in teams, I ask the students to complete the Motivated Strategies for Learning Questionnaire (see later) and I also ask for their college entrance examination grades. Using such information I form teams of three using the criteria previously mentioned. The few students who do not provide their grades are distributed randomly among the teams. Before I announce the makeup of the teams, the students work on assignments with classmates seated next to them.

It is important to explain the reasons for using cooperative learning when the students are first told they will be working in teams. I describe the interpersonal skills developed by working in groups, and tell them that those skills will be vital in their professional careers where they will certainly have to work in teams. I then tell them that less class time will be spent on lectures and more will be devoted to solving problems in their teams, promising them that I will correct *every* problem they solve and give each of them suggestions for improvement.

Cooperative learning requires careful preparation and implementation because instructors must ensure that the five defining criteria are met. As teachers, we need to reinforce appropriate social behaviors and discourage inappropriate ones, as personalities clearly influence the way in which students interact when they work in teams (Bertucci et al. 2005). Instructors must also be prepared to deal with problematic situations such as hitchhikers, dominant students, and non-cooperative team members (Oakley et al. 2004). It is important to deal with relational conflicts, because they are not only unfavorable for learning but also have detrimental social effects (Damon et al. 2002). "The best way to prevent school violence is to replace disparagement with respect, exclusion with inclusion, and lonely isolation with collaborative community" (Kagan 2001).

It is important for instructors to remember that most students have never been taught how to work in groups, and teams sometimes do not work as well one would hope (O'Donnell and O'Kelly 1994). "Unfortunately, successful cooperative group does not just happen according to the formula. The ability, maturity, and discipline of the students are big factors regarding how well the strategy will work" (Chiappetta and Koballa 2006). "The most important advice I could give a teacher who is planning to use cooperative learning is *be prepared*" (Slavin 1995). Guides to managing teams and helping them cope with difficulties can be found in the literature (Open Teaching Toolkit 1999; Oakley et al. 2004; Felder and Brent 2007).

It is not always easy to develop social cohesion between group members. Participating students need to develop social skills and tolerance for peers when working in teams. To minimize the potential difficulties mentioned above, I give the students participating in the general chemistry course handouts about how to work successfully in teams in which I stress the necessity of genuinely respecting and valuing each other's contributions, resolving disagreements amicably, and fulfilling their responsibilities in different team roles (Millis and Cottell 1998; Sleet et al. 1996; Cardellini and Felder 1999).

The formation of teams can be problematic, because the teacher does not know the students' motivations toward learning the subject (Bertucci et al. 2006), and if motivations are too diverse team dysfunctionalities can result. This problem can be addressed by forming new teams midway through the course unless every member of a team requests to stay together (Oakley et al. 2004).

Cooperative Problem Solving

It is well known that chemistry is for many students difficult, not well liked, and sometimes boring (Herron 1986; Nakhleh 1992; Johnstone 1993; Herron 1996; Childs and Sheehan 2009). According to Johnstone, the difficulties may lie in the both the intrinsic nature of the subject and the quality of its instruction. "The more I have studied chemistry, chemical education and the psychology of learning, the more I have become aware that we are trying to share our beautiful subject with young people in an apparently 'logical' way and, at the same time conflicting with what we know about the way people learn ('psychological')" (Johnstone 2000).

In the usual approach to chemistry instruction, the solution of problems is reduced to rote execution of some procedure, without any real cognitive gain for the students. Cooperative learning has been shown to have positive impact on students' problem-solving skills (Johnson et al. 1980; Qin et al. 1995; Millis and Cottell 1998). My teaching experiences support that conclusion. When I used the method for the first time, I started with a few questions and very short exercises in a lesson (I still use them) and then I increased the exercises and the time spent on them as I gained confidence with the method. At this point, about half of my class time is spent on group problem solving. The approach I use was developed by Johnson et al. (2006). The goal is to solve the problem correctly in a cooperative framework. The students have to develop and agree on one solution, and every team member must be able to explain the strategy used to solve the problem and to verify their solution. Positive interdependence is promoted by asking students to write their name on the solution sheet and the role they assumed, with the roles rotating in each new exercise. The students know that one of them will be randomly called to the blackboard to present and explain the solution, which assures individual accountability. I inform the students that many of the problems they will solve in teams or as homework will be included in the tests, which provides a high level of motivation to solve the problems.

Students participating in the general chemistry course are asked to solve problems related to every stoichiometric topic without explaining beforehand how the problems should be attacked, so sometimes teams go wrong. For example, the students participating in the general chemistry course solve the problem of Na_2CO_3 reacting with hydrochloric acid in the very first lesson, working in pairs. Several groups solve the problem correctly, but normally few or none of them can verify the correctness of the result. My goal in this task is to make them aware that they do not know how to approach and solve problems systematically. "Textbook solutions to problems and solutions presented by instructors on the blackboard are always efficient, well-organized paths to correct answers" (Herron 1986), that "provide no indication of the false starts, dead ends, illogical attempts, and wrong solutions that characterize the efforts of students when they work in problem solving" (Herron 1990).

After that initial experience, I can easily convince the students of the necessity of a different approach to problem solving. After some instruction, I present them with the same stoichiometric calculation. While the groups solve the problem, I wander around and look over the shoulders of some teams, making comments or suggestions, and also control the time spent on the task. As the course unfolds, more and more students ask for explanations. I never explain how to solve the problem, but I give clues for helping them reason and continue to cooperate. Then I collect the solutions and call someone to the blackboard to solve the problem and explain the solution. At times I ask the class how to determine whether a solution step is right or wrong. After the class agrees with a solution, I ask if there are more questions, and then proceed with the lecture or give another exercise.

Before starting each lecture, I collect the students' homework problem solutions. I subsequently correct each solution, noting the solution times and whether the students explained their steps, used proper units and the correct number of significant digits, and verified the results. The correctness of the calculations and the numeric result are important: in my General Chemistry course, the relative error allowed is 1 %. I give feedback on the students' performance and never miss an occasion to praise students by e-mail or in class who excel in something related to learning or problem solving.

An important issue is how to deal with the errors made by students while solving problems, particularly problems on new topics. The key is not the error in itself but understanding what went wrong. "When students make what the teacher considers to be an error, the teacher should try to find out what train of thought led the student to make that statement" (Cardellini 2006a). "Everyone has to learn starting from his/her own actual repertoire. This is why errors are not bad, but good in the educational enterprise: They tell every learner about the biases in his/her own repertoire of schemes. For this reason teachers should avoid associating learners' errors with negative feelings, emotions, or punishments" (Cardellini and Pascual-Leone 2004). Dealing in this way with errors is productive: as the course proceeds, I find fewer and fewer errors in the homework problem solutions, and when I examine the solutions I find increasing evidence of students correcting themselves.

One final consideration is about the use of extrinsic rewards as part of the cooperative learning method (Slavin 1995). Significantly positive effects of rewards on achievement were found in elementary and secondary schools (Slavin 1996). I choose not to give rewards because I want all students to contribute to the solution of problems and to maximize their participation in the group's discussion. It may be true that if there is a reward every member will make their best effort to contribute to the success of the group, but there is also a risk that the better students will do the work, discounting the contributions of less able group members. The only indirect reward for working in teams is the assurance that they will learn more and more meaningfully; in this way they will get something useful also for subsequent courses, and better scores on the exams. Students participating in the general chemistry course can get a bonus if they are able to solve problems in a way that are judged appropriate, original or new (Cardellini 2006b).

Reflection on the Practice

Students' motivation in academic tasks is influenced by their personal beliefs and by the learning environment (Ames 1992). The nature of the environment can be critical. "In supportive environments teachers expressed enthusiasm for learning, were respectful, used humor, and voiced expectations that all students would learn" (Patrick et al. 2003). The first days of a class are important for establishing a supportive environment. Our enthusiasm for the subject and our interest in the students' learning it can make chemistry interesting and relevant for them. If we are able to motivate some of them early in the course, they will lead and make more probable the engagement of their classmates. A number of authors offer suggestions for establishing a supportive learning environment early in a course (Hardy and Kirkwood 1994; Felder and Brent 2008, 2009a).

Motivation is more a process than a product: every class session should involve a variety of stimulating activities in class. A positive learning environment "…engages students in some higher-order intellectual activity: encouraging them to compare, apply, evaluate, analyze, and synthesize, but never only to listen and remember" (Bain 2004). A study investigated how students' level of motivation and use of specific cognitive and self-regulatory strategies changed over time in a course. It was found that their confidence that they would do well in class decreased over time, and they were decreasingly likely to believe that chemistry was important or useful to them (Zusho and Pintrich 2003). According to Richard Shavelson, in order to *engage* the students and making them *exert effort* in their learning, "they must relate new information to existing ideas. To this end, the content of education must be conceptually rich and challenging. Engaged and effortful learning occurs when students, confronted with challenging-but-within-reach-material *choose* to cognitively reorganize that material by modifying their prior knowledge to accommodate the new knowledge" (Novak 2010, Foreword).

Student motivation has to do with students' desire to participate in the learning process. Scholars distinguish between intrinsic and extrinsic motivation (Ryan and Deci 2000). A student who is intrinsically motivated undertakes an activity for its own sake, for the enjoyment it provides, the learning it permits, or the feelings of accomplishment it evokes. Research has shown that intrinsically motivated students tend to use strategies that require more effort and that allow them to process information more deeply than their extrinsically motivated colleagues (Lepper 1988). An extrinsically motivated student undertakes activities with the goal of obtain some reward or avoid some punishment external to the activity itself, such as grades or parents and teacher approval. An instructor may do the difference in motivating students to learn, because "stimulating students' motivation to learn includes encouraging them to use thoughtful information-processing and skill-building strategies when they are learning. This is quite different from merely offering them incentives for good performance later" (Brophy 2004).

Such an active learning environment is certainly very favorable for students because they have a variety of learning styles, according to the Index of Learning Style (Soloman and Felder 1988). This environment can also be very suitable for the development of self-regulated learning (Boekaerts 1997). The majority of students participating in the general chemistry course arrive at the university with great confidence in their capacities and very motivated toward the study, according to the Motivated Strategies for Learning Questionnaire (MSLQ), (Pintrich and De Groot 1990; Pintrich et al. 1993). But in such a learning environment, the individual response of students is also different (Vermetten et al. 2002): as with other pedagogical interventions, not all students like it.

This study examined a group of engineering students (9 females and 145 males, aged 19–22) in the second term of their first year at university. Three psychological measurements were applied to the group to see if there was any relationship between the results and the quality of the creative problem solving resulting from this approach. These were Formal Operational Reasoning, measured using the Group Assessment of Logical Thinking (GALT) test (Roadrangka et al. 1983). For $N = 54$ students, the scores ranged from 10 to 24 (out of 24) with a mean of 20.6 and standard deviation of 2.6. Disembedding ability, was measured using the Field Dependence/Field Independence test devised and calibrated by El-Banna (1987) based upon the original work of Witkin (Witkin 1974; Witkin et al. 1977; Witkin and Goodenough 1981). Out of a possible score of 20, for $N = 54$ students, the range achieved was 2–18, with a mean value of 12.8 and a standard deviation of 3.8. MSLQ: for $N = 148$ students, the scores ranged from 134 to 249 (out of 280), mean value of 200.8 and a standard deviation of 21.0.

The number of solutions of problems was about 13,000 from 89 students (mean value: 144.7; standard deviation 75.5); 20 students solved one or more problems in a creative way (Cardellini 2006b). After 6 months, 71 students passed the exam (mean mark: 25.7; standard deviation 4.6) and the majority of them handed over the material used for studying the general chemistry exam: 321 concept maps and 637 résumés were collected. The number of solutions of problems solved by the best students (final mark equal or greater than 27 out of 30) was about 5,500: mean value, 166.2; standard deviation 74.3 (from 37 to 335).

I set the stage for cooperative learning on the first day of class, when I explain to the students that we will be spending relatively little time on lectures and considerable time on problem solving in teams, and I briefly summarize the research showing that this approach will lead to more learning and better grades for most of them (Towns 1998). I also emphasize that we have a mutual goal, for all students to get interested in chemistry and to pass the exam, and that we should work cooperatively to achieve it. I then form teams of three and assign distinct roles to each team member that will rotate over the course of the semester, and I give them some challenging non-technical problems to get them accustomed to the way the class will be run (Cardellini 2006b).

As the course proceeds, the problems call for an increasing range of knowledge and problem-solving skills. While some students are initially resistant, their continuing success helps most of them develop growing confidence in their abilities, and by the midpoint of the course almost all of them express satisfaction with the class and in some cases strong enthusiasm. Most importantly, their problem-solving skills and interest in the course subject are significantly greater than they ever were when I taught more traditionally.

At the end of the course I ask the students to evaluate my teaching and to offer suggestions for improving the course. With the aim of improving my teaching, I use an action research approach, because "The fundamental aim of action research is to improve the practice rather then to produce knowledge. The production and utilization of knowledge is subordinate to, and conditioned by, this fundamental aim" (Elliott 1991). From the students' suggestions and from my observations I reflect about the improvements I can use in the next course: if my knowledge grows in teaching, the students will benefit (Shulman 2004). The teacher can know about the right direction of her/his teaching considering some indicators: students' attitudes and interest toward the subject must increase (Goldman et al. 1998).

A modification I plan to make in the future is to incorporate peer ratings into my evaluation of the students' performance. Some students may be able to cheat a teacher, but they cannot hide from their peers. A well-constructed peer rating protocol can promote individual accountability and can also give students valuable feedback on what they are doing well in their teams and which areas might need improvement (Brown 1995; Millis and Cottell 1998; Kaufman et al. 2000). In performing the latter function, peer ratings help address the fourth criterion of cooperative learning, which requires that the students be helped to develop the interpersonal skills required for high-performance teamwork, and the fifth criterion, which calls for teams to reflect on how well they are performing and to contemplate changes that will lead to improved performance. A free, powerful, and well-validated online peer rating system called CATME (Comprehensive Assessment of Team Member Effectiveness, http://www.catme.org) makes collecting peer ratings easy for instructors, uses the ratings to adjust team project grades for individual team members, gives the students feedback on their performance without compromising the confidentiality of the ratings, and gives instructors information about teams and individual students whose performance might require instructor intervention.

Conclusions

Teaching cannot be reduced to formulaic methods because many variables affect learning (Herron 1996; Bransford et al. 2000), including self-efficacy, utility and relevance of the material, and goal orientation (Ames 1992; Zusho and Pintrich 2003). Psychological factors and previous knowledge also play a role (Ausubel et al. 1978; Reid and Yang 2002). According to Shulman (2002), learning begins with students engagement and motivation. Because motivation to learn has affective components, we have to embody what we believe or preach: we need to show to our students the values we hope to see in our students' behavior. Students want professors who are knowledgeable and excited about the material and who care about their learning (Richlin 2006). Conversely, teachers who lack passion for the subject matter of their courses, are unable to connect students' interest to that subject matter, and convey indifference or hostility toward students, are likely to be ineffective (Carson 1999; Felder and Brent 2009a).

For learning chemistry with understanding we might need to take into account the human element. Learning is a human endeavor, so teachers can make a difference in the perception, motivation, and maybe in the lives of many students if we are able to interest them in our subject. We take learning seriously if we take their learning seriously, which can require a considerable effort. In my last course, about 100 of the 154 students enrolled attended the lectures regularly, and I received hundreds of e-mails and sent just as many. Students participating in the general chemistry course were quite happy to work in this engaged way and to be fully involved.

Cooperative learning refers to work done by student teams under conditions that assure positive interdependence, individual accountability, face-to-face interaction, development of team skills, and self-assessment of team functioning. Extensive research has shown that relative to the traditional instructional approach that emphasizes individual and competitive work, properly implemented cooperative learning leads to greater learning, greater confidence and self-esteem as problem solvers, higher student retention, and superior development of communication and social skills, such as leadership, project management, and conflict resolution skills (Dougherty et al. 1995; Johnson et al. 2000; Felder and Brent 2007). The technique has been widely used with considerable success in chemistry (Felder and Brent 2007). However, the benefits of cooperative learning are not automatic, and if not properly implemented, the method can create more difficulties for teachers than benefits for students. Instructors who undertake it should make sure they have read the literature on the method, understand the potential pitfalls (including student resistance to the method and possible team disfunctionalities), and know proven strategies for minimizing or eliminating those pitfalls.

Acknowledgments I am grateful to Richard Felder of North Carolina State University, who like a father helped me with advice and encouragement in my first hesitating steps in using cooperative learning, and who also offered suggestions for improving an early draft of this chapter.

References

Ames, C. (1992). Classrooms: Goals, structures, and student motivation. *Journal of Educational Psychology, 84*(3), 261–271.

Ausubel, D. P., Novak, J. D., & Hanesian, H. (1978). *Educational Psychology: A Cognitive View* (2nd ed.). New York: Holt, Rinehart and Winston. In J. D. Novak & D. B. Gowin (1995), *Learning how to learn*. New York: Cambridge University Press.

Bain, K. (2004). *What the best college teachers do*. Cambridge (MA): Harvard University Press.

Bertucci, A., Meloni, C., Conte, S., & Cardellini, L. (2005). The role of personality, gender and interaction in a cooperative and in a computer supported collaborative learning task. *Journal of Science Education*, *6*(Special issue), 32–35.

Bertucci, A., Conte, S., & Cardellini, L. (2006). Cooperative Learning. Una metodologia al servizio dei contesti educativi e del "brainframe" del nuovo millennio. *Orientamenti Pedagogici*, *53*(3), 517–536.

Boekaerts, M. (1997). Self-regulated learning: A new concept embraced by researchers, policy makers, educator, teachers, and students. *Learning and Instruction, 7*(2), 161–186.

Bransford, J. D., Brown, A. L., Cocking, R. R., Donovan, M. S., & Pellegrino, J. W. (2000). *How people learn. Brain, mind, experience, and school* (Expanded ed.). Washington, D.C.: National Academy Press.

Brophy, J. (2004). *Motivating students lo learn* (2nd ed.). Mahwah, NJ: Lawrence Erlbaum.

Brown, R. W. (1995). *Auto rating: Getting individual marks from team marks and enhancing teamwork.* Paper presented at the Frontiers in Education Conference, IEEE/ASEE, Pittsburgh.

Cardellini, L. (2006a). The foundations of radical constructivism: An interview with Ernst von Glasersfeld. *Foundations of Chemistry, 8*(2), 177–187.

Cardellini, L. (2006b). Fostering creative problem solving in chemistry through group work. *Chemistry Education Research and Practice, 7*(2), 131–140.

Cardellini, L., & Felder, R. M. (1999). L'apprendimento cooperativo: Un metodo per migliorare la preparazione e l'acquisizione di abilità cognitive negli studenti. *La Chimica nella Scuola, 21*(1), 18–25.

Cardellini, L., & Pascual-Leone, J. (2004). On mentors, cognitive development, education, and constructivism: An interview with Juan Pascual-Leone. *Journal of Cognitive Education and Psychology [online], 4*(2), 199–219.

Carson, B. H. (1999). Bad news in the service of good teaching: Students remember ineffective professors. *Journal of Excellence in College Teaching, 10*(1), 91–105.

Chiappetta, E. L., & Koballa, T. R. (2006). *Science instruction in the middle and secondary schools. Developing fundamental knowledge and skills for teaching* (6th ed.). Upper Saddle River, NJ: Pearson.

Childs, P. E., & Sheehan, M. (2009). What's difficult about chemistry? An Irish perspective. *Chemistry Education Research and Practice, 10*(3), 204–218.

Cooper, M. M. (1995). Cooperative learning. An approach for large enrolment courses. *Journal of Chemical Education, 72*(2), 162–164.

Damon, C., Buchs, C., & Butera, F. (2002). Epistemic and relational conflicts in sharing identical vs. complementary information during cooperative learning. *Swiss Journal of Psychology, 61*(3), 139–151.

Dougherty, R. C., Bowen, C. W., Berger, T., Rees, W., Mellon, E. K., & Pulliam, E. (1995). Cooperative learning and enhanced communication. Effect on student performance, retention, and attitudes in general chemistry. *Journal of Chemical Education, 72*(9), 793–797.

El-Banna, H. (1987). *The development of a predictive theory of science education based upon information processing theory*. Ph.D. thesis, University of Glasgow.

Elliott, J. (1991). *Action research for educational change*. Milton Keynes: Open University Press.

Felder, R. M. (1996). Active-inductive-cooperative learning: An instructional model for chemistry? *Journal of Chemical Education, 73*(9), 832–836.

Felder, R. M., & Brent, R. (1994). *Cooperative Learning in Technical Courses: Procedures, Pitfalls, and Payoffs*. (ERIC Document Reproduction Service ED 377038, October)

Felder, R. M., & Brent, R. (2007). Cooperative learning. In P. A. Mabrouk (Ed.), *Active learning: Models from the Analytical Sciences* (pp. 34–53). Washington, DC: American Chemical Society.

Felder, R. M., & Brent, R. (2008). The ten worst teaching mistakes I. Mistakes 5–10. *Chemical Engineering Education, 42*(4), 201–202.

Felder, R. M., & Brent, R. (2009a). The ten worst teaching mistakes I. Mistakes 1–4. *Chemical Engineering Education, 43*(1), 15–16.

Felder, R. M., & Brent, R. (2009b). Active Learning: An Introduction. *ASQ Higher Education Brief, 2*(4). Retrieved from http://www.ncsu.edu/felder-public/Papers/ALpaper(ASQ).pdf

Gillies, R. M., & Boyle, M. (2009). *Teachers' reflections on cooperative learning: Issues of implementation*. Teacher Education: Teaching and.

Goldman, S. R., Mayfield-Stewart, C., Bateman, H. V., Pellegrino, J. W., & Cognition and Technology Group at Vanderbilt (1998). Environments that support meaningful learning. In L. Hoffmann, A. Krapp, K. A. Renninger, & J. Baumert (Eds.), *Interest and learning* (pp. 184–196). Kiel: IPN.

Hardy, T., & Kirkwood, V. (1994). Toward creating effective learning environments for science teachers: The role of a science educator in the tertiary setting. *International Journal of Science Education, 16*(2), 231–251.

Heller, P., & Hollabaugh, M. (1992). Teaching problem solving through cooperative grouping. Part 2: Designing problems and structuring groups. *American Journal of Physics, 60*(7), 637–644.

Herron, J. D. (1986). What can we do about sue: A case study of competence. *Journal of Chemical Education, 63*, 528–531.

Herron, J. D. (1990). Research in chemical education: Results and directions. In M. Gardner, J. G. Greeno, F. Reif, A. H. Schoenfeld, A. Disessa, & E. Stage (Eds.), *Toward a scientific practice of science education* (pp. 31–54). Hillsdale, NJ: Erlbaum.

Herron, J. D. (1996). *The chemistry classroom. Formulas for successful teaching*. Washington, DC: American Chemical Society.

Johnson, D. W., Skon, L., & Johnson, R. T. (1980). Effects of cooperative, competitive, and individualistic conditions on children's problem-solving performance. *American Educational Research Journal, 17*(1), 83–93.

Johnson, D. W., Maruyama, G., Johnson, R., Nelson, D., & Skon, L. (1981). Effects of cooperative, competitive, and individualistic goal structures on achievement: A meta-analysis. *Psychological Bulletin, 89*(1), 47–62.

Johnson, D. W., Johnson, R. T., & Smith, K. A. (1998). Cooperative learning returns to college: What evidence is there that it works? *Change, 30*(4), 27–35.

Johnson, D. W., Johnson, R. T., & Stanne, M. E. (2000). *Cooperative Learning Methods: A meta-analysis*. Minneapolis: University of Minnesota, Cooperative Learning Center. Retrieved October 25, 2009, from http://www.co-operation.org/pages/cl-methods.html

Johnson, D. W., Johnson, R. T., & Smith, K. A. (2006). *Active Learning: Cooperation in the College Classroom* (3rd ed.). Edina, MN: Interaction Book.

Johnson, D. W., & Johnson, R. T. (2009). An educational psychology success story: social interdependence theory and cooperative learning. *Educational Researcher, 38*(5), 365–379.
Johnstone, A. H. (1993). The development of chemistry teaching. A changing response to changing demand. *Journal of Chemical Education, 70*(9), 701–705.
Johnstone, A. H. (1997). Chemistry teaching—science or alchemy? *Journal of Chemical Education, 74*(3), 262–268.
Johnstone, A. H. (2000). Teaching of chemistry—logical or psychological? *Chemistry Education Research and Practice, 1*(1), 9–15.
Kagan, S. (2001). Teaching for character and community. *Educational Leadership, 59*(2), 50–55.
Kaufman, D. B., Felder, R. M., & Fuller, H. (2000). Accounting for individual effort in cooperative learning teams. *Journal of Engineering Education, 89*(2), 133–140.
Laughlin, P. R., Hatch, E. C., Silver, J. S., & Boh, L. (2006). Groups perform better than the best individuals on letter-to-numbers problems: effects of group size. *Journal of Personality and Social Psychology, 90*(4), 644–651.
Lazarowitz, R., Hertz, R. L., Baird, J. H., & Bowlden, V. (1988). Academic achievement and on-task behaviour instructed in a cooperative students instructed in a cooperative small investigative group. *Science Education, 72*(4), 475–487.
Lazarowitz, R., Hertz-Lazarowitz, R., & Baird, J. H. (1994). Learning science in a cooperative setting: Academic achievement and affective outcomes. *Journal of Research in Science Teaching, 31*(10), 1121–1131.
Ledlow, S. (2001). An interview with Richard Felder. Retrieved October 30, 2009, from http://clte.asu.edu/active/feldertranscript.htm
Lepper, M. R. (1988). Motivational considerations in the study of instruction. *Cognition and Instruction, 5*(4), 289–309.
Mazur, E. (1997). *Peer instruction. A user's manual.* Upper Saddle River, NJ: Pearson.
McKeachie, W. J. (1994). *Teaching tips. Strategies, research, and theory for college and university teachers* (9th ed.). Lexington, MA: D. C. Heath and Co.
Millis, B. J., & Cottell, P. G, Jr. (1998). *Cooperative learning for higher education faculty.* Phoenix, AZ: Oryx Press.
Nakhleh, M. B. (1992). Why some students don't learn chemistry. Chemical misconceptions. *Journal of Chemical Education, 69*(3), 191–196.
Novak, J. D. (2010). *Learning, creating, and using knowledge: Concept maps as facilitative tools in schools and corporations* (2nd ed.). New York: Routledge.
Nurrenbern, S. C. (Ed.). (1995). *Experiences in cooperative learning: A collection for chemistry teachers.* Madison, WI: Institute for Chemical Education.
Oakley, B., Felder, R. M., Brent, R., & Elhajj, I. (2004). Turning student groups into effective teams. *Journal of Student Centered Learning, 2*(1), 9–34.
Obaya, A. (1999). Getting cooperative learning. *Science Education International, 10*(2), 25–27.
O'Donnell, A. M., & O'Kelly, J. (1994). Learning from peers: Beyond the rhetoric of positive results. *Educational Psychology Review, 6*(4), 321–349.
Okebukola, P. A., & Ogunniyi, M. B. (1984). Cooperative, competitive, and individualistic science laboratory interaction patterns—Effects on students' achievement and acquisition of practical skills. *Journal of Research in Science Teaching, 21*(9), 875–884.
Open Teaching Toolkit. (1999). *Effective group work.* Milton Keynes: Open University Press.
Patrick, H., Turner, J. C., Meyer, D. K., & Midgley, C. (2003). How teachers establish psychological environments during the first days of school: associations with avoidance in mathematics. *Teachers College Record, 105*(8), 1521–1558.
Pintrich, P. R., & De Groot, E. V. (1990). Motivational and self-regulated learning components of classroom academic performance. *Journal of Educational Psychology, 82*(1), 33–40.
Pintrich, P. R., Smith, D. A. F., Garcia, T., & McKeachie, W. J. (1993). Reliability and predictive validity of the motivated strategies for learning questionnaire (MSLQ). *Educational and Psychological Measurement, 53*(3), 801–813.
Prince, M. (2004). Does active learning work? A review of the research. *Journal of Engineering Education, 93*(3), 223–231.

Qin, Z., Johnson, D. W., & Johnson, R. T. (1995). Cooperative versus competitive efforts and problem solving. *Review of Educational Research, 65*(2), 129–143.
Raven, B. H., & Shaw, J. I. (1970). Interdependence and group problem-solving in the triad. *Journal of Personality and Social Psychology, 14*(2), 157–165.
Reid, N., & Yang, M. J. (2002). Open-ended problem solving in school chemistry: A preliminary investigation. *International Journal Science Education, 24*(12), 1313–1332.
Richlin, L. (2006). *Blueprint for learning. Constructing college courses to facilitate, assess, and document learning*. Sterling: VA, Stylus.
Roadrangka, V., Yeany, R. H., & Padilla, M. J. (1983). *The construction and validation of Group Assessment of Logical Thinking (GALT)*, Paper presented at the Annual Meeting of the National Association for Research in Science Teaching, Dallas.
Ryan, R. M., & Deci, E. L. (2000). Intrinsic and extrinsic motivations: Classic definitions and new directions. *Contemporary Educational Psychology, 25*(1), 54–67.
Robinson, W. R. (1995). Groups-of-three study. In S. C. Nurrenbern (Ed.), *Experiences in cooperative learning: A collection for chemistry teachers* (pp. 148–149). Madison, WI: Institute for Chemical Education.
Shulman, L. S. (2002). Making differences: A table of learning. *Change, 34*(6), 36–44.
Shulman, L. S. (2004). Those who understand: Knowledge growth in teaching. In S. M. Wilson (Ed.), *The wisdom of practice. Essays on teaching, learning, and learning to teach* (pp. 189–215). San Francisco, CA: Jossey-Bass.
Slavin, R. E. (1980). Cooperative learning. *Review of Educational Research, 50*(2), 315–342.
Slavin, R. E. (1995). *Cooperative learning* (2nd ed.). Needham Heights, MA: Allyn & Bacon.
Slavin, R. E. (1996). Research on cooperative learning and achievement: What we know, what we need to know. *Contemporary Educational Psychology, 21*(1), 43–69.
Sleet, R. J., Hager, P., Logan, P., & Hopper, M. (1996). *Broader Skill Requirements of Science Graduates* (pp. 128-130). Sydney: University of Technology.
Smith, K., Johnson, D. W., & Johnson, R. T. (1981). Can conflict be constructive? Controversy versus concurrence seeking in learning groups. *Journal of Educational Psychology, 73*(5), 651–663.
Smith, K. A., Sheppard, S. D., Johnson, D. W., & Johnson, R. T. (2005). Pedagogies of engagement: Classroom-based practices. *Journal of Engineering Education, 94*(1), 1–15.
Soloman, B. A., & Felder, R. M. (1988). Index of Learning Style. Retrieved October 20, 2009, from http://www.engr.ncsu.edu/learningstyles/ilsweb.html
Springer, L., Stanne, M. E., & Donovan, S. (1997). Measuring the Success of Small-Group Learning in College-Level SMET Teaching: A Meta-Analysis. Retrieved November 15, 2009, from http://www.wcer.wisc.edu/archive/CL1/CL/resource/scismet.pdf
Towns, M. H. (1998). How do i get my students to work together? Getting cooperative learning started. *Journal of Chemical Education, 75*(1), 67–69.
Vermetten, Y. J., Vermunt, J. D., & Lodewijks, H. G. (2002). Powerful learning environments? How university students differ in their response to instructional measures. *Learning and Instruction, 12*(3), 263–284.
Vygotsky, L. S. (1962). *Thought and language*. Cambridge, MA: MIT Press.
Witkin, H. A. (1974). *Psychological differentiation: studies of development*. New York: Wiley.
Witkin, H. A., Moore, C. A., Goodenough, D. R., & Cox, P. W. (1977). Field-dependent and field-independent cognitive styles and their educational implications. *Review of Educational Research, 47*(1), 1–64.
Witkin, H. A., & Goodenough, D. R. (1981). *Cognitive styles: essence and origins. Field dependence and field independence*. New York: International Universities Press.
Ziegler, S. (1981). The effectiveness of cooperative learning teams for increasing cross-ethnic friendship: Additional evidence. *Human Organization, 40*(3), 264–268.
Zusho, A., & Pintrich, P. R. (2003). Skill and will: the role of motivation and cognition in the learning of college chemistry. *International Journal of Science Education, 25*(9), 1081–1094.

Chapter 9
The Learning Company Approach to Promote Active Chemistry Learning: Examples and Experiences from Lower Secondary Education in Germany

Torsten Witteck, Katharina Beck, Bettina Most, Stephan Kienast and Ingo Eilks

Introduction and Pedagogical Justification

There seems to be no question that laboratory work is a natural component of secondary school chemistry teaching (Nakhleh et al. 2002). In the chemistry education literature over the last decades, laboratory work has repeatedly been described as essential for teaching the scientific method and for learning chemical content (Blosser 1983), or for understanding the nature of science (Duschl 1990). Nevertheless, there have also been very cautious remarks that laboratory work's positive role in learning chemistry is not self-evident (e.g., Hofstein and Lunetta 1982; Tobin 1990; Hofstein 2004). This position is increasingly supported by empirical research evidence, which shows that including laboratory work into our classrooms does not automatically lead to positive results in either cognitive achievement or learning about the scientific method and the nature of science (Lunetta 1998; Nakhleh et al. 2002; Hofstein 2004). Lunetta (1998, p. 250) with

T. Witteck
Westfalen-Kolleg (Grammar School), Bielefeld, Germany

K. Beck
KGS Kirchweyhe (Comprehensive School), Brinkum, Germany

B. Most
Konrad-Adenauer-Realschule (Middle School), Hamm, Germany

S. Kienast
Ursulinengymnasium (Grammar School), Werl, Germany

I. Eilks (✉)
Department of Biology and Chemistry, Institute for Didactics of Science (IDN), Didactics of Chemistry, University of Bremen, Bremen, Germany
e-mail: ingo.eilks@uni-bremen.de

I. Devetak and S. A. Glažar (eds.), *Learning with Understanding in the Chemistry Classroom*, DOI: 10.1007/978-94-007-4366-3_9,

reference to Champagne, Gunstone and Klopfer (1985), Eylon and Linn (1988), Tasker (1981) described the situation like this:

> Students often fail to understand the relationship between the purpose of the investigation and the design of the experiment which they had conducted, they do not connect the experiment with what they have done earlier, and they seldom note the discrepancies between their own concepts, the concepts of their peers, and those of the science community. […] To many students, a 'lab' means manipulating equipment but not manipulating ideas.

From these less-than-promising results in recent practices, Bates (1978), Tobin (1990), Gunstone and Champagne (1990), Herrington and Nakleh (2003) and Hofstein (2004), among others, worked out why carrying out laboratory work in school chemistry lessons is quite often not very successful. Out of this emerge several suggestions for changes in the common practice of school chemistry experimentation. One—and perhaps the most often suggested change—is that school laboratory work should go beyond doing "cookbook recipe" experiments (Tamir and Lunetta 1981; Tobin 1990). Pleas for opening up experimental tasks with an eye toward more inquiry-oriented modes, increased student self-regulation, and the inclusion of planning, evaluation and documentation of experiments into students' activities have been repeatedly made (Hofstein and Lunetta 1982; Gunstone and Champagne 1990; Kipnis and Hofstein 2007). However, it has also been suggested that experiments should be more carefully connected to both content learning and meaningful contexts (Lunetta 1998).

Coming from the distributed cognition framework, Nakleh et al. (2002) have additionally suggested the construction of a more cooperative laboratory learning environment in order to recognize the dynamic and interactive aspect of knowledge generation. This aspect was recently worked out further, e.g., by Witteck and Eilks (2005) in the application of the pairs-to-share method during the protocolling of experiments. Lunetta (1998) also argued for taking the aspect of communication during laboratory activities more thoroughly into account based on cooperative learning factors such as those described in detail by Johnson and Johnson (1999) and the social constructivist approach (Hodson and Hodson 1998).

Positive developments in achievement, including growing skills and self-esteem, were reported in laboratory work settings where such cooperative learning functions well (Lunetta 1990; Quin et al. 1995). The same has been described for cooperative learning in science education in general (Lazarowitz and Hertz-Lazarowitz 1998). Furthermore, Nakleh et al. (2002) suggested that we should also think about different forms of assessment, with the focus on good group performance. They also referred directly back to the method of poster presentations.

This chapter summarizes the development and application of three different lesson plans for secondary chemistry education in Germany, which explicitly follow the above-mentioned suggestions for organizing lab-work, by applying the learning company approach—a cooperative learning method coupled with strategies leading classroom practice into a different style of experimentation—to chemistry education. The lesson plans deal with methods of separating matter in initial lower secondary chemistry lessons, first phenomena of chemical reaction in the second year, and introducing acid–base chemistry in advanced lower secondary chemistry education. The lesson plans are described briefly and summarize the experiences given in teacher and student feedback. An overall outlook is also provided.

The Path Toward the Learning Company Approach in Chemistry Teaching

The idea for a "learning company" (or "learning office") was developed in the field of didactics of vocational education. According to Pätzold and Lang (1999), the learning company is a didactically-constructed classroom structure, analogous to existing or "ideal" companies. The learning environment should allow for simulating practical, profession-oriented tasks in business. The students are supposed to learn through a model based on already-existing or idealized companies, how processes in a company occur. The aim is for students to recognize how businesses are structured and how differing tasks within the company are linked by a cause–effect relationship to one another, to the economy and to the environment. This learning also incorporates aspects of functional cooperation within and between different departments or individuals.

The learning company idea obviously is an interesting and authentic method for vocational training. But, one might think that teaching how business and industrial structures function is not a goal of lower secondary chemistry education in compulsory schools. However, the above-mentioned thoughts might offer possibilities to promote motivation, the encouragement of cooperative learning and the framing of experimentation in a different style also in chemistry teaching. This is why a group of teachers in a roughly 10-year-old Participatory Action Research project (Eilks and Ralle 2002; Eilks 2003; Eilks and Markic 2011) decided to engage themselves with the question of whether, how, and with what effect the learning company approach might be an interesting innovation for compulsory lower secondary chemistry education. Some seven years ago the group started to develop lesson plans connecting the learning company idea with new approaches toward more openness and student self-regulation, i.e., concerning laboratory work tasks. The group members were all familiar with the creation and evaluation of lesson plans for lower secondary chemistry education using the cyclical process of Participatory Action Research. They were also very experienced in applying different forms of cooperative learning in such lesson plans, e.g., jigsaw classrooms (Eilks 2005), inside-outside-circle (Witteck et al. 2004), or pairs-to-share (Witteck and Eilks 2005).

The initial idea was provoked by a project awarded by the CEFIC Science Education Award in 1999 (CEFIC 1999), which made the group both aware of and curious about the learning company approach. Unfortunately, this special project seemed to be highly unique and hardly transferable into regular chemistry teaching in most cases. However, the group emphasized developing further examples—including the relevant teaching materials—which might potentially be useful for every German secondary school under normal teaching conditions and fulfill parts of the required governmental syllabus.

A first learning company lesson plan based on acid–base chemistry was cyclically designed and tested for tenth grade chemistry pupils. The objective was to include all relevant aspects of the respective syllabus parts in one learning company lesson plan in both the theoretical and hands-on aspects. The teaching methods selected intended to aid the pupils in performing all necessary learning steps on their own, based on small learning groups. Lessons started with open-ended tasks (goal-oriented "work orders" sent from the manager in the learning company (the teacher) to his departments (the student groups)), instead of simply using prescribed "cookbook recipes." Each lesson was based on experimental work (see below; Witteck and Eilks 2006a).

Classroom observations and teachers' reported initial experiences (stemming from communal self-reflection exercises during the action research group meetings) showed that the acid–base learning company was an amazing idea with great potential. In the teachers' opinion, their students had achieved unanticipatedly good results as compared to the teachers' predictions before testing the lesson—and pupils had managed this on their own. The teachers described very high levels of student motivation, enormously self-regulated and successful student activity, and admirable cognitive achievement. The first set of feedback from the pupils supported the teachers' view. This led the teacher group to the question of whether such an open approach is also applicable to other domains of secondary chemistry teaching, i.e., for younger students. This is why two further examples were developed and all lesson plans intensely reflected upon using the basis of student and teacher points-of-view.

The Process of Development

The lesson plans described below were developed by a team of about 15 teachers in a Participatory Action Research project accompanied by the University of Bremen (Eilks and Ralle 2002; Fig. 9.1). The action research group had already existed for about 6 years prior to working on the learning company approach (Eilks 2003; Eilks and Markic 2011). This group meets roughly every 4 weeks for one whole afternoon. During the meetings, lesson plans are developed and feedback is discussed. The entire process of structuring the lesson plan and respective materials is cyclical. Each cycle consists of development, testing, and reflection between university researchers in chemistry education and classroom practitioners. Structuring of each of the lesson plans took place over a period of about a year. The main steps in structuring were led by one practitioner (TW) from the action research group.

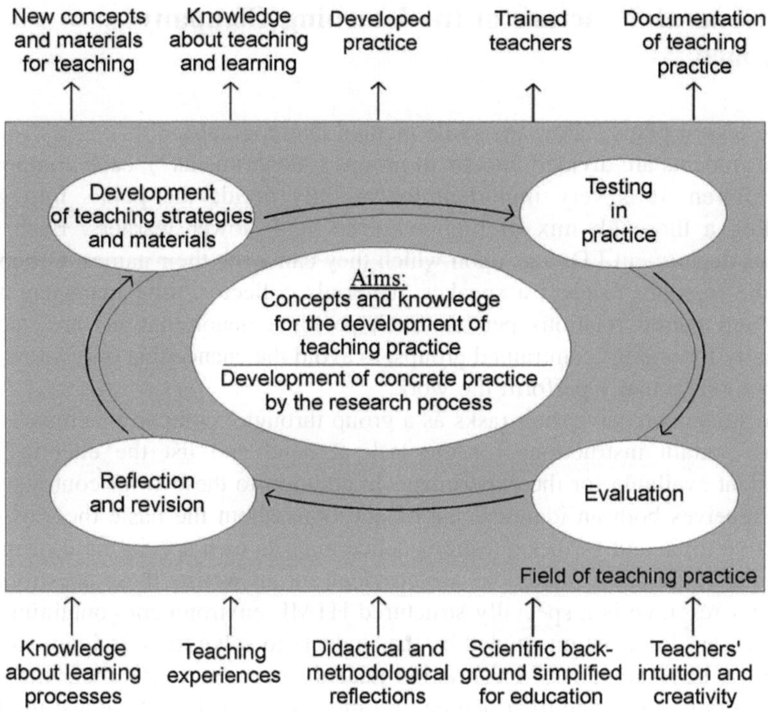

Fig. 9.1 The Model of Participatory Action Research in chemistry education by Eilks and Ralle (2002)

All the lesson plans were tested by different research group practitioners in a sample of about ten different learning groups located in different grammar, middle, and comprehensive schools. The first cycle of testing always accompanied the last steps of structuring the lesson plan. Later cycles with other learning groups then were carried out with a time lag ranging from a few months up to about a year later. Nevertheless, all groups were taught using almost an identical lesson plan and working materials. More details about the testing process can be reviewed in Witteck and Eilks (2006a), Witteck et al. (2007), Beck et al. (2010).

In all cases, the considerations of the teachers were collected in open group discussions during the regular Participatory Action Research group meetings. Additional data came from written student questionnaires, which asked for the students' personal experiences and criticisms. A combination of an open- and a Likert-type questionnaire was used for this purpose. The students were first asked in an open questionnaire to evaluate which aspects of the lesson plan were important enough to be mentioned (from the students' viewpoint)—either in a positive or a negative sense. After this, a Likert questionnaire was used to gather information on those points considered important by the teachers and researchers. The questionnaires were structured similarly to those used in, e.g., Witteck et al. (2004), or Witteck and Eilks (2005).

The General Structure of the Learning Company Approach

All the lesson plans follow the same method choreography:

The students are divided into small groups ("departments"), each composed of 4–5 children. It is very important to carefully divide the pupils into groups including a thorough mix of high-achievers and slower learners. Each group receives department I.D. tags upon which they can write their names. Groups also have the freedom to elect a speaker, materials collector, time manager, minute taker, and public relations person. Such personal denominations are valuable, especially for young or untrained groups, to avoid the chance that only a few of the students will actually perform the work.

The students receive their tasks as a group through "company memos." These memos contain instructions for the task at hand and list the chemicals and equipment available for the experiment. In addition to these work contracts, each group receives both an identical list of questions about the basic theories of the respective topic and further questions concerning its own special field of inquiry. Computer resources or textbooks are provided for answering these questions. The computer resource is a specially structured HTML-environment containing all of the necessary information needed by the students to solve their assignments. This includes a large amount of information related to everyday life about the topic.[1] The list of questions and their degree of difficulty can be used to adjust the lesson plan to a specific group's learning capabilities and a priori knowledge. For further learning, there are several experiments available in the computer resource, which are accessible via video clips. If applicable, small animations at the particle level of matter are presented.

After this, pupils are provided with 1–2 lesson periods (45 min each) of preparation time for their experiments. If electronic aids are intended, the students can spend this time learning on the computer and use the learning environment created for this purpose. Should the students already be polished enough to find potential solutions without any help, they can initially be asked to make their proposal without additional material. If no computer resources are available, hard copies of the learning environment, textbooks, and relevant working materials can be provided to the students. It is entirely possible that some of the pupils' plans might end up having to be scrapped, since their execution might not lead to the learners' desired goal. Therefore, the independent planning phase of the experiment becomes more important than the actual experimentation itself. The students can work creatively and freely develop themselves and their ideas. It is a very special event, if and when their independent planning actually leads to successful results.

---

[1] To get an idea about the HTML-learning environment a German version is available at http://www.idn.uni-bremen.de/chemiedidaktik/material/Teilchen/SaeurenBasen/SaeurenBasen/index.html.

Before starting the experiment, the students must explain their plans toward the teacher and to ask for a "green light" to start. This negotiation also has to cover any relevant questions of safety regulations and risk assessment. After discussing and planning the procedures, the learners have another 1–2 lesson periods to carry out their experiments, including careful documentation of all their activities. It is helpful at this point if access to the computer-based learning environment is also available during this phase. If a department cannot find a workable solution, the teacher can provide the pupils with ideas or—in a worst-case scenario—give them a descriptive procedure for the experiment. The laboratory work should be carefully diagrammed on a poster, so that students in the various other departments can later understand and absorb both the contents and the experimental results presented. Other forms of presentation might be also used, e.g., presentations using transparencies or PowerPoint.

In the end phase, the pupils' experimental results are presented to the whole class as unit. All students receive a worksheet, upon which they must document the results of the other departments. Furthermore, the pupils must fill out an evaluation form, which asks them to evaluate and critique the results and poster presentations from the other groups. In the final stage after the presentation, the students can secure their knowledge of the various experimental works from other groups. They have the opportunity to actively review those contents which they either did not understand the first time around, or where they still have questions about the experimental procedures or end results. This takes place through the multimedia learning environment.

Three Examples from Lower Secondary Chemistry Teaching in Germany

The following section gives a short overview of the three lesson plans mentioned above.

Dr. Taste—Introducing methods of separating matter (Witteck et al. 2007).

One typical issue found in Germany's introductory lower secondary science education is an examination of the properties of matter. This includes using these properties as a basis for choosing different methods when attempting to separate matter. In Germany, this issue is mostly conducted in grade 6 or 7 (age range 11–13). Typical methods the students must become familiar with are distillation, filtration, and centrifugation. Theoretical explanations using a particle model are not always an integrated part of the respective teaching unit. In most cases, explanations at the particle level are dealt with later in the same school year.

The action research group decided to apply the learning company approach to this part of the syllabus. The learning company "*Dr. Taste*" (in German: *Dr. Schmeck*) was constructed. This fictional company mimics an analytical institute focusing on the analysis of food and drinks, and it consists of a managing director (the teacher) and different departments (student groups). The departments are responsible for different operations within the analytical institute. Each of the groups is asked to become experts on a particular method of separating matter. *Dr. Taste* consists of seven departments: distillation, filtration, chromatography, extraction, adsorption, centrifugation, and decantation (or a subset of these with respect to the teacher's goals and group size).

As typical for the learning company approach in the interpretation described here, the students do not get a "recipe" for what to do. The departments are only supplied with goal-oriented "work orders" from the managing director (the teacher). These work orders describe a more-or-less open problem. This task to be solved is, however, embedded in the very clear-cut context of everyday life, in particular the field of food and drink production. A prototypical work order structure and layout is shown in Fig. 9.2.

The work orders do not cover prescribed procedures or sketches of the setup. This enables the openness described in the pedagogical justification above. The orders are organized so that no experimental direction is explicitly given. Thus, the assigned experimental problems have to be conquered through self-dictated, self-organized, self-responsible learning, including interpersonal communication and negotiation within the group. Nevertheless, the multimedia learning environment, textbooks, or a respective folder of information materials can act as a backup, so that the exercise can be solved without resorting to a prescribed path. These resources are handled by the teacher and given to the students with respect to their cognitive abilities, level of creativity, and foreknowledge.

Aside from controlling the level of openness and the degree of self-regulation through electronic aids, a second tool for fine-tuning openness is build into the learning company structure. This tool consists of varying the available equipment for solving the problem. In general, the more materials the students are provided with, the more they are able to invent creative solutions for their problem. The smaller the list of available materials is, the more the students are forced to seek a specifically targeted procedure (premeditated by the instructor).

Within this framework, the pupils are asked to inform themselves about the open task, its background, and any possible strategies for solving the problem. In the case of Dr. Taste, all groups are purposely offered the same equipment: beakers, Erlenmeyer flasks, test tubes, funnels, filter paper, pipettes, mortars and pestles, tripods, Bunsen burners, lighters, crystallization or Petri dishes, thermometers, curved glass rods, evaporization dishes, perforated stoppers, and paper towels. This equipment covers most of the standard tools used in early secondary chemistry lessons in Germany. In this early phase of overall chemistry education, students are asked to familiarize themselves with these tools and their functions. To allow for openness, learners are given more laboratory tools than are really necessary to solve the individual problem. This strategy follows Miller et al. (2004), who described how important it is for pupils to learn about the function of laboratory equipment for any subsequent learning in the laboratory. Playing with these tools and trying out their uses is in addition to the objectives stated within the open experimental tasks within this unit.

Dr. Taste

Institute for Foodstuff Research

Foodstuff Institute Dr. Taste

To:
Chromatography Department

- in house document-

Task

Customer	Number	Reference	Date
	257894	S-15/07	01.04.06

Food colours are frequently used in the manufacture of foodstuffs, especially cakes and sweets. Smarties brand candy is a good example. Different packages containing various colours can be purchased in stores for cooking and baking purposes. Many of these colours contain only a single colouring agent, however, many are made up of mixtures of two or more colouring agents. Find out which colours appear in different colours of Smarties.

The warehouse has provided the following chemicals and equipment for you to use:

Equipment: beakers, Erlenmeyer flasks, test tubes, funnels, filter paper, pipettes, mortar and pestle, tripod, Bunsen burner, lighter, crystallization or Petri dishes, thermometer, glass rod, evaporization dish

Chemicals: Smarties, water, alcohol, nail polish remover

Good Luck!

The Management

Fig. 9.2 Work order toward the department of chromatography

Having received their tasks and equipment, the pupils are then asked to develop their own ideas of how to solve the given problem. Each student has to search for theoretical information and come up with ideas for solving the problem in an experimental way. A specially constructed multimedia learning environment (Dr. Taste's intranet) is offered to them as an aid, as well as the use of a textbook where necessary. In each case, we used a multimedia learning environment offering information on methods of separating matter and sparking ideas for how to structure the experiments. Nevertheless, neither the textbook nor the multimedia tool explicitly describes how to conduct the experiment based on the equipment in the learners' possession. Only the teacher has such a description as a fallback plan, should the teacher decide that a group is floundering and needs assistance. Initial ideas can be located found on the Internet, but must be independently found by the pupils themselves.

Starting from the information located, the learners have to negotiate their group's strategy and are free to try out different approaches. They therefore can attune the method and pace of their work to their own personal capabilities. Strategies of structuring the groups' cooperative activities can be introduced to help the students in their self-organization, if required. One possibility, for example, is to give the students individual roles within their group: speaker, time manager, protocoller, etc. The final objective of the groups' work is the presentation of the activities, results, and background in poster form. Table 9.1 gives a short overview of the open tasks and potential solutions. The complete teaching materials in German are given in Witteck and Eilks (2006c). Figure 9.3 gives some impressions of the students' work.

Sabine Sweet & Co.—Introducing the basic phenomena of chemical reactions (Beck et al. 2010).

At the end of the first—or early in the second-year of chemistry teaching in Germany (mostly grade 6–7, age range 11–13) initial chemical reactions are introduced. Students should learn that chemical reactions are the change from one kind of substance into another. Within this framework, pupils need to learn to recognize chemical reactions, to be able to differentiate them from purely physical changes, and to learn about different attributes connected with chemical reactions.

For introducing the chemical reaction, the standard example employed is burning a candle. Different reactions between wax, wood, and different metals with oxygen are also discussed in most cases. Alternate reactions can be selected using different substances from the household. One very motivating and multifunctional compound in this respect is sugar. Sugar undergoes many different reactions, thus allowing for a broad range of different phenomena connected to chemical change.

Starting from this point, the Participatory Action Research group developed a learning company named *Sabine Sweet & Co.* (in German *Sabine Süß & Co.*). Its structure exactly follows the pattern described above. *Sabine Sweet & Co.* is a company addicted to any form of business or trading which deals with regular sugar (saccharine). Table 9.2 gives an overview of the six different departments, their tasks, and some potential solutions. During testing, some of the learning groups implemented a small shift away from the final poster product, substituting instead PowerPoint presentations in the final group work exchange phase. The complete teaching materials in German can be found in Beck et al. (2009).

Max Sour Ltd.—Introducing acids and bases and the Brønsted theory (Witteck and Eilks 2006a)

In Germany, the chemistry of acids and bases is one of the common topics covered by advanced lower secondary chemistry education. Their properties and macroscopic behavior are essential for understanding many processes in our

Table 9.1 Overview of the 'departments' in Dr. Taste's Foodstuff Institute

Department	Open task	Potential solution
Distillation	Alcoholic drinks, e.g., wine, beer, champagne, schnapps, grain alcohol, and sherry contain alcohol. But what does alcohol look like? Separate the alcohol from a sample of red wine. Describe the properties of the alcohol that you extract	One distillation apparatus should be set up by having an Erlenmeyer flask, stopper, and glass tube with a wet paper towel on it. The distillate is caught in a beaker
Extraction	Potato chips taste good, but are viewed as an unhealthy sort of food because they contain so much fat. But what does this fat actually look like? Your task is to separate the fat out of a potato chip sample and describe its properties	A fresh portion of potato chips is ground in the mortar. In an Erlenmeyer flask the pieces are stirred in acetone for 10 min. The mixture is filtered and both components are dried
Filtration	If you pulp an apple, you are left with a mixture which doesn't leave much juice to drink. If you press this mixture through a cloth, you are left with cloudy apple juice. The apple juice which you can purchase in the store is no longer cloudy. It is clear. Such juice is the most popular type of apple juice and sells better than other sorts. Grind up an apple and press out the juice. Make clear apple juice from the cloudy juice and describe its properties	The naturally cloudy apple juice is filtered very carefully. An alternative would be centrifuging
Chromatography	Smarties brand candy packages contain Smarties of various colors. Many of these colors contain only a single coloring agent, however, many are made up of mixtures of two or more coloring agents. Find out which colors appear in different colors of Smarties, which of them contain only one single coloring agent, which do consist of more	A Smarty is placed in the middle of a piece of filter paper. A single drop of water is placed very slowly and carefully onto the Smarty with a pipette or glass rod. A further drop of water is added after the coloring of the candy dissolved in the first drop. By adding 1–2 more drops singly until the colored solution drips onto the filter paper and starts to expand outwards
Adsorption	Nowadays you are able to buy soft drinks with very intensive and unusual colors, for example blue. We have purchased a bottle of the blue soft drink "Powerade." Remove the color from the blue soft drink and describe its properties	"Powerade" is placed into a beaker. Kernels of active charcoal are added to the solution and the mixture is stirred for a while. The solution is filtered
Centrifuging	Orange juice is a popular drink which appears in many varieties, for example, with or without orange pulp. Your task is to make pulp-free orange juice out of orange juice with pulp. How much pulp does 100 mL of orange juice with pulp contain?	Orange juice with pulp is placed into a centrifuge and the juice is separated from the pulp. The juice is decanted from the pulp after the separation is complete

Fig. 9.3 Students preparing, experimenting, making posters, and exchanging information

Table 9.2 Overview of the 'departments' in Sabine Sweet & Co

Department	Open task	Potential solution
Producing artificial honey	Our stock of artificial honey has shrunk significantly. Our customers are already complaining about shortages, but they have also criticized the standard product because it contains different preservatives. Your department's task is to develop a recipe for producing artificial honey made exclusively from sugar and a limited number of other natural compounds	5 g sugar is dissolved in 10 mL water. The mixture is heated to 80 °C. 1 mL of citric acid solution is added and the mixture stirred for 15 min. After the mixture has cooled, the product of glucose, fructose and water smells, tastes, and appears like honey
Producing caramel	The janitors have told you that a water leak in your caramel production department has led to corrosion. Your task is to research the formation of caramel to determine whether water is set free during the reaction. Examine the relationship between the masses of reactants and products found in caramel formation. Additionally, please find a way to produce 'caramel coloring' a brown agent used to coloring foodstuffs	10 g of sugar is heated in a beaker until the mass starts to turn brown. The evaporating vapors are condensed on a cold glass and tested for water content. The Law of Conservation of Mass is checked by repeating the experiment in a stoppered Erlenmeyer flask with a glass tube connected to a balloon. The entire apparatus is weighed before and after the reaction
Getting heat energy from sugar	Rising energy prices have the management wondering if sugar can be burned for heat energy. Unfortunately, sugar does not burn easily. Your department's task is to find a way to burn sugar with a flame to extract heat energy from it	Sugar can only be ignited by a flame if the right catalyst is present. Iron salts in the ashes of plants work well as a catalyst
Fermentation of sugar	The company needs alcohol to fill its candies. Due to cost effectiveness we intend to produce our own alcohol from sugar. Research and develop a procedure for showing the formation of alcohol	10 g of sugar and 20 g baker's yeast are added to 100 mL of water at 30 °C. The products formed are tested for carbon dioxide using lime water and a digital alcohol test is applied
Using bio-catalysis	Regular sugar (saccharose) is a problem for diabetics. Other forms of sugar (i.e., fructose) are less problematic. Fructose can be made from saccharine. Research and develop a procedure for showing the formation of fructose from saccharose	Some acid is added to a solution of sugar. After a short time, fructose formation can be proven using a fructose test strip
A mirror with the help of sugar	Classical procedures for producing mirrors have made use of special kinds of sugar. Evaluate the procedure. Research and develop a procedure showing the process of making a sugar mirror	2.5 mL of silver nitrate solution is placed in a glass tube. A drop of ammonia is added. Then sodium hydroxide and glucose are added and slightly heated. Silver deposits on the glass

everyday lives, e.g., how and why household cleaners function. Topics covered are the behavior of acids and bases, the dangers associated with them, the pH-scale and indicators, and the most important reactions of acids and bases. In addition to the phenomenological level, a submicroscopic understanding is introduced at the end of lower secondary chemistry education in middle and grammar schools. Here, the Brønsted acid–base concept is the preferred model of explanation: an acid is a proton donor and a base is a proton acceptor.

The objective of the *Max Sour Ltd.* learning company (in German: *Max Sauer GmbH*) is to deal with all relevant aspects of acid–base chemistry using a learning company lesson plan. This covers the hands-on as well as the theoretically aspects, up to and including a submicroscopic understanding of the Brønsted acid–base theory.

The lesson plan follows the single steps described previously. *Max Sour* Ltd. contains seven different departments: two research divisions ("Synthetic Indicators" and "Plant-based Indicators"), two analytical divisions ("Cosmetics" and "Pharmaceuticals"), the company's canteen, a janitorial department, and an environmental department.

Whereas the use of additional resources was optional in the other examples, in this case such multimedia learning aids become essential. The students must learn and apply a complete new model of explanation when it comes to chemical reactions. The Brønsted model cannot simply be created by the students on their own, even if the entire group consists of extreme high-achievers. This is why additional resources must be provided. They allow the learners to accustom themselves to the role of oxonium and hydroxide ions. Such aids also help in explaining the theory of proton transfer when clarifying acidic and basic behavior and neutralization.

Table 9.3 gives an overview of the departments, their tasks, and potential solutions. The teaching materials in German are described in Witteck and Eilks (2006b).

Experiences and Evaluation

The Teachers' View

In their self-reflections, the teachers considered the learning company to be a highly motivating form of learning in all the three examples described above. The students had shown themselves to be extremely curious, even during the initial presentation of the learning company idea. The pupils quickly became engaged in a competition among themselves, beginning their work with a clear focus on the problems to be solved. The teachers interpreted this to mean that the framework offered learners a quasi-authentic and very challenging situation. According to the

Table 9.3 Overview of the 'departments' in Max Sour Ltd

Department	Open task	Potential solution
Research division "Synthetic Indicators"	The standard universal indicator of the company is used up. Invent a new one from different single indicators so that the new universal indicator can differentiate between a preset range of pH-values (1, 4, 7, 10, 14)	A large number of indicators are provided for the task. The pupils must discover a good combination of the solutions by finding out the pH-ranges and colors, inductive mixing, and deductive thinking
Research division "Plant-based Indicators"	The head of the company heard of the potential of using plant extracts as indicators. Produce a new, natural indicator from radish and write an instruction manual for this indicator, including a color scale which makes predetermined pH-change points (1, 4, 7, 10, 14) visible	The students have to find out how to use the peel and extract it using alcohol. The extraction should be as concentrated as possible. The color scale can be made using a digital camera
Analytical division "Cosmetics"	Research the behavior of a "pH-skin neutral" hand-cleaning lotion (pH 5.5) and compare it with an already-existing, common soap. Furthermore, you must test an acne product and an anti-peeling product for their characteristics	The pH-values and constituents of these products must also be determined and the effect has to be researched on the Internet
Analytical division "Pharmaceuticals"	Due to the levels of stress in the company, the doctor suggests using an antacid. Research the composition and behavior of antacids and compare them with sodium hydroxide	The students must uncover the properties of the antacid with respect to the process of neutralization, either as a tablet or in a liquid form. They must also describe how both forms function
Canteen	The canteen is required to prepare its usual "pot roast with red cabbage". The bosses emphatically express their wish for red cabbage, since the red cabbage turned blue the last time. The kitchen workers must figure out why this happened and avoid repeating the mistake	The students must uncover the behavior of the coloring agents in red cabbage and their behavior under different conditions

(continued)

Table 9.3 (continued)

Department	Open task	Potential solution
Janitors	Find a way for the company's canteen to clean calcified heating-elements in the dishwasher and to free a plugged drainpipe	The students must find out how acidic cleaners work concerning the decomposition of calcification and how basic drainpipe cleaners function
Environmental department	Examine, neutralize, and dispose of the growing quantities of waste from the other six divisions from last month. Neutralize the solutions, evaporate them, and determine the mass of their constituent parts for cost-effective disposal	The students have to apply the concept of neutralization for environmentally sound disposal of acidic and basic liquids

teachers, their students seemed to directly identify themselves with their group or department.

One of the most important impressions mentioned repeatedly by the teachers were the intense, content-focused discussions taking place among their pupils. The learners were very concerned with the question of how to effectively structure promising experimental activities to solve their given task. Each of the groups found a way to solve its problem. However, their strategies differed widely, ranging from purely trial-and-error approaches to well-thought-out, meticulously planned procedures. With respect to the students' foreknowledge and cognitive abilities, inductive and, especially, deductive strategies of problem-solving were applied. Some even mixed both approaches and purposely shifted their methods due to discussion and reflection on both approaches. The offer of worksheets with example solutions by the teachers was only used in some middle school classes as a control. They were almost never used in grammar school classes.

Another important issue stemming from the teachers' reflections concerned the combination of different methodological elements within the learning environments. The teachers saw large advantages in the students' openness concerning the selection, sequencing, and weighing of their activities. Pupils were forced to repeatedly shift their focus between hands-on activities and the search for additional information in the multimedia learning environment, Internet, or textbooks. The same held true for the negotiations within the group. Such networked activity dealing with theoretical information, practical work, communication, and negotiation within the group, was described as a total turnabout from the teachers' past experiences. The teachers were amazed at the completely different atmosphere during laboratory work exercises and the increased levels of pupil self-reflection. Such a self-regulated combination of different activities while focusing on a content problem did seem to be very challenging and demanding. In the teachers' opinion, however, these hurdles were taken in stride with high levels of success and unanticipatedly few problems.

All in all, this method seemed fruitful in provoking the discussions claimed by Nakleh et al. (2002). With reference to Gunstone and Champagne (1990), this approach helps to place more emphasis on student activity in the planning and evaluating phases of experiments. This aspect was also strengthened by the final presentation containing the actions and end-results of each department. Students stepped up to the plate when it came to preparing and executing their summaries at the end of the lesson. The results from the learning groups—their minutes, posters, slides, and presentations—more than fulfilled the pre-testing phase expectations stated by the teachers for that age level. Difficulties were only observed concerning single steps and in some of the learning group compositions.

The Students' Views

The positive conclusions of the teachers for all three learning companies were supported by the students' views recorded in the open and Likert-questionnaires. The students mentioned both the fun they had had and the openness (freedom to follow their own interests, ideas, and pathways) which they felt while developing and conducting the experiments in a group especially positively.

> The biggest difference for me was that we had to do everything ourselves and that we weren't as strictly controlled as in other experiments in class. I especially liked the fact that we had to get to the results all by ourselves and were allowed to simply forge ahead as we liked. That was really fun.

> I liked the work with the computer learning environment because we could work more independently than normal. In addition we had to create the experiment by ourselves. I really liked having to work independently and having to carefully think out how to perform the experiment.

The teaching approach was described as demanding, but also challenging. The lessons were seen as making the pupils more active and making chemistry lessons more interesting.

> With this method of teaching I could be a lot more active and think and act more freely. I understood most everything better than in "normal" lessons. I find that independent work is much more demanding and more interesting than normal teaching methods.

The high levels of self-reliance, the cooperative atmosphere (being allowed to do things together as a group independently from the teacher), and the experimental activities without the teachers' close guidance and control were mentioned as being especially positive.

> I liked the group work. It was independent work. We had to do all the work ourselves without the teacher helping us (well, maybe a little)… I really liked the learning company because we could perform experiments. You could do experiments with the things and materials that were given to us and some were pretty cool.

Some of the students also recognized the importance of first making themselves familiar with the intended hands-on activities before they carried them out.

> I liked the learning company because we first developed the write-up of the experiment and later on we had to conduct the experiment (exactly in the way we planned it).

Table 9.4 gives an overview of the frequency of selected aspects as mentioned by the students in the open questionnaire. This particular bunch comes from four learning groups in Example 1 Dr. Taste's Institute (Witteck et al. 2007).

The learners' opinion was also very positive for the Likert-items, as illustrated by the third example, Max Sour Ltd. The students mentioned that they liked the learning company a lot better than normal classroom teaching. The alternative teaching methods were classified by the students as less boring and more fun (82 % "agreed" or "pretty much agreed" to this). They especially liked the fact

Table 9.4 Frequency of aspects mentioned in the open questionnaire ($N = 86$)

Positive comments	Frequency
More intense and effective learning	18
Liked the cooperative learning atmosphere	39
Liked being more self-sufficient	
• Because of being more active	44
• Because being responsible for my own actions	2
• Because of being allowed to make my own decisions about the best path	31
• Because of the possibility for self-regulated and self-organized experimentation	24
Considered the lesson plan explicitly attractive and described having more fun in class	38
Negative comments	
Too little control by the teacher	4
Too high demands due to a limited time frame	1
Problems within the individual groups	4
Being disturbed by too much noise	2

($N = 86$; 3 students did not answer the open questionnaire, most students gave comments in more than one category)

that they could work on the content with their fellow students (88 %), carry out experiments without being given a "recipe" (80 %), and perform experiments in self-directed, small groups (95 %). At the same time, the students expressed feelings that they had learned a lot (84 %), had worked more intensively (76 %), and had been more independent (93 %) than they normally experienced in a chemistry classroom situation. The students did not perceive the interdependence between self-organization, problem-solving, and interpersonal relationships to solve the work orders as a negative factor. Nevertheless, in this particular case nearly half the learners expressed a lack of adequate control by the teacher (Fig. 9.4). In all three lesson plans, the opinions about the cooperative and laboratory work phases were much more positive than those regarding work on the computer. Although computers themselves have been described as positively affecting outcomes, opinions about the multimedia phase of the plan were split, despite the fact that the computer work had been carefully devised to complement the rest of the setting. Similar results for all items were also revealed for the other two examples.

Conclusions

The three lesson plans described above attempted to refine pathways to cooperative, inquiry-oriented learning by adopting the learning company approach. Each teaching unit showed that it had high potential for promoting active chemistry learning in its respective setting (see also Witteck and Eilks 2006a; Witteck et al. 2007 and Beck et al. 2010). All three examples successfully led to a different

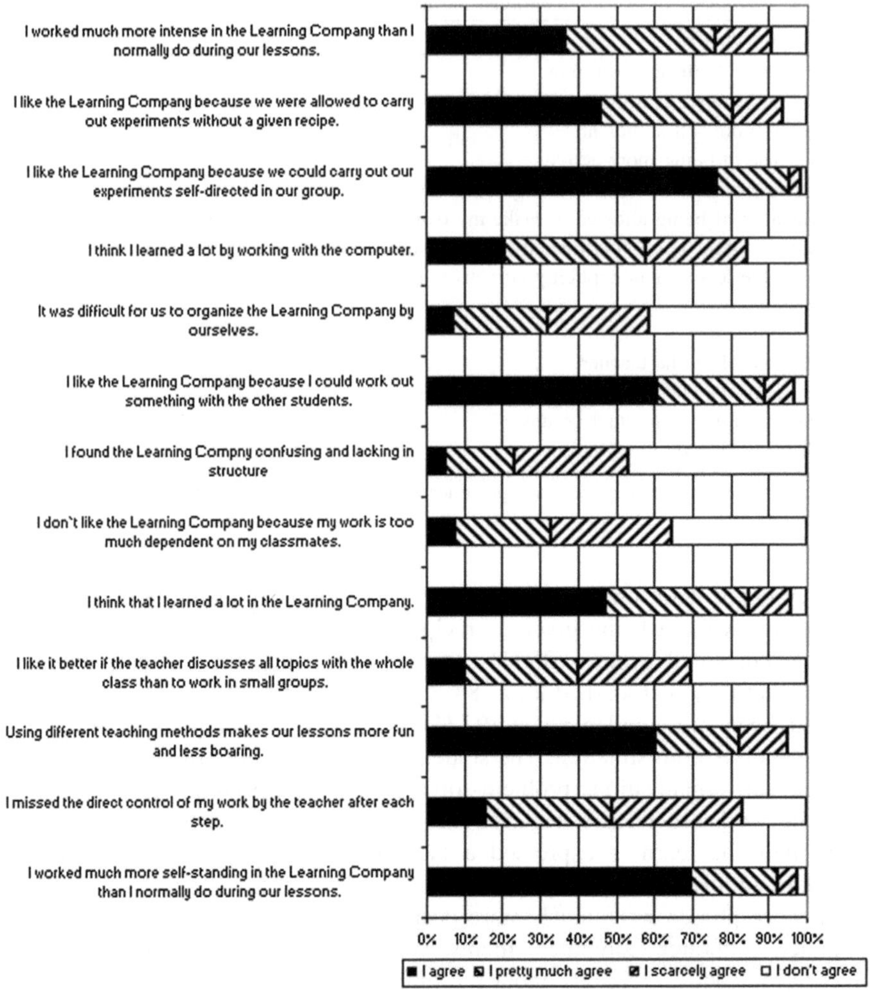

Fig. 9.4 Responses to the Likert-items concerning example 3 Max Sour Ltd. (Witteck and Eilks 2006a)

learning culture within chemistry lessons. This new culture was characterized by self-dictated, self-organized, self-responsible learning, according to repeatedly stated opinions during the instructors' self-reflection periods. From our point of view, we believe that a cooperative learning environment approach in order to solve open experimental tasks holds great promise for overcoming the lack of student motivation which is often reported in chemistry education. Additionally, the studies suggest that such learning forms not only do not decrease cognitive achievement, but also clearly evidence great methodological potential for increasing student skills in various strategies of problem-solving, negotiation, and presentation relevant to scientific inquiry. Based on this, we consider this method to evidence considerable potential in aiding students to learn typical paths of scientific inquiry as a part of their growing knowledge about the Nature of Science. In our considered opinion, this aspect leads us to the conclusion that educators should more often use inquiry-based experiments and (as suggested repeatedly in past research efforts, e.g., Hofstein and Lunetta 1982; Gunstone and Champagne 1990; Kipnis and Hofstein 2007) to implement them systematically in cooperative modes of learning and instruction (Nakleh et al. 2002).

References

Bates, G. R. (1978). The role of the laboratory in secondary school science programs. In M. B. Rowe (Ed.), *What research says to the science teacher* (pp. 55–82). Washington: NSTA.

Beck, K., Witteck, T., & Eilks, I. (2009). Sabine Süß & Co—Eine Lernfirma rund um chemische Reaktionen des Zuckers für den Chemieanfangsunterricht [Sabine Sweet & co—A learning company around chemical reactions of sugar for introudctory chemistry teaching]. *RAABits Chemie Sekundarstufe I.* I B/8, 1–12.

Beck, K., Witteck, T., & Eilks, I. (2010). Open experimentation on phenomena of chemical reactions via the learning company approach in early secondary chemistry education. *Eurasia Journal of Mathematics, Science and Technology Education, 6*(2), 163–171.

Blosser, P. E. (1983). What research says: The role of the laboratory in science teaching. *School, Science and Mathematics, 83*(2), 165–169.

CEFIC (1999). Retrieved July 01, 2006, from www.cefic.be/activities/science-edu/award/99/03.htm.

Champagne, A. B., Gunstone, R. F., & Klopfer, L. E. (1985). Instructional consequences of students' knowledge about physical phenomena. In L. H. T. West & A. L. Pines (Eds.), *Cognitive structure and conceptual change* (pp. 61–68). New York: Academic Press.

Duschl, R. A. (1990). *Restructuring science education: The importance of theories and their development*. New York: Teachers College Press.

Eilks, I. (2003). Co-operative curriculum development in a project of Participatory Action Research within chemical education: Teachers' reflections. *Science Education International, 14*(4), 41–49.

Eilks, I. (2005). Experiences and reflections about teaching atomic structure in a jigsaw classroom in lower secondary school chemistry lessons. *Journal of Chemical Education, 82*(2), 313–320.

Eilks, I., & Markic, S. (2011). Effects of a long-term Participatory Action Research project on science teachers' professional development. *Eurasia Journal of Mathematics, Science and Technology Education, 7*(3), 149–160.

Eilks, I., & Ralle, B. (2002). Participatory Action Research within chemical education. In B. Ralle & I. Eilks (Eds.), *Research in chemical education—What does this mean?* (pp. 87–98). Aachen: Shaker.

Eylon, B. S., & Linn, M. C. (1988). Learning and instruction: An examination of four research perspectives in science education. *Review of Educational Research, 58*(3), 251–301.

Gunstone, R. F., & Champagne, A. B. (1990). Promoting conceptual change in the laboratory. In E. Hegarty-Hazel (Ed.), *The student laboratory and the science curriculum* (pp. 159–182). London: Routledge.

Herrington, D. G., & Nakhleh, M. B. (2003). What defines effective chemistry laboratory instruction? Teaching assistant and student perspectives. *Journal of Chemical Education, 80*(10), 1197–1205.

Hodson, D., & Hodson, J. (1998). From constructivism to social constructivism: A Vygotskyan perspective on teaching and learning science. *School Science Review, 79*(289), 33–41.
Hofstein, A. (2004). The laboratory in chemistry education: Thirty years of experience with developments, implementation, and research. *Chemistry Education Research and Practice, 5*(2), 247–264.
Hofstein, A., & Lunetta, V. N. (1982). The role of the laboratory in science teaching: Neglected aspects of research. *Review of Educational Research, 52*(2), 201–217.
Johnson, D. W., & Johnson, R. T. (1999). *Learning together and alone: Cooperative, competitive, and individualistic learning* (5th ed.). Boston: Allyn and Bacon.
Kipnis, M., & Hofstein, A. (2007). The inquiry laboratory as a source for development of metacognitive skills. *International Journal of Science and Mathematics Education, 6*(3), 601–627.
Lazarowitz, R., & Hertz-Lazarowitz, R. (1998). Cooperative learning in the science curriculum. In B. J. Fraser & K. G. Tobin (Eds.), *International handbook of science education* (pp. 623–640). Dordrecht: Kluwer.
Lunetta, V. N. (1990). Cooperative learning in science, mathematics and computer problem solving. In M. Gardner, J. Greeno, F. Reif, A. Schoenfeld, A. Disessa, & E. Stage (Eds.), *Toward a scientific practice of science education* (pp. 235–249). Hillsdale: Lawrence Erlbaum.
Lunetta, V. N. (1998). The school science laboratory: Historical perspectives and contexts for contemporary teaching. In B. J. Fraser & K. G. Tobin (Eds.), *International handbook of science education* (pp. 249–268). Dordrecht: Kluwer.
Miller, L. S., Nakhleh, M. B., Nash, J. J., & Meyer, J. A. (2004). Students' attitudes toward and conceptual understanding of chemical instrumentation. *Journal of Chemical Education, 81*(12), 1801–1808.
Nakhleh, M. B., Polles, J., & Malina, K. (2002). Learning chemistry in a laboratory environment. In J. K. Gilbert, O. De Jong, R. S. Justi, D. F. Treagust, & J. H. Van Driel (Eds.), *Chemical education: Towards research-based practice* (pp. 69–94). Dordrecht: Kluwer.
Pätzold, G., & Lang, M. (1999). *Lernkulturen im Wandel [Learning cultures in change]*. Bielefeld: Bertelsmann.
Qin, Z., Johnson, D. W., & Johnson, R. T. (1995). Cooperative versus competitive efforts and problem solving. *Review of Educational Research, 65*(2), 129–143.
Tamir, P., & Lunetta, V. N. (1981). Inquiry related tasks in high school science laboratory handbooks. *Science Education, 65*(5), 477–484.
Tasker, R. (1981). Children's views and classroom experiences. *Australian Science Teachers Journal, 27*(3), 33–37.
Tobin, K. G. (1990). Research on science laboratory activities: In pursuit of better questions and answers to improve learning. *School Science and Mathematics, 90*(5), 403–418.
Witteck, T., & Eilks, I. (2005). Writing up an experiment cooperatively. *School Science Review, 86*(317), 107–110.
Witteck, T., & Eilks, I. (2006a). Max Sour Ltd.—Open experimentation and problem solving in a cooperative learning company. *School Science Review, 88*(323), 95–102.
Witteck, T., & Eilks, I. (2006b). Die Max Sauer GmbH—Materialien für eine Lernfirma zu Säuren und Basen in der Sekundarstufe I [Max Sour Ltd—Materials for a learning company on acids and bases for lower secondary education]. *RAABits Chemie Sekundarstufe I, I/E, 9*, 1–26.
Witteck, T., & Eilks, I. (2006c). Lernumgebung und Materialien zu einer Lernfirma zu Methoden der Stofftrennung für den Anfangsunterricht (Dr. Schmeck's Lebensmitteluntersuchungen) [Learning environemnt and materials for a learning company on methods of seperating matter for initial chemistry education (Dr. Taste's investigations of food)]. Retrieved March 01, 2013 from http://www.chemiedidaktik.uni-bremen.de/material/Teilchen/Stofftrennung/Stofftrennung.htm.

Witteck, T., Most, B., Kienast, S., & Eilks, I. (2007). A lesson plan on separating matter based on the learning company approach—A motivating frame for self-regulated and open lab-work in introductory chemistry lessons. *Chemistry Education Research and Practice, 7*(2), 108–119.
Witteck, T., Most, B., Leerhoff, G., & Eilks, I. (2004). Cooperative learning on the internet using the ball bearing method (Inside-outside-circle). *Science Education International, 15*(3), 209–223.

Section II
Approaches in Chemistry Teaching for Learning with Understanding

Teaching Strategies

To ensure that students engage themselves mentally active in learning, science teachers at all levels of education try to use different teaching strategies or approaches. If learning is to take place, students should think about the content presented by the teacher, textbook, online or otherwise. The most important problem that science teachers face is how to motivate students to learn for their future lives as active citizens. It is important to present to the students that they are not learning just to pass exams, but to become scientifically literate adults, who will make important and correct decisions. To achieve this, teachers and science education researchers try to find the ways to make students learn science concepts with understanding and for life. This usually involves experimental work, using different pictorial material, context-based approaches, multimedia environments... Some of these aspects are presented in second part of Section II "Approaches in Chemistry Teaching for Learning with Understanding".

Chapter 10 by Williamson presents teaching chemical concepts through implementing three levels of chemical concepts. This chapter upgrades Chaps. 1–3 and 6. Williamson concludes that traditionally, chemistry at all educational levels has been taught as a mathematical course that emphasised algorithmic problem-solving almost exclusively. Because research showed that students at all levels have trouble with conceptual understanding of chemistry new approaches to teach chemistry had to emerge. Some chemistry teachers at all levels of education intuitively teach chemistry conceptually, many still have difficulty knowing how to do this and what teaching strategies are available to them. Conceptual teaching, as a teaching strategy emphasises students' ability to explain relationships, to predict outcomes, to visualise/explain particle behaviour and to understand the macroscopic, particulate, symbolic and mathematical levels of chemical concepts presentations. In this chapter the author highlights different teaching strategies to make chemistry teaching more conceptual and less mathematical when it is not really necessary to deeply understand chemical concepts. These strategies can be used with large or small classes and they include the application of macroscopic representations, particulate representations (both dynamic and static models), group problem-solving, algorithmic and conceptual assessments, etc.

Students' achievement in learning chemistry through the design and construction approach to laboratory activity and its relation with their prior achievements and motivation to learn is discussed by Vrtačnik, Sodja and Juriševič in Chap. 11. The authors claim that the design and construction approach to activities in chemistry lessons for middle school students is regarded as an authentic science activity, and that this approach to learn chemistry is rarely practised in science classes. In this approach students were asked to design their own experiments and control variables. The results suggest that students' success in the design and construction approach depended upon the complexity of a particular task. A significant drop off in achievements and motivation scores was found with tasks based on more abstract thinking, e.g. analysing data and setting up hypotheses. In evaluating the design and construction approach, students expressed the highest appreciation for a positive classroom atmosphere and their active participation in the laboratory activity. The research findings revealed that students with higher achievement in chemistry are also highly extrinsically and intrinsically motivated for learning chemistry and have a higher academic self-concept.

In Chap. 12, Parchmann, Dunker and Endres look at the value of the contexts as chemistry learning catalysts for students and teachers. They presented two approaches; *Chemie im Kontext* and CHEMOL. The authors emphasised that the teaching tradition in German chemistry classrooms uses experiments as stimuli for chemistry learning. However, these experiments were often not connected to the experiences of students and to relevant topics. The projects *Chemie im Kontext* for secondary level and CHEMOL for primary level therefore use contexts from the students' daily-life or contexts connected to important socio-scientific issues to raise questions which can then be researched by students. The active learning of the students is supported by scaffolding material and a variety of teaching and learning methods, involving different roles for teachers and students. The authors also presented that active learning did not take place only in the classroom but also within the meetings of teachers of different schools and chemistry educators from university. The chapter describes the structure of both projects, gives two examples of teaching and learning processes and discusses results from qualitative and quantitative research studies.

In Chap. 13 entitled "How Does Level of Guidance Affect Understanding When Students Use a Dynamic Simulation of Liquid–Vapor Equilibrium?" Akaygün and Jones present research on visualisations of molecular structure and dynamics being powerful learning tools. The authors emphasised that students often need guidance to understand what they are seeing in simulations, animations and static visualisations of chemical phenomena, particularly at the submicroscopic level. Scientific visualisations that are enlightening for experts may not only be difficult for novices to interpret, they may also not address misconceptions commonly held by novices. The chapter explores students' learning using either a worksheet with a high level of guidance or a more open-ended worksheet with a minimal level of guidance. Students also completed a pre-test and post-test of conceptual understanding and an attitude survey. Results showed that many students were able to correct their understanding after learning with simulation. No difference in conceptual understanding was found between the groups using worksheets of different guidance levels. However, comments about both simulation and worksheets on the evaluation questionnaire were more positive for students who had used the open-ended version. Students who had used the open-ended worksheet were also more likely to focus on the content of the lesson in their remarks, while students who used the more guided worksheets were more likely to focus on the structure of the lesson.

Treagust, Mthembu and Chandrasegaran addressed an evaluation of the Predict-Observe-Explain instructional strategy to enhance students' understanding of redox reactions in Chap. 14. Following a teacher in-service programme on the use of the Predict-Observe-Explain (POE) instructional strategy to enhance students' understanding of redox reaction concepts, its efficacy was evaluated in a study involving South African students. Eight hands-on activities involving redox reactions were conducted over a four-week period by a teacher who had participated in the programme. Instruction was evaluated using multiple methods, including laboratory observations, interviews with students, questionnaires to assess students' attitudes concerning the use of POEs and a pre- and post-test on redox reactions. The findings of the study indicated that students improved their understanding of redox reactions and held positive attitudes on the use of POEs in facilitating this understanding. However, despite the overall positive outcomes of the instruction, several students were unable to differentiate between specific concepts. The authors stress that the success of the presented teaching strategy has the potential to enhance constructivist teaching and learning practices of science teachers.

In Chap. 15 by Maciejowska, Wietecha-Posłuszny, Woźniakiewicz and Kościelniak, which concludes this section, a case study and role-playing in forensic chemistry and analytical chemistry is the focus of attention. In this chapter, role-playing and case study are presented as students' active learning approach to foster chemical concepts understanding. Role-playing topics are often related to current and controversial issues such as environmental and forensic ones. The authors introduce role-playing in a university-level forensic class that is based on a real story, well known from Polish newspapers. Another presented example of a role-playing class was introduced into the compulsory analytical chemistry laboratory course. When comparing the opinions of students, graduates, doctoral students, professionals and the authors of classes, it can be stated that the authors are most careful about judging the impact of their classes on the further vocational career of graduates. They all agree that the classes conducted using the role-playing teaching approach are interesting and motivating for both students and lecturers.

Chapter 10
Teaching Chemistry Conceptually

Vickie M. Williamson

Introduction

Traditionally, precollege and college-level chemistry have been taught as a mathematical course that emphasized algorithmic problem solving almost exclusively, with the belief that understanding of particle behavior would necessarily be derived from the ability to solve mathematical problems. Nonetheless, early research in the area has shown that chemistry students from middle school to college have little scientific understanding of and many misconceptions about the particulate nature of matter (PNM) (Abraham et al. 1994). Abraham et al. found that subjects who had completed a middle school physical science course, a high-school chemistry course, or a two-semester college chemistry course had the same high number of misconceptions about particle behavior, although the specific type of misconceptions did change with age. Other researchers had similar findings showing that students of all ages have difficulty with conceptual understanding about particles (Gabel et al. 1987; Novick and Nussbaum 1981). What contributes to these difficulties?

The results from a number of studies have shown that many of these difficulties are caused by students' application of macroscopic explanations, those derived from their everyday experience, to particles (de Vos and Verdonk 1987; Haidar and Abraham 1991) or by students' inability to visualize, diagram, or depict the behavior of particles (Ben-Zvi et al. 1986; Gabel et al. 1987). Conceptual understanding of the behavior of particles in chemical processes is very different from algorithmic or mathematical problem solving. Students can often solve chemistry problems requiring algorithmic or "plug and chug" strategies without an understanding of the underlying chemistry concepts (Niaz and Robinson 1993; Nurrenbern and Pickering 1987). This is at odds with the fact that chemists explain most natural phenomena in terms of particle behavior.

V. M. Williamson (✉)
Texas A and M University, College Station, TX, USA
e-mail: williamson@tamu.edu

I. Devetak and S. A. Glažar (eds.), *Learning with Understanding in the Chemistry Classroom*, DOI: 10.1007/978-94-007-4366-3_10,

In all levels of college first-year chemistry courses, researchers have shown that students are consistently more successful with problems that can be solved using an algorithm than with problems involving conceptual thinking (Nakhleh 1993; Nurrenbern and Pickering 1987; Sawrey 1990). While the gas laws have been the topic most often used, the conceptual algorithmic gap has been observed with other topics such as stoichiometry, equations, limiting reagent, empirical formulas, and density. Johnstone (1993) proposed that chemistry is actually three basic representations: observable evidence (macroscopic), mathematical and chemical symbols (symbolic), and atomic/molecular/particle behavior (submicroscopic). Further, he proposed that chemists could seamlessly move between representations, while novices cannot.

The research studies showing that students have trouble with conceptual understanding resulted in a call to teach chemistry both algorithmically (mathematically) and conceptually (Pickering 1990). While some instructors intuitively teach conceptually, a number of instructors have difficulty with what it means to teach chemistry conceptually. Conceptual teaching puts an emphasis on students' ability to explain relationships, to predict outcomes, to visualize/explain particle behavior, and to understand the macroscopic, particulate, symbolic, and mathematical levels. Current research findings have focused on teaching strategies that can be used to help students understand chemical phenomena on these different levels, particularly the conceptual particle understanding. These strategies can be used with large or small classes and focus on active student engagement in the learning process. This article will outline a number of these methods.

Macroscopic Representations

The use of macroscopic representations in class helps student understanding by promoting the formation of macroscopic mental models in the students' mind. These can include laboratories, demonstrations, videos of a demonstration or of actual phenomenon, and computer simulations of a "real-life" laboratory. Macroscopic representations show students views of the phenomena that can be seen with their eyes.

Laboratories have been the traditional area of the course that allowed for active participation of students and have given students a visual experience with chemical phenomena. The level of student activity may vary with the type of laboratory. Laboratories can be categorized as verification, guided-inquiry, or open-inquiry (Abraham 2004). In a traditional verification laboratory, the lecture on the topic has already been given and students often know the outcome of the laboratory, as they are just verifying that the relationships given in the lecture are correct. Students have a much more active role in the inquiry-based laboratories. In guided-inquiry laboratories, students are directed to collect variables on phenomena new to them, and then are asked to look for meaning via patterns or relationships in their own data. In open-inquiry laboratories, students design their own procedures to answer a question, which asks about a new setting or new aspect to a relationship established in a previous guided-inquiry laboratory.

Chatterjee et al. (2009) found that students had a more positive attitude toward guided-inquiry over open-inquiry laboratories and that students perceived that they learned more with guided-inquiry laboratories. These authors believed that this difference may be accounted for in the longer type of report usually required by the open-inquiry laboratories and cautioned that instructors should be aware of these feelings, but still incorporate both types of laboratories into their courses. Instructors can find a number of sources of inquiry-based laboratories in prepared laboratory manuals from publishers. Alternatively, there are a number of published 'recipes' for converting traditional laboratories into inquiry ones (e.g., Allen et al. 1986). Conversion can be done if you have experience with inquiry or if you have a collaborator who has experience with inquiry. Conversion can be very difficult for someone without these things.

A demonstration can be performed in an "inquiry" manner, such that the demonstrator is handling the chemicals and directing students to record data, but not lecturing on the concept. Students then use the data and work in groups to construct chemical concepts by analyzing the data for relationships between variables, patterns, etc., as in a guided-inquiry laboratory. McKee et al. (2007) found that inquiry demonstrations were just as effective as guided-inquiry laboratories, at least with the high reasoning ability students in their sample. The students were not actively involved with manipulating the glassware, but were actively involved in making sense of the data as they developed the relationship. The authors suggested that inquiry demonstrations could be an alternative when laboratories could not be done, but warned that the same effect may not exist for students with lower reasoning abilities that their college-level, general chemistry subjects. A video of a demonstration could work much in the same way if it preformed in an inquiry manner as described above or if it were paused at key points to allow students to make predictions and hypotheses.

There are computer laboratory simulations at the macroscopic level that allow students to perform a virtual laboratory. Students manipulate equipment and chemicals on-screen. Similarly, these computer versions can approach the laboratory in a verification or inquiry mode. There is not a clear advantage for the virtual laboratories over the actual ones that this author could find in the literature, but some researchers suggest that virtual laboratories used with a hand-on laboratory may provide the best experience (Bourque and Carlson 1987; Martinez-Jimenez et al. 2003). It may be that the virtual laboratories are best when they are in an "inquiry" manner, where students are being guided to collect certain variables to develop a concept or are being allowed to design their own experiments to ask a question. Further, virtual laboratories may be best in conjunction with a hands-on laboratory or when a hands-on laboratory is not possible due to safety or physical constraints. These can be used in a computer laboratory, with groups of students, or assigned for out-of-class work. Computer laboratory simulations are available on the Internet, from publishers, or from commercial companies. Instructors should investigate sites for free materials such as the *Journal of*

Chemical Education Digital Library (http://jchemed.chem.wisc.edu/JCEDLib/index.html), which is part of the National Science Digital Library (http://nsdl.org/), Multimedia Educational Resource for Learning and Online Teaching (MERLOT at http://www.merlot.org/merlot/index.htm), and the laboratory simulations is housed at Iowa State University (http://www.chem.iastate.edu/group/Greenbowe/sections/projectfolder/animationsindex.htm).

Particulate Representations

Particulate representations help students to visualize the particle nature of matter. Johnstone (1993) called this level of understanding "submicroscopic;" however, particulate may be a better term since this includes atoms, molecules, etc. As previously discussed in the Introduction, students have difficulties understanding the behavior of particles. Since particles are not visible, students must rely on their mental models of the particles and particle action. For some students, it is difficult to visualize particles. Instructors should use techniques to promote the formation of mental models of particles in their students. Physical models, student-generated drawings and computer animation are a few techniques that may be easily used in the classroom. For a more complete list of techniques to promote visualization, see Williamson and Jose (2009).

Physical models (such as model kits, play-doh, magnets, or gumdrops and toothpicks) are easy to incorporate into the classroom and have been found to benefit student understanding. In their study of 326 high-school chemistry students, Gabel and Sherwood (1980) found that students who manipulated space-filling molecular models performed significantly better on solving general chemistry problems that those who only saw teacher demonstrations with the models. Students who manipulate physical models construct more understanding between the models and underlying chemistry concepts (Friedel et al. 1990). Physical models can be used with more that just atomic structure and VSPER theory. It is important to ask students to use these models to show both structure and interaction of particles throughout the semester.

There are computer animation programs that simply depict fixed pictures of molecules with either no movement or simple rotation. Students are not allowed to make mistakes in the drawing of the molecules, only to request certain molecules for viewing. Dori and Barak (2001) suggested that there might be an advantage to these computer models, especially when used with physical molecular models with inquiry-based learning tasks. In their study, 154 high-school students who used these computer and physical molecular models gave better explanations of structure than did the 122 students in a traditional group, where the teacher used physical models only for demonstrations. The authors attributed the increase in understanding to the active learning techniques with virtual and physical molecular models. Prices vary for these computer programs. With this type of program, the molecules are often simply shown to the class by the instructor, but research shows that this diminishes possible results. If facilities are available, a more student active use of this type of program is to give students a set of questions to guide them to investigate the molecule either individually or in groups. Questions could include those about the atoms attached, bond angles, length of bonds, overall shape, shape at various rotation points, etc.

Asking students to create a picture of particles and their behavior has also been shown to help understanding. These pictures can be drawn by hand in the form of a series of pictures (story boards) or flipbooks or by computer program. Milne (1999) described a flipbook activity to help students visualize both the kinetic and the stoichiometric nature of chemical reactions at the particle level. A number of computer programs can be used to draw molecules. Some are low cost or free (e.g., *Chem Sketch Freeware* at http://www.acdlabs.com/download/), while others are expensive. The free programs change quickly. Instructors wishing for their students to create electronic particulate drawings should search on the Internet for the current-free drawing programs. Also using computer programs, animations or a series of moving pictures can be created by the students. Software to create these student-generated animations can range from free (e.g., *ChemSense* at http://chemsense.org/) to expense.

Williamson et al. (2013) found that students' content understanding of equilibrium and mental rotation ability was increased with the creation of storyboards and ChemSense computer animations, but that there was no difference between the treatments in the population of college chemistry students used. Their students were required to create one storyboard or animation for each of 2 weeks during a unit on equilibrium. One assignment dealt with physical equilibrium (changes between phases), while the other dealt with chemical equilibrium (a chemical reaction in equilibrium). Authors propose that their students approached the assignments as a series of frames or storyboards, regardless of whether the assignment was for the paper-and-pencil version or the computer version. The good news for instructors may be that if the benefits are the same, the paper-and-pencil storyboards may easier to assign, especially for small classes. Animations and storyboards work better with dynamic processes and as out of class or laboratory assignments. Instructors can require that student drawings, storyboards, or animations be included with reports and homework. Student misconceptions can be easily detected from their own creations.

There are a number of particulate animations that are already constructed by professionals and show a series of images to depict a dynamic process. These can be used by the instructor in class or assigned to the student for viewing during laboratory or as homework. We know that dynamic particulate animations increase students' conceptual understanding over static particle pictures (Williamson and Abraham 1995). The conceptual understanding of particles was increased for college chemistry students who were exposed to short 1- or 2-min animations daily over a 2-week unit. The conceptual understanding was evaluated on new topics, not those shown in the animations. Something about the dynamic quality seems to promote the formation of superior mental models of particulate behavior that will transfer to new phenomena. This finding has been echoed by other research. For example, Sanger and Badger (2001) found that animations and electron density plots were superior to static particulate drawings and wooden models, and Yezierski and Birk (2006) found that animations helped improve conceptual understanding and close an initial, pre-test gender gap in their study of students from middle school to college level. Sources for these professional animations include the Internet (see all of the sites referenced at the end of the previous section on Macroscopic Models) and publishers (as almost every textbook now comes with a set of particulate animations).

Since allowing students to "experience" the particulate level involves computers or other projection technologies, it is important to realize that there are a number of factors that facilitate or impede the use of visualizations in the classroom. In a study with high-level, high-school chemistry teachers, teachers upon leaving a 3-week summer workshop believed they could implement computer molecular visualization programs and animations at their schools. Williamson et al. (2005) found a number of factors which emerged during the school year that did impact the instructors' use of these programs and animations. These factors included the: (1) availability of computers in the classroom for student and instructor use (many classroom had no computers), (2) type of access or safety controls on the school Internet access, (3) school computer lab availability (many times the school computer laboratory is already scheduled by other courses), (4) preparation time for the instructor to integrate computer applications into the curriculum, (5) technical support availability (the length of time it took to get help), and (6) moral/monetary support from administration.

These were factors even with instructors who fully intended to use molecular visualizations programs and animations as they left the 3-week workshop. Any of these factors can impede the utilization of any technique involving computers. Before instructors plan to use computers, they should take these factors into account in their own institution. The factors may temper whether the technique can be used at all or may impact the assignment place (during lecture, laboratory, or homework), grouping (individual or group), or frequency (once during the term or with every report/unit of study).

There can also be other interesting outcomes when using techniques to encourage particulate understanding. For example, we know that an extensive use of particulate visualizations (both animations and programs) can increase subjects' mental rotation ability (Williamson and Jose 2008). These authors found that teachers significantly increased in their mental rotation ability over a 3-week workshop. This increase in spatial ability occurred for both sexes, but was lost with lack of use, only to be gained again during a 3-week workshop the next year. Authors describe the increase as a "use it or lose it" quantity. Williamson et al. (2013) also found that college general chemistry students increased their mental rotation ability after a 2-week unit in which they were either drawing particles on storyboards or drawing them with a computer program (ChemSense). This is an area that we do not fully understand and which needs more research.

Symbolic Representations

Symbolic representations include both the chemical symbols and mathematical representations for the phenomena. Student difficulties with writing formulas and chemical equations have long been documented at both the high school and beginning college levels (e.g., Bennett 1925). Symbolic representations traditionally were presented first to students (the chemical formulas and mathematics concerning a phenomena). Next, students would work on problems and do a verification laboratory activity (macroscopic representation). It was assumed that conceptual understanding of the particulate level would follow. This traditional method is counter to the more inquiry-based strategies previously discussed, which usually begin with data from macroscopic investigations (laboratory or field work), then moved on to particulate and symbolic representations. Whether the symbolic representation is given initially or not, it is still an area that can give students difficulties.

Methods to help students with chemical formulas include gaming activities. Chimeno et al. (2006) evaluated three groups of students (Traditional Learning, Rainbow Wheel, and Rainbow Matrix) on their ability to name and write ionic compounds. A pre-test established that the groups were equivalent. The Rainbow Wheel is an educational game, while the Rainbow Matrix is its computerized version. Both games dealt with combining cations and anions to achieve the correct formula and naming the compound according to the Stock method. All groups were provided with a periodic table and a list of common ions for a 2.5-h practice session, in which each group had the same number of problems converting chemical names to formula and vice versa. Students were given a post-test and had a later exam covering the material. Students from both the game treatments significantly outscored the traditional group on both the post-test and on the course exam, leading the authors to assert that students may be more receptive to learning methods involving a game. Authors gave three possible reasons for the outcome: the quality and timing of feedback given to the student in the games, the use of ion cutouts in the games, and the motivational issues associated with games. One method touted to help with balancing equations is the use of analogy, usually with cooking or food examples. Haim et al. (2003) proposed the use of analogy to understand formula, equations, conserving mass, limiting reagent, and yield. In their example, students were told that two types of sandwiches had been ordered, but that this was the wrong order. Received sandwiches needed to be disassembled and remade. Authors acknowledged that the use of understandable, everyday problems allowed students to realize the mathematical relationships.

Methods to help students with mathematical representations include the use of factor analysis and cross multiplication (cross proportional). In fact, each of these two methods has been proposed as the "best" method at different times in history. Currently, most textbooks use the factor analysis methods. However, Cook and Cook (2005) contend that the cross proportional method promotes student learning and helps students visualize the connections between chemical concepts, increasing their conceptual understanding. Authors believe that students can spot errors more easily using the cross proportional method than when using the factor analysis method, which allows for misconceptions to be more readily addressed. The use of the factor analysis method should never be used as an algorithm, which students blindly follow. Instructors, who choose to use the factor analysis method, can use conceptual questions as previously discussed to help insure that students can verbalize their understanding of each step in a factor analysis.

Relating the Representations

Multiple representations may be needed to increase student learning. Russell et al. (1997) found improvement in students' conceptual understanding and ability to create dynamic mental models when macroscopic, particulate, and symbolic representations are used. Additionally, students were able to better correlate or move between the three levels of representation. In a summary of the literature on the development and use of animations, Burke et al. (1998) also noted that animations of short duration and the use of demonstrations with the animations could be effective.

Velázquez-Marcano et al. (2004) found that both macroscopic and particulate visualizations were needed to gain the best predictive ability, at least in their study of gases and liquids. Students in this study were shown a video of a demonstration and an animation of the particles for the same phenomena. Although treatment groups varied the order of presentation, both the macroscopic and the particulate representation were needed for students to best predict the outcome of the demonstrations. While order did not matter for predicting the outcome, it seems that there may be a preferred order to insure maximum student understanding for the phenomena (Williamson et al. 2012). Macroscopic representations used prior to particulate ones give the best results when students were asked for the reasons for the outcome of the phenomena. Research findings show this order (macroscopic followed by particulate) is preferred at least with concepts of the study. This means that students need to experience the laboratory, see a demonstration or a video of a demonstration prior to moving on to the particulate level. In a way, this seems intuitive to go from concrete to abstract. This has similarities with Piaget's idea that one must go through the concrete level first before going to his formal level (Piaget 1977). The movement from the macroscopic level to particulate level could then be followed by a move to the mathematical level; however, more research needs to be done in this area to see if this pattern holds for all topics.

Rappoport and Ashkenazi (2008) found that experts thought the macroscopic and symbolic aspects of the phenomena emerged or came from the interactions of the particles (emergent perspective), while students either failed to link the levels or thought that the macroscopic and symbolic aspects controlled or guided the particle behavior (submergent perspective). This difference in view of the role of the particles between the experts and students (particles in control or particles being controlled) could impact conceptual understanding. The authors proposed that the gap between algorithmic and conceptual problem solving could be due to this submergent perspective of students and suggested that teaching should include the directionality when connecting the particle behavior. More research needs to be done in this area. Instructors should consider first presenting the macroscopic aspect of the phenomena followed by the particle aspect, making sure to emphasize the directionality of the relationship between the two. Symbolic and mathematical representations could then be linked to the particles and macroscopic levels.

Group Problem Solving

Another technique to actively involve students is to replace at least some lecture time with group problem solving. There are a number of methods that can be used with cooperative groups, including assigning students to groups versus allowing students to form their own groups, static versus changing groups throughout the course duration, assigning roles versus allowing the group to simply work together, etc. Williamson and Rowe (2002) found a significantly lower dropout rate when group problem solving was used exclusively in a college, quantitative analysis class as compared to a traditional lecture class. For this study, groups were assigned to achieve heterogeneous groups based on reasoning ability and gender. Groups contained four members, one of high reasoning ability, two of medium ability, and one of lower ability. The finding of a lower drop out rate for the section that used group work versus the section that used traditional lecture seems to imply that the involvement of students actively solving problems was superior to simply watching the professor solve problems. The authors formulated four assertions from their quantitative and qualitative study:

Assertion #1 Even bright students are more self-assured when they have more opportunities to exercise/verbalize their understandings and abilities.

Assertion #2 It seems that students of slower abilities can be brought "up to speed" more quickly by a peer, often due to the reluctance to see the instructor. Students are less intimidated when a peer points out errors (More-intimidated students withdraw from the class).

Assertion #3 The feelings of comradeship will enable students to persist, while feelings of isolation lead to withdrawals.

Assertion #4 Students and instructors resist new methods.

There are a variety of methods that can be used by the instructor to incorporate group problem solving into the classroom. The most high-tech version would be the use of a personal response system or clicker. Students' individual answers can be recorded, and class data can be tabulated. MacArthur and Jones (2008) reviewed the literature on the use of clickers in college chemistry classes. The authors found that publications reported improved student attitudes, and while most reported improved student learning, others were inconclusive. They also suggested that the collaborative work and formative assessment are the largest benefits. The drawbacks include student adjustment, time limitations, and technology issues. It was also noted that often multiple-choice, conceptual questions can be used, although many clicker systems also allow for numeric answers. Bunce et al. (2006) warn that there must be an opportunity for reflection and review of clicker questions, in addition to the practice clickers give students during lecture, in order to attain maximum benefit.

There is also a low-tech version of the clickers, used in many college and high-school classes. For multiple-choice questions, bright cardboard squares that are lettered can be used. Alternately, the letter A can be printed on one color of paper, with the process repeated for letters B–D on their own color. These can be inserted into a clear sheet protector, such that the A and B show on 1 side, each on its own color of paper, with C and D on the backside. After a few minutes of lecture, a conceptual question can be posed and a timer begun, at the end of which students will hold up the folded sheet protector to display the letter of their choice. This allows the instructor to know that a sea of pink indicate that most of the class has chosen a letter. This low-tech option still allows for formative assessment, but does not allow for recording of student response. Instructors can simply put a box of these out at the beginning of class and ask students pick them up at the beginning of class, with return at the end of the period. Most instructors ask students to sit with a partner or in groups depending on the room arrangement, which will force a group decision and cut down the number of sheet proctors required. Most instructors use between two and five of these conceptual questions throughout an hour lecture, although the topic and the length of time dictate the exact number of questions allowed. Most questions require 1 min, although some harder numeric questions may require more time. There are a number of sources of concept questions on the Internet. Three of these include: http://people.brandeis.edu/~herzfeld/conceptests.html, http://chem.pdx.edu/~wamserc/ConcepTests/default.htm, and http://www.jce.divched.org/JCEDLib/QBank/collection/ConcepTests/.

Assessment: Use of Both Algorithmic and Conceptual Questions

With so many aspects of teaching, it will not help to teach in a more student active manner with more emphasis on conceptual understanding if we do not also assess with both mathematical (algorithmic) and conceptual questions. Students learn very quickly to only pay attention to what is going to be on the test. If we are serious about developing both algorithmic and conceptual student understanding, then we must assess in both areas. The three web sources given for conceptual questions in the section above can also be used to help generate conceptual questions for your examinations and assessments.

If you need a standardized examination, the American Chemical Society (ACS) through its Division of Chemical Education and Examinations Institute has a conceptual general chemistry examination that can be used. They also have a first- and second-term paired question examination, which offers both algorithmic and conceptual questions on the same topics. These examinations are available for purchase at http://chemexams.chem.iastate.edu/. Like all standardized examinations, security is an issue, which results in extra diligence required of the instructor. If a standardized examination is not required, it is less trouble to model your own assessments after those from the websites previously given.

Other Techniques

Many instructors have found that assigning a creative project increases student interest and retention (Lerman 1986). These projects can be as simple as role-playing molecules transitioning between phases or in a reaction to more complex multimedia reports. These projects can focus on the macroscopic, particulate, or symbolic representation of the phenomenon under investigation. Topics for projects should be generated by the students, but approved by the instructor before students proceed (Lerman 1986). There is not an exhaustive list of techniques in this manuscript, but rather the techniques given here are intended to represent a few common methods to help an instructor incorporate active learning and a few conceptual methodologies into his/her classroom.

Summary

So how might this work for a specific topic? The gas laws or specifically Charles' Law is usually taught in both high-school and college chemistry classes. This is the concept that there is a direct relationship between temperature and volume of a gas (as the temperature in absolute increases, so does the volume). Traditionally, an instructor would lecture on the topic, giving the students the equation that volume divided by temperature is equal to a constant. The students would then work problems and go to laboratory to verify that the instructor was correct.

An alternate method would be to have students obtain data about the volume and temperature in an inquiry manner, which could be done by a laboratory or a demonstration of heating balloons attached to flasks, or by using a computer laboratory simulation. Good data are needed if students are going to use the data to find a relationship (that volume divided by temperature is equal to a constant). The consistency of data may be an advantage for the computer simulation.

After the data are analyzed, a discussion could consist of having the groups report their findings and looking for patterns in the data. After the class realizes that the data show that volume divided by temperature is equal to a constant, the instructor could give the a number of concept questions, allowing groups to respond via clickers or cards, about different gases, cooling a gas, etc. Graphical treatment of the data will allow students to generate absolute zero. (Students love to think about the temperature at which volume is zero. How can matter disappear?) The instructor then might want to give the historical background on Charles' Law and ask students to portray the particles of air in the flask/balloon system. This portrayal could be done with drawings, storyboards, computer animation programs, or other creative projects. After eliciting the students' understanding, the instructor might want to show some professional animations of particle behavior. Next, use some clicker or card questions that focus more on the algorithmic aspect of the relationship (the traditional homework), mixed in with conceptual questions. Students should be ready now for algorithmic and conceptual homework problems! This method is based on the guided-inquiry approach discussed by Abraham (2004), in which concepts are inductively generated from data to be later applied deductively via an open-inquiry activity.

A second example is the concept of synthesis reactions and the idea that compounds have definite proportions of weight and numbers of atoms. In an inquiry mode, students would first go to the laboratory to investigate the heating of magnesium metal ribbon in an evaporating dish. To help insure good data, a small amount of distilled water is added after the initial heating and cooling, then gently boiled away. By comparing the beginning and end weights, student will realize that there has been a weight gain. The instructor can then direct the students to compare the weight of the magnesium to the weight change, by encouraging students to try various mathematical operations on these numbers (add, subtract, multiply, and divide).

Through a discussion, students establish that a chemical change has occurred due to the weight gain and the difference in properties of the beginning and ending materials. The question can then be posed to ask student, what combined with the magnesium to give the weight gain. Possible substances can be generated that have combined with the magnesium. Since the magnesium has only been in contact with the air and the water, possible gases listed by students might include water vapor, oxygen gas, nitrogen gas, carbon dioxide, hydrogen, and other gases in the atmosphere. The instructor can then direct students to consider the possible products of each when combined with magnesium. Depending on the prior knowledge of the class, the instructor may have to guide the class to write these possible products. Usually, these products are suggested once the noble gases are eliminated–MgO, Mg_3N_2, $Mg(OH)_2$, while MgH_2 and $MgCO_3$ might be suggested. Next students can pool their data. When looking for a pattern in the data, the weight of the magnesium divided by the weight change will give a constant value of about 1.5. The instructor can ask the students to compare the data ratio to the ratio given by the possible products from the periodic table. Only the MgO will give the same ratio. The terms "synthesis reaction" and "law of constant proportions" should be discussed.

At this point, ask the student to portray the particle behavior with drawings, storyboards, computer animation programs, or other creative projects to elicit the student's understanding. Focus now is on relating particles to the balanced equation. Computer animation of the particles involved in a reaction could be used in class to help students relate what is happening at the particle level to the chemical reaction. Students can also relate these prepared computer animations to their own drawings, storyboards, or animations. Depending on the class, the instructor may wish to go through all of the methods using only synthesis reactions by continuing on to the algorithmic and conceptual problems in class and for homework, then repeat the process for decomposition and other types of equations. Alternatively, the instructor can introduce decomposition and other types of reactions via particle animations, laboratory work, or problems. Investigation of a decomposition reaction like the heating of a hydrated compound will help to consolidate understanding.

Next, use some clicker or card questions with student groups that have both conceptual problems and problems focused on the more algorithmic aspect of the stoichiometry using simple balanced equations like mole-to-mole and gram-to-gram problems. The use of games and analogies can help solidify student's understanding. Students should be ready now for algorithmic (the traditional homework) and conceptual homework problems!

Conclusion

Research findings indicate that it is important to teach chemistry both conceptually and mathematically. Instructors can add conceptual teaching techniques to their classrooms by making small changes. Individual instructors should choose a small number of strategies to implement, perhaps even just one. Once this strategy has been incorporated into the classroom, another can be attempted. The benefits of adding conceptual methods to one's teaching are many and include improved student understanding and attitudes, as the students begin to see the macroscopic, particulate, symbolic, and mathematical sides to chemistry.

This chapter is meant to give instructors a set of proven conceptual strategies from which to choose. All of the strategies previously discussed have cited references, which give evidence that the strategy is successful in promoting conceptual understanding. There are few comparative studies of conceptual strategies in the literature, so most of the cited references did not compare the strategy to other methods, only examining if the strategy resulted in desired outcomes, with the exception of the comparative studies previously discussed. The chapter gives a limited list of common strategies and does not rank the strategies or imply that one conceptual strategy creates better learning than another. The goal of the chapter is to acquaint instructors with a list of conceptual strategies so that they may implement one or more strategy of their choice. The basic strategies presented here include the use of macroscopic representations, particulate representations (both dynamic and static models), symbolic representations, relating the three representations, group problem solving, both algorithmic and conceptual assessments, plus others.

References

Abraham, M. R., Williamson, V. M., & Westbrook, S. L. (1994). A cross age study of five chemistry concepts. *Journal of Research in Science Teaching, 31*(2), 147–165.

Abraham, M. R. (2004). Inquiry and the learning cycle approach. In N. J. Pienta, M. M. Cooper, & T. J. Greenbowe (Eds.), *Chemists' guide to effective teaching* (pp. 41–52). Upper Saddle River, NJ: Pearson.

Allen, J. B., Baker, L. N., & Ramsden, J. H. (1986). Guided inquiry laboratory. *Journal of Chemical Education, 63*(6), 533–534.

Ben-Zvi, R., Eylon, B. R., & Silberstein, J. (1986). Is an atom of copper malleable? *Journal of Chemical Education, 63*(1), 64–66.

Bennett, J. C. (1925). A study of pupil errors in chemistry. *Journal of Chemical Education, 2*(9), 760–769.

Bourque, D. R., & Carlson, G. R. (1987). Hands-on versus computer simulation methods in chemistry. *Journal of Chemical Education, 64*(3), 232–234.

Bunce, D. M., VandenPlas, J. R., & Havanki, K. L. (2006). Comparing the effectiveness on student achievement of a student response system versus online WebCT quizzes. *Journal of Chemical Education, 83*(3), 488–493.

Burke, K. A., Greenbowe, T. J., & Windschitl, M. A. (1998). Developing and using conceptual computer animations for chemistry instruction. *Journal of Chemical Education, 75*(12), 1658–1661.

Chatterjee, S., Williamson, V. M., McCann, K., & Peck, M. L. (2009). Attitude and perception of students towards guided and open inquiry laboratories. *Journal of Chemical Education, 86*(12), 1427–1432.

Chimeno, J. S., Wulfsberg, G. P., Sanger, M. J., & Melton, T. J. (2006). The rainbow wheel and rainbow matrix: Two effective tools for learning ionic nomenclature. *Journal of Chemical Education, 83*(4), 651–654.

Cook, E., & Cook, R. L. (2005). Cross-proportions: A conceptual method for developing quantitative problem-solving skills. *Journal of Chemical Education, 82*(8), 1187–1189.

de Vos, W., & Verdonk, A. H. (1987). A new road to reaction: Part 4. The substance and its molecules. *Journal of Chemical Education, 64*(8), 692–694.

Dori, Y. J., & Barak, M. (2001). Virtual and physical molecular modeling: Fostering model perception and spatial understanding. *Educational Technology & Society, 4*(1), 61–74.

Friedel, A. W., Gabel, D. L., & Samuel, J. (1990). Using analogs for chemistry problem solving: Does it increase understanding? *School Science and Mathematics, 90*(8), 674–682.

Gabel, D. L., Samuel, K. V., & Hunn, D. (1987). Understanding the particulate nature of matter. *Journal of Chemical Education, 64*(8), 695–697.

Gabel, D., & Sherwood, R. (1980). The effect of student manipulation of molecular models on chemistry achievement according to Piagetian level. *Journal of Research in Science Teaching, 17*(1), 75–81.

Haidar, A. H., & Abraham, M. R. (1991). A comparison of applied and theoretical knowledge of concepts based on the particulate nature of matter. *Journal of Research in Science Teaching, 28*(10), 919–938.

Haim, L., Corton, E., Kocmur, S., & Galagovsky, L. (2003). Learning stoichiometry with hamburger sandwiches. *Journal of Chemical Education, 80*(9), 1021–1022.

Johnstone, A. H. (1993). Development of chemistry teaching. *Journal of Chemical Education, 70*(9), 701–705.

Lerman, Z. (1986). Chemistry for art and communication students. *Journal of Chemical Education, 63*(2), 142–143.

MacArthur, J. R., & Jones, L. L. (2008). A review of literature reports of clickers applicable to college chemistry classrooms. *Chemistry Education Research and Practice, 9*(3), 187–195.

Martínez-Jimenez, P., Pontes-Pedrajas, A., Polo, J., & Climent-Bellido, M. S. (2003). Learning in chemistry with virtual laboratories. *Journal of Chemical Education, 80*(3), 346–352.

McKee, E., Williamson, V. M., & Ruebush, L. E. (2007). Effects of a demonstration laboratory on student learning. *Journal of Science Education and Technology, 16*(5), 395–400.

Milne, R. W. (1999). A low-cost activity for particle conceptualization at the secondary level. *Journal of Chemical Education, 76*(1), 50–51.

Nakhleh, M. B. (1993). Are our students conceptual thinkers or algorithmic problem solvers?: Identifying conceptual students in general chemistry. *Journal of Chemical Education, 70*(1), 52–55.

Niaz, M., & Robinson, W. R. (1993). Teaching algorithmic problem solving or conceptual understanding: Role of developmental level, mental capacity, and cognitive style. *Journal of Science Education and Technology, 2*(2), 407–416.

Novick, S., & Nussbaum, J. (1981). Pupil's understanding of the particulate nature of matter: A cross-age study. *Science Education, 65*(2), 187–196.

Nurrenbern, S. C., & Pickering, M. (1987). Concept learning versus problem solving is there a difference? *Journal of Chemical Education, 64*(6), 508–510.

Piaget, J. (1977). *The development of thought: Equilibrium of cognitve structures*. Newyork, NY: Viking.

Pickering, M. (1990). Further studies on concept learning versus problem solving. *Journal of Chemical Education, 67*(3), 254–255.

Rappoport, L. T., & Ashkenazi, G. (2008). Connecting levels of representation: Emergent versus submergent perspective. *International Journal of Science Education, 30*(12), 1585–1603.

Russell, J., Kozma, R. B., Jones, T., Wykoff, J., Marx, N., & Davis, J. (1997). Use of simultaneous-synchronized macroscopic, microscopic, and symbolic representations to enhance the teaching and learning of chemical concepts. *Journal of Chemical Education, 74*(3), 330–334.

Sanger, M. J., & Badger, S. M. (2001). Using computer-based visualization strategies to improve students' understanding of molecular polarity and miscibility. *Journal of Chemical Education, 78*(10), 1412–1416.

Sawrey, B. A. (1990). Concept learning versus problem solving: Revisited. *Journal of Chemical Education, 67*(3), 253–254.

Williamson, V. M., & Abraham, M. R. (1995). The effects of computer animation on the particulate mental models of college chemistry students. *Journal of Research in Science Teaching, 32*(5), 521–534.

Williamson, V. M., & Jose, T. J. (2008). Effects of a two-year molecular visualization experience on teachers' attitudes, content knowledge, and spatial ability. *Journal of Chemical Education, 85*(5), 718–723.

Williamson, V. M., & Jose, T. J. (2009). Using visualization techniques in chemistry teaching. In N. J. Pienta, M. M. Cooper, & T. J. Greenbowe (Eds.), *Chemists' guide to effective teaching* (Vol. 2), (pp 71–88). Upper Saddle River, NJ: Prentice Hall.

Williamson, V. M., & Rowe, M. W. (2002). Group problem solving versus lecture in college-level quantitative analysis: The good, the bad, and the ugly. *Journal of Chemical Education, 79*(9), 1131–1134.

Williamson, V. M., Brown, L. M., Peck, M. L., & Simpson, M. (2005). Facilitators and barriers to teacher implementation of molecular visualization. *The Texas Science Teacher, 34*(2), 12–16.

Williamson, V. M., Lane, S., Gilbreath, T., Tasker, R., Ashkenazi, G., Williamson, K. C., et al. (2012). The effect of macroscopic and particulate visualizations on students' particulate explanations. *Journal of Chemical Education, 89*(8), 979–987.

Williamson, V. M., Watkins, J. T., & Williamson, K. C. (2013). The effect of student-constructed animations versus storyboards on students' mental rotation ability, equilibrium content knowledge and attitudes. Peer-reviewed chapter. In J. Suits & M. Sanger (Eds.), *Pedagogic roles of animations and simulations in chemistry courses* (pp. 293-311). Washington, DC: American Chemical Society.

Velázquez-Marcano, A., Williamson, V. M., Ashkenazi, G., Tasker, R., & Williamson, K. C. (2004). The use of video demonstrations and particulate animation in the general chemistry. *Journal of Science Education and Technology, 13*(3), 315–323.

Yezierski, E. J., & Birk, J. P. (2006). Misconceptions about the particulate nature of matter: Using animations to close the gender gap. *Journal of Chemical Education, 83*(6), 954–960.

Chapter 11
Students' Achievement in Learning Chemistry Through the Design and Construction Approach to Laboratory Activity and the Relation with their Prior Achievements and Motivation to Learn

Margareta Vrtačnik, Kristina Sodja and Mojca Juriševič

Theoretical Framework

The majority of emerging active teaching strategies or student-centered strategies are rooted in constructivist learning theory. Their common feature is challenging open-ended investigations in a realistic, meaningful context which allows learners to explore and generate many possibilities, both affirmative and contradictory (Fosnot and Perry 2005). Thus the constructivist approach emphasizes teacher's role in mediating learners to construct their own scientific models and to explore their domains of applicability (Matthews 2008). The relation of the constructivist learning model with Piaget's theory of intellectual development and its implication for better understanding of some problems in teaching and learning chemistry is given by Bodner (1986). Research results that focused on laboratory activities which are inquiry-based and on an instructional technique (The Science Writing Heuristic) that combines inquiry, collaborative learning, and writing proved that these approaches have potential for improving the pedagogical value of laboratory work by changing the nature of the chemistry laboratory (Burke et al. 2006; Cacciatore et al. 2008; Furlan 2009; Rudd et al. 2001; Tarhan and Sezen 2010).

The hands-on/minds-on approach toward teaching and learning is therefore one of the active strategies which most science educators advocate, in spite of severe criticism toward constructivism derived from empirical studies on the effectiveness of this approach in comparison with the guided approach in teaching science

M. Vrtačnik (✉) · K. Sodja
Faculty of Natural Sciences and Engineering, University of Ljubljana,
Vegova 4, Ljubljana, Slovenia
e-mail: metka.vrtacnik@guest.arnes.si

M. Juriševič
Faculty of Education, University of Ljubljana, Kardeljeva pl 16, Ljubljana, Slovenia

I. Devetak and S. A. Glažar (eds.), *Learning with Understanding in the Chemistry Classroom*, DOI: 10.1007/978-94-007-4366-3_11,

(Kirschner et al. 2006; Kroesbergen et al. 2004; Mayer 2004). But according to current neurophysiological research findings, the doubt about and the criticism of the constructivist paradigm is not totally justified. Current neuroscience research has shown that information and knowledge are assimilated with different degrees of effectiveness, depending on the mood and tendency of emotions (Cohen and Magen 2004; Cozolino 2006; Erk et al. 2003). There are neurological reasons why learning contents should not be presented neutrally but in an emotional, interesting, and exciting manner (Thiel et al. 2002).

Design and construction of laboratory activity evokes a series of positive emotions in the students, since it is conducted in a relaxed atmosphere, without fear and pressure. Students design and plan their activities according to their own pace, knowing that they will support each other in achieving the mutual goal.

Unfortunately, many of the activities students perform in chemistry classrooms are usually related to listening to teacher's explanations, following demonstrations, or conducting experiments in small groups according to a carefully prescribed procedure, and answering questions related to the experimental activity. Seldom are opportunities available to carry out more authentic science activities. However, when asked to design their own experiments and control variables, students are more likely to think like a scientist and apply science competencies to solving problems (Jones 1999). Construction and design activity therefore offers the possibility to develop some basic science competencies such as: define a problem, design experiments and/or observations, control variables, conduct experiments, take notes, analyze the results, set up hypotheses, check hypotheses, and report and communicate findings (Vrtačnik 2011).

It can be concluded that motivational initiatives which teachers apply while teaching have an important role for the neurophysiological processes in learning (Byrnes and Fox 1998; Jang 2008; Schunk 1998; Urdan and Schoenfelder 2006). They could be divided into two broader categories, didactic and psychological (Juriševič 2006). Didactic motivational initiatives represent the organizational side of the learning surroundings and learning process, type of instructional methods and resources used, while the psychological ones represent the mediation role of the teacher through the student's learning process (i.e., coaching, scaffolding, and modeling, see Brophy 1999). Both of the initiatives have an important impact on the student's motivation to learn and indirectly on the student's performance and achievement. Especially when both of them are congruent at the relational and content level simultaneously (e.g., interest learning units and instructional methods, positive classroom climate, accepting and stimulating teacher, etc.), it is possible that students—beside responding to the external motivations—step by step develop also more intrinsic motivations for learning,—since in an acceptable and stimulating environment students feel psychologically safe, develop positive academic self-concept, interests, and curiosity (Eccles et al. 1998; Stipek 1998).

Learning motivation could be defined as a mediation variable of academic achievements, as it affects the academic performance through various quantitative and qualitative indicators of the learning process; it is also connected with some other personality traits of students as well as demographic characteristics, such as

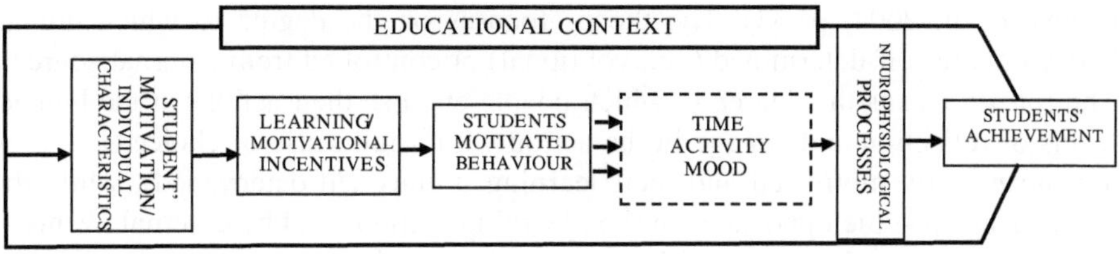

Fig. 11.1 The dynamics of motivation to learn (adapted from Rheinberg et al. 2000)

the nature of the temperament, anxiety, needs, abilities or nationality (Alexander and Murphy 1998; Jarvela and Niemivirta 2001; Juriševič 2006; Pintrich and Schunk 1996; Rheinberg et al. 2000; Rothbart and Hwang 2005). Research shows that motivation is connected with storing information into the long-term memory and with its recognition and retrieval (Schiefele and Rheinberg 1997). According to Corno (1994) and other contemporary authors motivation is attributed the key role in the decision-taking processes for certain learning behaviors.

In the opinion of Stipek (1998) learning motivation is mainly expressed in the attitudes of students toward learning and in their different approaches to learning. Jarvela and Niemivirta (2001) point out the fact that learning motivation encourages higher forms of learning, and consequently contributes to higher quality knowledge.

Rheinberg et al. (2000) provided for a detailed definition of the relationships among motivation, learning processes, and academic achievements. To their belief the influence of motivation upon learning is exercised on three different levels, namely: (1) in the duration and frequency of learning activities; (2) in the form of learning activities pursued; and (3) in functional disposition of the student during the learning activity. The first level of influence means active learning time (ALT)ù in which the relation between motivation and academic achievement is a rather complex one. It is not necessarily positive in all cases, as it is interwoven in the network of other learning variables (e.g., abilities, learning strategies, previous knowledge). Motivation influences the form or nature of learning activities in a variety of ways: on the one hand it balances the effort invested by the pupil into learning (in proportion to the level of difficulty of the learning task), whereas on the other hand it influences the application of learning strategies, encouraging the student to learn and enabling the successful reaching of goals. The third level of motivational influence on the learning outcomes is related to the optimal psychological state of the student during learning (Fig. 11.1).

According to Ryan and Deci (2000), intrinsic motivation is an individual's inherent inclination from which stems his/her tendency to learn about particular areas of life regardless of the presence of external enticements. In their Self-determination theory (*SDT*) the authors "assumed that all students, no matter how unskilled or how impoverished their backgrounds, possess inherent growth tendencies and innate psychological needs that provide a motivational foundation for their autonomous motivation and healthy psychological development"

(Reeve et al. 2004, p. 33). The theory focuses on the degree to which human behaviors are self-determined (i.e., volitional) or controlled from external sources. The former means the degree to which people endorse their actions at the highest levels of reflection and engage in the action with a full sense of choice (i.e., they are intrinsically motivated and their learning is thus self-determined), while the latter means just the opposite side (i.e., learning is motivated by external events—rewards, praise, punishment, without internal regulation).

According to this theory, learning activities in the chemistry classroom should be designed in such a way that students would value and self-regulate these activities without or with a minimum of external pressure. This process is realized through internalization (the process of taking in a value or regulation) and integration (a process by which individuals transform the regulation into their own so that it will emanate from their sense of self) (Ryan and Deci 2000). Namely, research shows that learners with internalized, integrated, or pure internal motivation achieve better results in knowledge tests, get higher achievement scores, and have a highly positive learning self-concept. In comparison with their peers with more extrinsic motivation, they show also less academic anxiety, and are less dependent on external motivational stimuli (Green et al. 2007; Gottfried et al. 2001). Personal satisfaction experienced through learning is also linked to higher creativity (Amabile 1985, cited in Csikszentmihalyi and Nakamura 1989; Shachar and Fischer 2004). Highly intrinsically motivated students are more successful in learning new concepts and show better understanding of the learning matter (Stipek 1998). Rennie (1990), on the basis of the research study on science learning, also concluded that higher results in science are related to the learner's active engagement in learning tasks, to his/her positive attitude toward the subject and to a highly positive self-concept in science, all of which imply the learner's intrinsic motivation to learn.

Approaches to chemistry teaching and learning, based on theories of scientific literacy, motivation, and situated learning yielded positive results regarding students' interest, achievements, and motivation for learning chemistry concepts (Bobich 2008; Nentwig et al. 2007; Chimeno et al. 2006).

Combining information technologies with the intent of improving the science-learning environment in terms of student motivation and learning efficiency, additionally increased students' positive perception of their learning and confidence, Charlesworth and Vician (2003).

Providing direction for students to review topics from previous chemistry classes, designing courses for early introduction to current research topics, using applied chemistry examples for solving problems, and analogies to teach chemistry, are approaches reported in the literature, which aim to tackle motivation problems for learning chemistry (Rieck 1998; Holme 1994; Thiele and Treagust 1994; Woodburn 1977).

The model of expertise in chemistry problem solving based on Anderson's Adaptive Control of Thought-Rational (ACT-R) theory, which was tested by Taasoobshirazi and Glynn (2009) showed how conceptualization, self-efficacy, and strategy interacted and contributed to the successful solution of quantitative, well-defined chemistry problems. The impact of self-concepts, self-efficiency, usefulness of science study, and interest in chemistry and physics, on students' academic performance was also revealed in the study by Lavonen and Laaksonen (2009).

Research results of Glynn et al. (2009) provided evidence that the students conceptualized their motivation to learn science in terms of five dimensions: intrinsic motivation and personal relevance, self-efficacy and assessment anxiety, self-determination career motivation, and grade motivation, and especially the belief in the relevance of science to students' careers was found by Glynn et al. (2007) as a strong predictor of students' motivation for learning science. Palmer (2009) investigated situational interest as a short-term form of motivation which occurred during a inquiry-based science lesson. The results indicated that interest arousal was substantial but did fluctuate throughout the lesson, according to the types of activities in which students were involved. The main source of interest was novelty, although choice, physical activity, and social involvement were also implicated.

Reseach Problem and Research Questions

This study aimed to investigate the correlation between students' academic achievements obtained through the design and construct approach to laboratory activities and their motivation to learn chemistry. The research questions were:

1. Did the design and construct approach to laboratory activities in learning concepts related to foam in the 9th grade, enable the understanding of the concepts selected?
2. How did the knowledge achieved through this approach correlate with students' prior knowledge (chemistry achievements from the 8th grade, and science achievements from the 7th grade) and their motivation to learn chemistry (i.e., controlled motivation, regulated motivation, intrinsic motivation, and academic self-concept)?
3. How did students evaluate the design and construct approach to laboratory activities in comparison with other teaching/learning strategies usually experienced in their chemistry classes.

We assume that this information will be useful to science (chemistry) teachers and science education researchers in applying more appropriate instructional methods, and thus fostering motivation for learning science (chemistry) with the objective of attaining a deeper understanding of chemical concepts and higher achievements.

Method

Participants

A total of 132 9th Grade students (58 males and 74 females) from four different Slovenian schools participated in the study. Their average age was 14.4 years. The sample represented an urban and rural population with mixed socioeconomic status and was randomly selected.

Instruments

Student's Handouts

A handout composed of five segments was designed for collecting feed-back on student achievements in designing and carrying out experiments of the teaching unit "Foam, foam" (Vrtačnik 2009). The structure of the handout followed hierarchically ordered steps by the process complexity of the design and construction approach. At the beginning of each segment there was a short explanation of the purpose of the segment. The first segment was dedicated to designing the experiments; students had to write down all possible pairs of salt solutions they could form from four different salt solutions (con. 0.5 mol/L) they were given on the tray. The second segment involved to carrying out wet experiments and taking notes of observations. For the purpose of the evaluation of students' results, a sketch of an empty table was attached as a hint for colleting data and a legend for uniform marking of the amount of foam produced in mixing pairs of salt solutions. The third segment was dedicated to setting up reasons for abundant and stable foam formation. A table was included for marking the pH of salt solutions, and a short guideline for setting up the hypothesis was also added. The fourth segment was intended to find out the role of detergent in foam formation and the nature of the gas trapped in the foam. The fifth segment was intended to link the macroscopic findings in foam formation with submicroscopic presentation of the role of detergent and water molecules in foam stabilization. This part was also connected with a short animation, which shows how detergent molecules are oriented around the bubble of gas and how water molecules surround the polar heads of detergent molecules. Scoring of the handouts was done for each task of the experimental procedure separately; for the first task (each combination of reagents 1 point), for the second task (see Fig. 11.3 for the combinations of salt solutions Al/Zn, Al/Na, Zn/Na, and Na/NaHC assigned as no foam or very little foam 0.5 points, for Al/NaHC and Zn/NaHC very abundant or abundant 1.5 points), for the third task (each correct determination of pH of salt solution 0.5 points, and for correct statement of the hypothesis 3 points, for partially correct 1.5 points), for the fourth task 6 points (1 point for each correct observation and 1 point for each correct explanation), and for the fifth task 5 points (for correct orientations of detergent molecules and water molecules around gas bubbles, and partially correct, meaning that the majority of the presented molecules were oriented correctly, 2.5 points). The total score achieved by each student (score maximum 31 points) was defined as student achievement. In order to draw a distinction between correct and partial correct answers, 20 % of students' handouts were collected and their answers analyzed. A list of accepted correct and partially correct answers was prepared and used in the further scoring procedure.

Students' Motivation for Learning Chemistry

A 37-item questionnaire for assessment of students' motivation was constructed on the basis of two questionnaires used in previous research (Black and Deci 2000; Juriševič et al. 2008) with the theoretical background from educational psychology research on motivation and self-concept (Ryan and Deci 2000; Marsh 1990). Specifically, the instrument was designed to assess (1) different components of students' motivation for learning chemistry (i.e., controlled motivation based on extrinsic motivational stimuli, regulated motivation based on internalized and integrated motivational stimuli, intrinsic motivation, and academic self-concept), (2) students' reasons for preference regarding the instructional method used in the study, and (3) students' preferences for different learning methods usually applied in chemistry classrooms.

Administration of the instrument took approximately 15 min in the classroom; students were asked to respond to a simple declarative sentence on a 5-point Likert scale ranging from 1—not at all true to 5—very true for me.

Reseach Design

The teacher brought some foam products into the classroom and initiated discussion on foams and their usage. Afterward he demonstrated an experiment in which very abundant and stable foam was formed. The teacher poured into 200 mL beaker 50 mL of 0.5 mol/L aqueous solution of $NaHCO_3$, added 2–3 drops of liquid detergent and 50 mL of 0.5 mol/L aqueous solution of $Al_2(SO_4)_3$, but the students were not told which salt solutions were mixed. This experiment was the starting point of the experimental design and construction approach. From the teacher's demonstration, the students had to observe that equal volumes of two salt solutions were mixed and that only a few drops of liquid detergent were added to one of the salt solutions. These observations were crucial for their own experimental design, which followed the demonstration. Students worked in pairs, each pair of students was given reagents on a plastic tray: 0.5 mol/L solutions of the following salts: $NaHCO_3$, $Al_2(SO_4)_3$, $ZnSO_4$, Na_2SO_4, beakers, measuring cylinders, pH papers, liquid detergent, and other materials. On the handouts they were presented with the following problems: (1) to find out the combination of two salt solutions which would upon mixing form the most stable and abundant foam, (2) to find out which gas was trapped in the foam, and (3) to find out the role of detergent in the experiments. The teaching unit lasted 40 min; during the lesson students were filling in the handouts. After one week, during regular chemistry class, students' motivation to learn chemistry was assessed.

Data were statistically analyzed with the SPPS package, version 17.0 on descriptive and bivariate levels of analysis. On the descriptive level, the basic statistics of variables were calculated. Correlations between variables were calculated based on Pearson correlation coefficients.

Results with Discussion

Students' Achievement in the Design and Construction Approach Used in Teaching the Unit » Foam, Foam «

Task: Combinations of Pairs of Salt Solutions

Students had to predict pairs of all possible combinations of salt solutions which they were given on plastic trays in 250 mL reagent flasks. As an example of how to write the combinations, $Al_2(SO_4)_3 + ZnSO_4$ was already written on the handouts. The results are presented in Fig. 11.2.

The majority of students (90.4 %) found and correctly wrote the formulae of all five combinations of salt solutions. Mistakes were due to presenting the same combination of salt solutions several times (6.3 %) or writing the incorrect formula of the salt (4.3 %).

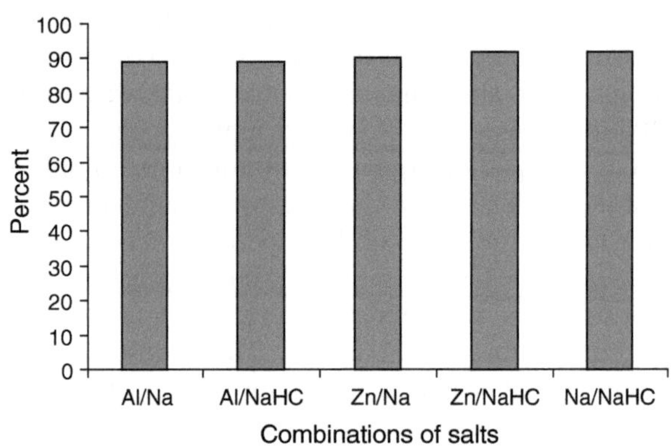

Fig. 11.2 Students' achievements in finding out pairs of salts. Legend: $Al_2(SO_4)_3 + Na_2SO_4$ (Al/Na); $Al_2(SO_4)_3 + NaHCO_3$ (Al/NaHC); $ZnSO_4 + Na_2SO_4$ (Zn/Na); $ZnSO_4 + NaHCO_3$ (ZnNaHC); $Na_2SO_4 + NaHCO_3$ (Na/NaHC)

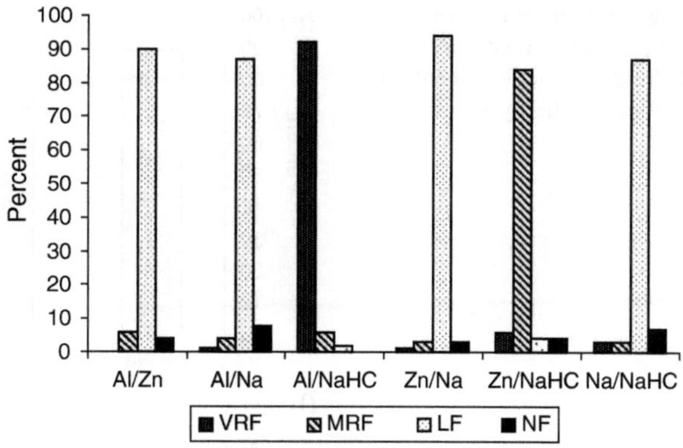

Fig. 11.3 Students' achievements in determining the amount of foam. *Legend* VRF—very rich foam; MRF—medium rich foam; LF—little foam; NF—no foam

Task: Carrying Out Experiments—Collecting Data

Planning experiments was followed by carrying out wet-experiments according to the plan. As a hint for collecting data an empty layout of the table, and a legend for assigning the amount of foam were included in the handout. The students had to calculate in advance the amount of salt solutions they should use for one experiment in order to be able to complete the task successfully. Their achievements are presented in Fig. 11.3.

Over 90 % (92 %) of students correctly observed that the maximum amount of stable foam was formed upon mixing the water solution of aluminum sulfate and sodium hydrogen carbonate. Further, 84 % of students found out correctly that medium rich foam was formed upon mixing the solution of zinc sulfate with sodium hydrogen carbonate. During carrying out the experiments, careful observation and precision were necessary. If students were careless and did not pay attention to the amount of foam formed, or forgot which salt solutions they were mixing, their results were incorrect. These observations are supported with the findings that 16 % of students additionally found that the medium amount of foam was formed also in cases where no foam could be expected, and 8 % of students found for the same combination of salts that very rich foam was formed. Some students (4 %), in spite of the guidelines, used nearly all solutions of salts for only a few experiments, and 4 % of students mixed the same salt solutions twice.

Task: Setting up the Hypothesis on the Correlation Between the pH of the Salt Solution and the Amount of Foam

Students had to measure the pH of salt solutions and find out the relation between pH of salts and the amount of foam formation. Teacher helped them by focusing their attention on pH papers. This teacher intervention acted as a support for directing students' thinking. The results are presented in Fig. 11.4a.

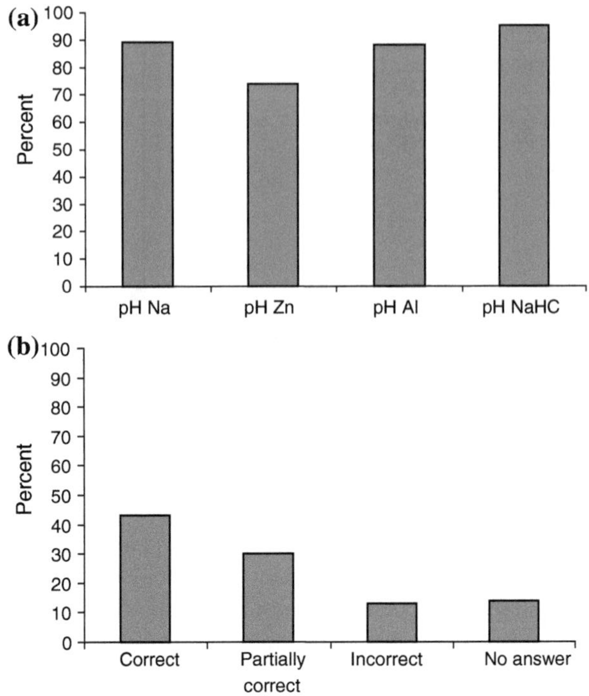

Fig. 11.4 **a** Percent of correct determinations of pH. **b** Students' achievements in setting up the hypothesis

On average 90.7 % of students found correctly the pH of three salt solutions, the only exception being zinc sulfate, for which only 74 % of answers were correct. Observation of students' work showed that mistakes were due to exchanging the names of solutions or using of the same pH paper several times.

In the second part of this task, students had to set up the hypothesis about the correlation between the pH of salt solutions and the amount of foam formed upon mixing two salt solutions. For the students this part was much more difficult than the previous one, Fig. 11.4b.

Less than half of the students (43 %) stated the hypothesis correctly: » The greater the difference between pH values of two salts in combination, the greater is the amount of foam formed «. Less than one third of students (30 %) gave a partially correct answer, meaning that the hypothesis was not correctly formulated, e.g., » Because there is the greatest difference in pH. «, or » The greater the difference in water solutions, the more foam is formed «. 13 % of answers were totally incorrect and 14 % of students did not state the hypothesis. These results proved our assumption that in Slovenian schools chemistry is mostly taught in the traditional way, and that teachers are not paying enough attention to science process skills and hands-on activities. Consequently, students are not used to formulating their own opinions during the school experimental work. In addition, the level of student chemistry literacy is rather low, therefore the majority of students were not able to formulate meaningful sentences from their observations, because they are lacking opportunities to discuss and express openly their own opinions about the concept taught.

Task: The Role of Detergent and the Nature of Gas

According to the instructions, students had to repeat the experiment (mixing solutions of aluminum sulfate and sodium hydrogen carbonate with detergent and without detergent) in order to find out the role of detergent and the nature of the gas trapped in the foam they obtained by using the burning splint. Their task was to describe in a coherent way the results of both experiments. The results are presented in Fig. 11.5.

Expected observations and explanations:

1. Foam is formed also when detergent is not added, but is not very stable or rich.
2. Gas which evolves upon mixing, extinguishes the flame of the burning splint when the splint reaches the rim of the beaker filled with gas. The gas is carbon dioxide.
3. Detergent traps the gas bubbles, thus preventing them from escaping from the beaker.

One fifth (21 %) of student observations and explanations were in line with our expectations, while two thirds (62 %) of students were able to describe correctly only one observation, 7 % of students did not describe correctly any of the observations and 10 % of students did not give answers. The majority (62 %) focused their attention only on one experimental observation e.g., results of the reaction without detergent, or the experiment with a burning splint, or describing the role of detergent. Only one fifth of them linked correctly all observations into a coherent set of explanations.

Fig. 11.5 Distribution of answers

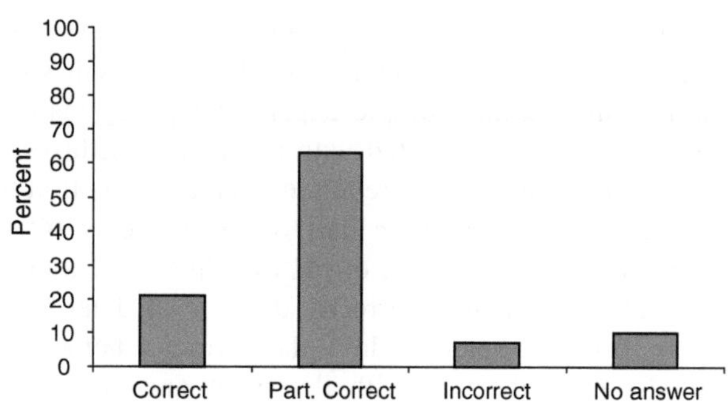

Table 11.1 Schematic presentations

Model of gas bubble	Model of water molecule	Model of detergent molecule
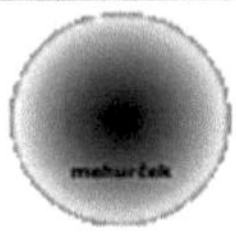		

Explaining the Role of Detergent at Submicroscopic Level

Students had to follow a short animation which showed at the molecular level how detergent and water molecules stabilize gas bubbles. Then they had to use models of particles (Table 11.1) and draw their own presentation of the stabilization process.

Nearly two thirds of students (60 %) were able to draw correctly the orientation of detergent and water molecules around the bubble of gas, 40 % of students oriented the models incorrectly or else they did not draw the scheme, because they probably did not understand the meaning of the animation.

Synthesis of Results: Steps of the Design and Construction Procedure and Students' Achievements

Figure 11.6 shows how the percent of correct answers differs according to different steps of the design and construction procedure. Students' achievements on the learning method used depend on the complexity of the thinking process required for finding correct answers. 90.4 % of students were able to predict all possible combinations of salt solutions, 92 % of students carried out experiments precisely enough that without difficulty they found the combination of salts which upon mixing gave the most stable and rich foam. 86.5 % of students determined correctly, within experimental error, the pH of salt solutions. Students were able to design and conduct simple experiments, they also proved to be good observers, however when confronted with more difficult tasks, where chemical literacy and analytical thinking were required, a great drop off in the number of correct answers was observed. Less than half of the students (40 %) were able to set up the hypothesis on the impact of pH of salts on the amount and stability of foam, and only 20 % explained correctly the role of detergent in foam formation. Surprisingly, 60 % of students drew the correct scheme showing at molecular level the stabilization of foam with detergent and water molecules. The results might confirm that students proved to be good observers.

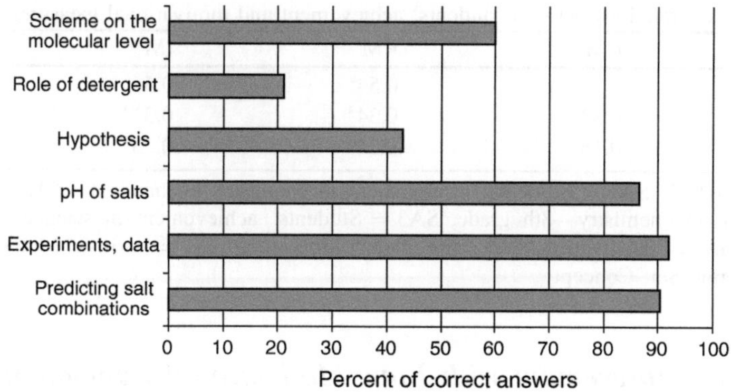

Fig. 11.6 Syntheses of results

Relation Between Students' Achievement in the Design and Construction Activity and their Prior Achievement

Table 11.2 summarizes the results related to student achievements in the design and construction approach with their prior achievements: science in 7th grade and chemistry in 8th grade. The values of the Pearson correlation coefficients are 0.49 ($p < .001$) for science grade and 0.55 ($p < 0.001$) for chemistry grade.

The results indicate the importance of pre-knowledge in learning chemistry concepts. New concepts could not be understood if pre-knowledge of the concepts in which they are rooted did not exist. According to the expectations, students with better pre-knowledge of chemistry and science were better in planning, controlling, and executing experiments as well as in analyzing data and correlating results with theory. These results are in line with the findings of Doppelt et al. (2008), Zangyuan (2003), who reported that students' prior knowledge and free exploration in teaching scientific concepts may have the advantage of engaging more students in the learning process and advancing their achievements. The relevance of prior knowledge on students' performance in science and more specifically chemistry, and its effect on the instructional design has also been proved in a series of studies e.g., Seery (2009), Hailikari et al. (2008), Chambers and Andre (1997), Hewson and Hewson (1983).

Table 11.2 The correlation between achievement in the design and construction approach and pre-knowledge in science and chemistry

	Prior achievement	
	Science 7th grade	Chemistry 8th grade
Achievement in the design and construction approach	0.49[a]	0.55[a]

[a] $p < 0.001$

Table 11.3 Correlations between students' achievement and motivational measures

	CM	RM	IM	AS-C
SA1	0.54[a]	0.54[a]	0.55[a]	0.65[a]
SA2	0.32[a]	0.34[a]	0.31[a]	0.64[a]
SA3	0.16	0.24[a]	0.23[a]	0.56[a]

Note [a] $p < 0.001$. SA1 = Students' achievement in chemistry on handouts, SA2 = Students' achievement in chemistry—8th grade, SA3 = Students' achievement in science—7th grade, CM = Controlled Motivation, RM = Regulated Motivation, IM = Intrinsic Motivation, AS-C = Academic Self-Concept

Students' Achievement with Four Motivational Components

The next set of analyses encompassed assessing relations between students' chemistry achievements and four motivational components. Target correlations are displayed in Table 11.3.

Correlations between students' achievement in handouts on the topic of foam, based on the design and construction approach to laboratory activities in Table 11.3 are bolded. All target correlations are of medium–high level and are significant, ranging between 0.54 and 0.65. Correlations between students' achievement in chemistry and science from grade 7 and 8 follow the same path, but they are slightly lower, ranging between 0.32 and 0.64. The strongest correlations are between students' current achievement and motivational measures, especially with academic self-concept in chemistry, the one highly correlated also with the other two measures of students' achievement in chemistry and science in grade 7 and grade 8. However, it is possible that the finding is in line with the benefits of the instructional method used in the study, under the assumption that the design and construction approach is more appropriate for students with a higher academic self-concept, or that it has a positive impact on its enhancement. Kaya and Rice (2010) investigated the effects of individual student factors, among them self-confidence, and classroom factors on elementary science achievement within and across five countries. At the student level, higher levels of home resources and self-confidence yielded higher science scores on the TIMSS 2003. Statistically significant correlation between students' science achievement and their self-confidence and interest in science as well as instructional design, was also revealed in the studies of Chang and Cheng (2008), Feltham and Downs (2002), Romance and Vitale (1992, 2001), and Tarhan and Sesen (2010). The results of the study by Nieswandt (2007) also revealed the importance of a strong and positive self-concept, the feeling of doing well in the chemistry class, for developing a meaningful understanding of scientific concepts.

Research shows that didactic and psychological motivations based on active learning methods encourage students to learn more confidently and autonomously,

Table 11.4 Correlations among motivational measures

	CM	RM	IM
RM	0.58[a]		
IM	0.40[a]	0.69[a]	
AS-C	0.49[a]	0.50[a]	0.58[a]

Note [a] $p < 0.001$. CM = Controlled Motivation, RM = Regulated Motivation, IM = Intrinsic Motivation, AS-C = Academic Self-Concept

making the learning context personal, interesting, and meaningful (Schunk and Pajers 2009; Reeve et al. 2004; Urdan and Schoenfelder 2006). On the other hand, it is important to take into consideration also two principles in motivation development—differentiation and multidimensionality. The first principle claims that the more the students develop different motivational components for the chemistry topics, the better is the result on their achievements (DeBacker and Nelson 2000). The second principle, multidimensionality, states that different motivational components are integrated in the motivational patterns that students develop through their schooling (Juriševič 2006) and "...allows us to understand the extent to which domain specificity might vary as a function of the construct under focus" (Green et al. 2007, p. 271). From this point of view it is important to note also the relation among motivational constructs as shown in Table 11.4. All correlations are significant and of medium–high level, indicating that different motivational constructs are correlated but still different enough to confirm the multidimensional framework.

On the basis of these results, it can be concluded that students' chemistry achievements and their motivation to learn chemistry are correlated; students with high achievements in chemistry have a higher academic self-concept and are also highly motivated—they have a strongly expressed extrinsic as well as intrinsic motivation to learn chemistry.

Students' Appreciation of the Design and Construction Approach to Laboratory Activities and Correlation with their Prior and Current Achievements

In evaluating the design and construction approach to laboratory activities, students had to estimate their opinion by specifying their levels of agreement with the five-level Likert item statements (*1*—meaning "I totally disagree." and *5*—meaning "I fully agree.") for the following attributes of the teaching units: (a) correlation of concepts with their prior experience with foam, (b) teacher guidance, (c) communication with peers, (b) help between and within groups, (d) relaxing and working atmosphere, (e) understanding new concepts, and (f) learning by doing experiments. The results of the descriptive statistics are presented in Table 11.5.

Table 11.5 Results of the descriptive statistics for students' opinions about the design and construction approach

Teaching unit attribute	Mean	Mode
Correlation with experience	2.04	1.00
Teacher guidance	2.38	1.00
Communication with peers	2.64	2.00
Help between and within groups	3.16	3.00
Relaxing and working atmosphere	3.58	4.00
Understanding new concept	2.54	1.00
Learning by doing experiments	3.42	5.00

The results show that students most appreciated the relaxed, yet working atmosphere which prevailed during the lesson, (mean = 3.58, mode = 4, frequency = 36), and learning through doing experiments (mean = 3.42, mode = 5, frequency = 41). Students' high appreciation of the relaxing atmosphere and learning through doing experiments, as experienced throughout the teaching unit, is in accordance with the findings of neurophysiologic research on the impact of emotions on motivation for learning. In such an atmosphere, positive emotions could be easily evoked, thus supporting a more positive attitude for fulfilling different tasks of the construction and design approach (Pecrun 2009). They expressed a neutral opinion about help between and among groups (mean = 3.16, mode = 3, frequency = 35). They did not have high opinions about the relation between the concepts presented in the teaching unit and their life experiences with foams, (mean = 2.04, mode = 1, frequency 53) and about the role of teacher who was, in this case, more a guide through different steps of the approach than a presenter of knowledge (mean = 2.38, mode = 1, frequency = 40). They also disagreed with the statement that the teaching unit contributes to a better understanding of the concept of foam (mean = 2.54, mode = 1, frequency = 36).

We were also interested in correlations between attributes of the teaching unit and students' achievement in the teaching unit, and their pre-knowledge. Results are summarized in Table 11.6.

A statistically significant medium–strong correlation at the level less then 0.001 was found only for one attribute of the teaching unit—relaxing and working atmosphere—with prior knowledge of science 7th grade and chemistry 8th grade ($r = 0.319$ and $r = 0.321$, respectively). Students with better grades in chemistry

Table 11.6 Correlations between attributes of the teaching unit and achievements

Attributes of teaching unit	Prior achievement		Achievement in the design and construction approach
Items	7th grade	8th grade	
Relaxing and working atmosphere	0.319[a]	0.321[a]	0.221[b]
Learning based on doing experiments	0.132	0.206[b]	0.143

[a] Correlation is significant at the <0.0001 level (2-tailed)
[b] Correlation is significant at the 0.05 level (2-tailed)

felt more self-confident and they appreciated more the relaxing and working atmosphere than did students with poorer prior achievements. Better students were also more successful at filling in handouts correctly and they appreciated more the relaxing atmosphere during the lesson, which evoked positive emotions. These results are in line with the study by Randler (2009), which also showed a positive association of emotion with achievement.

Students' Appreciation of Different Teaching/Learning Strategies

With the last set of the five-level Likert item statements, students had to estimate their level of agreement with selected teaching/learning strategies used in chemistry classes: teacher's lecturing, learning in pairs, independent learning with computer, independent learning with textbook, learning by doing experiments, Table 11.7.

The results reveal that in chemistry classrooms students most prefer learning by doing experiments (mean = 3.88, mode = 5, frequency = 55). This result is in accordance with students' estimation of the attribute (learning by doing experiments) of the design and construction approach. Regarding other learning strategies, students expressed rather neutral opinions about teacher's explanations, learning in pairs and independent learning with the computer (means = 2.69, 3.37, 2.97, modes = 3.00, frequencies = 40, 40, and 28), however the majority of students did not like to learn independently with the textbook (mean = 2.01, mode = 1, frequency = 31). The question is: why is the independent learning with the textbook so unpopular?

One possible answer is that teachers are not giving enough encouragement to their students to use more regularly the textbook as an important source of data and knowledge in chemistry classes. According to Harder (1989), science teachers, should be aware of the students' frustration when confronted with reading technical material in science textbooks. Acknowledging this problem by recommending possible solutions (e.g., model of a variety of comprehension strategies, guided discussions, small group discussions) can produce a positive change in attitude.

We were further interested in how student preferences toward different learning strategies correlate with students' pre-knowledge and their achievements in the design and construction approach, Table 11.8.

Statistically significant medium–high correlations at the 0.01 level were found only for learning based on doing experiments and pre-knowledge ($r = 0.389$, science—7th grade and $r = 0.405$, chemistry 8th grade). Statistically significant weak correlations at the 0.05 level were also found for learning in pairs and

Table 11.7 Results of the descriptive statistics about students' estimation of different teaching/learning strategies

Learning method	Mean	Mode
Teacher's explanations	2.69	3.00
Learning in pairs	3.37	3.00
Independent learning with computer	2.97	3.00
Independent learning with textbook	2.01	1.00
Learning by doing experiments	3.88	5.00

Table 11.8 Correlations between different learning/teaching methods, prior knowledge, and achievements in the design and construction approach

Learning/teaching method	Prior knowledge		Achievement in the design and construction approach
	Science 7th grade	Chemistry 8th grade	
Teachers' explanations			0.223[a]
Learning in pairs	0.207[a]		
Independent learning with textbook			0.200[a]
Learning based on doing experiments	0.389[b]	0.405[b]	

[a] Correlation is significant at the 0.05 level (two-tailed)
[b] Correlation is significant at the 0.01 level (two-tailed)

science grade (r = 0.207), teachers' explanations and student achievements in the design and construction approach (r = 0.223), and independent learning with the textbook and knowledge gained (r = 0.200).

Conclusions

The main problem of the present study was to evaluate the 9th graders' chemistry knowledge constructed through experimental work—design and construction approach to the topic "Foam, foam." It was presumed that an active instructional approach would enhance students' learning and thus the level of their chemistry knowledge. The results show that the majority of students (>90 %) did not have problems in resolving easier tasks within the learning activity, e.g., in predicting combinations of salt solutions, carrying out wet experiments, estimating the amount of reagents, and finding out the combination of two salts which upon mixing gave the most stable and rich foam (Fig. 11.6). However it is necessary to stress that only simple experimental skills (e.g., using measuring cylinders and beakers, pH papers) were expected for successful fulfillment of the mentioned tasks. But on those tasks of the approach where abstract thinking and higher order thinking skills were needed (e.g., analyzing data and setting up the hypothesis, determining the role of detergent, evaluating the results) less than half of the students were successful in completing the tasks (40 and 20 %, Fig. 11.6). Actually, students showed weakness in science literacy by having serious problems with formulation and verbalization of the hypothesis (Laugksch 2000). This is an important finding of the study itself, and at the same time also a contribution of the instructional method used for the purposes of this study. Namely, it is probably hard to discover and consequently also to deal with such kind of problems, although they are crucial for improving understanding of chemistry, if the prevailing teaching method is the classic (i.e., frontal method) without any possibility of detecting these obstacles while students are learning in school.

Another conclusion of the study is also valuable, as it is based on analyses of correlations between students' achievement in the design and construction approach to laboratory activity and pre-knowledge of science in 7th grade and chemistry in 8th grade, where modest correlations were found at the α level of 0.1 % (Table 11.2). This means that students with a better background did better also on the present learning tasks, so it can be concluded that the students' chemistry knowledge is upgrading or deepening throughout schooling, from middle to high school, and that in this case the grades have a relatively strong predictive value for students' enhancement and achievements.

Further statistical analyses revealed also that students' achievement through the design and construction approach is correlated with four motivational components: controlled and regulated motivation, and especially with intrinsic motivation and academic self-concept (Table 11.3). Only students with prior higher achievements and higher motivation for learning science/chemistry successfully accomplished the more demanding and complex tasks of the design and construction approach. This finding suggests that the pedagogical work in the chemistry classroom should focus more fully on internal motivation constructs (e.g., interest, self-concept) in order to empower students to use deep learning strategies, which leads to higher levels of knowledge (Green et al. 2007; Reeve et al. 2004; Zimmerman and Clearly 2009).

These findings should not discourage teachers from applying the design and construction approach to laboratory activity in their chemistry classes, since in evaluating different attributes of the approach, students reported that they most appreciate the relaxing and working atmosphere and the opportunity to learn chemistry by doing (Table 11.5). In other words, it means that more authentic learning tasks, together with the mediating role of the teacher within the supportive learning environment, create the best conditions for learning (Brophy 1999; DeBacker and Nelson 2000; Urdan and Schhoenfelder 2006). This is especially valuable when also the rationale for learning is mediated through teaching, so that students can understand immediately the usefulness of learning in everyday life; as Jang (2008) reports, the rationale enhances students' autonomously motivated learning behavior, which is needed to engage them actively and constructively in learning, regardless of its difficulty.

Acknowledgments The authors wish to thank the financers of this project and especially all participating students and teachers, without whose dedication and willingness to cooperate the realization of the project would not have been possible.

References

Alexander, P. A., & Murphy, P. K. (1998). The research base for APA's learner-centred psychological principles. In N. M. Lambert & B. L. McCombs (Eds.), *How students learn: Reforming schools through learner-centred education* (pp. 25–60). Washington, DC: APA.

Bobich, J. A. (2008). Active learning of biochemistry made easy (for the teacher). *Journal of Chemical Education, 85*(4), 234–236.

Bodner, G. M. (1986). Constructivism – A theory of knowledge. *Journal of Chemical Education, 63*(10), 873–878.

Black, A. E., & Deci, E. L. (2000). The effects of instructors' autonomy support and students' autonomous motivation on learning organic chemistry: A self-determination theory perspective. *Science Education, 84*(6), 740–756.

Brophy, J. (1999). Toward a model of the value aspects of motivation in education: Developing appreciation for particular learning domains and activities. *Educational Psychologist, 34*(2), 75–85.

Burke, K. A., Greenbowe, T. J., & Hand, B. M. (2006). Implementing the science writing heuristic in the chemistry laboratory. *Journal of Chemical Education, 83*(7), 1032–1038.

Byrnes, J. P., & Fox, N. A. (1998). The educational relevance of research in cognitive neuroscience. *Educational Psychology Review, 10*(3), 297–342.

Cacciatore, K. L., Amado, J., Evans, J. J., & Sevian, H. (2008). Connecting solubility, equilibrium, and periodicity in a green, inquiry experiment for the general chemistry laboratory. *Journal of Chemical Education, 85*(2), 251–253.

Chambers, S. K., & Andre, T. (1997). Gender, prior knowledge, interest, and experience in electricity and conceptual change text manipulations in learning about direct current. *Journal of Research in Science Teaching, 34*(2), 107–123.

Chang, C. Y., & Cheng, W. Y. (2008). Science achievement and students' self-confidence and interest in science: A Taiwanese representative sample study. *International Journal of Science Education, 30*(9), 1183–1200.

Charlesworth, P., & Vician, C. (2003). Leveraging technology for chemical sciences education: An early assessment of WebCT usage in first-year chemistry courses. *Journal of Chemical Education, 80*(11), 1333–1337.

Chimeno, J. S., Wulfsberg, G. P., Sanger, M. J., & Melton, T. J. (2006). The rainbow wheel and rainbow matrix: Two effective tools for learning ionic nomenclature. *Journal of Chemical Education, 83*(4), 651–654.

Cohen, A., & Magen, H. (2004). Hierarchical systems of attention and action. In G. W. Humphreys & M. J. Riddoch (Eds.), *Attention in action: Advances from cognitive neuroscience* (pp. 27–68). Howe: Taylor & Francis.

Corno, L. (1994). Student volition and education: outcomes, influences, and practices. In D. H. Schunk & B. J. Zimmerman (Eds.), *Self-regulation of learning and performance: Issues and educational applications* (pp. 229–251). Hillsdale, New Jersey: Erlbaum.

Cozolino, L. (2006). *The neuroscience of human relationships: Attachment and the developing social brain.* New York: Norton.

Csikszentmihalyi, M., & Nakamura, J. (1989). The dynamics of intrinsic motivation: A study of adolescents. In R. Ames & C. Ames (Eds.), *Research on motivation in education: Goals and cognitions* (Vol. 3, pp. 45–72). San Diego, CA: Academic Press.

DeBacker, T. K., & Nelson, R. M. (2000). Motivation to learn science: Differences related to gender, class type, and ability level. *Journal of Educational Research, 93*(4), 245–254.

Doppelt, Y., Mehalik, M. M., Schunn, C. D., Silk, E., & Krysinski, D. (2008). Engagement and achievements: A case study of design-based learning in a science context. *Journal of Technology Education, 19*(2), 22–39.

Eccles, J. S., Wigfield, A., & Schiefele, U. (1998). Motivation to succeed. In W. Damon & N. Eisenberg (Eds.), *Handbook of child psychology* (Vol. 3, pp. 1017–1095). New York, NY: Wiley.

Erk, S., Kiefer, M., Grothe, J., Wunderlich, A. P., Spitzer, M., & Walter, H. (2003). Emotional context modulates subsequent memory effect. *Neuroimage, 18*(2), 439–447.

Feltham, N. F., & Downs, C. T. (2002). Three forms of assessment of prior knowledge, and improved performance following an enrichment programme, of English second language biology students within the context of a marine theme. *International Journal of Science Education, 24*(2), 157–184.

Fosnot, C. T., & Perry, R. S. (2005). Constructivism: A psychological theory of learning. In C. T. Fosnot (Ed.), *Constructivism: Theory, perspectives and practice*. New York: Teachers College Press, Columbia University.

Furlan, P. Y. (2009). Engaging students in early exploration of nanoscience topics using hands-on activities and scanning tunneling microscopy. *Journal of Chemical Education, 86*(6), 705–711.

Glynn, S. M., Taasoobshirazi, G., & Brickman, P. (2009). Science motivation questionnaire: Construct validation with nonscience majors. *Journal of Research in Science Teaching, 46*(2), 127–146.

Glynn, S. M., Taasoobshirazi, G., & Brickman, P. (2007). Nonscience majors learning science: A theoretical model of motivation. *Journal of Research in Science Teaching, 44*(8), 1088–1107.

Gottfried, A. E., Fleming, J. S., & Gottfried, A. W. (2001). Continuity of academic intrinsic motivation from childhood through late adolescence: A longitudinal study. *Journal of Educational Psychology, 93*, 3–13.

Green, J., Martin, J. A., & Marsh, H. W. (2007). Motivation and engagement in english, mathematics and science high school subjects: Towards an understanding of multidimensional domain specificity. *Learning and Individual Differences, 17*(3), 269–279.

Hailikari, T., Katajavuori, N., & Lindblom-Ylanne, S. (2008). The relevance of prior knowledge in learning and instructional design. *American Journal of Pharmaceutical Education, 72*(5), 113.

Harder, A. K. (1989). Attitudes toward reading science textbooks. *The American Biology Teacher, 51*(4), 208–212.

Hewson, M. G., & Hewson, P. W. (1983). Effect of instruction using students' prior knowledge and conceptual change strategies on science learning. *Journal of Research in Science Teaching, 20*(8), 731–743.

Holme, T. A. (1994). Providing motivation for the general-chemistry course through early introduction of current research topics. *Journal of Chemical Education, 71*(11), 919–921.

Jang, H. (2008). Supporting students' motivation, engagement, and learning during an uninteresting activity. *Journal of Educational Psychology, 100*(4), 798–811.

Jarvela, S., & Niemivirta, M. (2001). Motivation in context: Challenges and possibilities in studying the role of motivation in new pedagogical cultures. In S. Volet & S. Jarvela (Eds.), *Motivation in learning context: Theoretical advances and methodological implications* (pp. 105–127). Amsterdam: Pergamon.

Jones, L. L. (1999). Learning chemistry through design and construction. *UniServe Science News, 14*, 3–7.

Juriševič, M. (2006). *Učna motivacija in razlike med učenci [Motivation to learn and individual differences]*. Ljubljana: Univerza v Ljubljani, Pedagoška fakulteta.

Juriševič, M., Razdevšek, Pučko C., Devetak, I., & Glažar, S. A. (2008). Intrinsic motivation of pre-service primary school teachers for learning chemistry in relation to their academic achievement. *International Journal of Science Education, 30*(1), 87–107.

Kaya, S., & Rice, D. C. (2010). Multilevel effects of student and classroom factors on elementary science achievement in five countries. *International Journal of Science Education, 32*(10), 1337–1363.

Kirschner, P. A., Sweller, J., & Clark, R. E. (2006). Why minimal guidance during instruction does not work: An analysis of the failure of constructivist, discovery, problem-based, experiential, and inquiry-based. *Educational Psychologist, 41*(2), 75–86.

Kroesbergen, E. H., Van Luit, J. E. H., & Maas, C. J. M. (2004). Effectiveness of explicit and constructivist mathematics instruction for low-achieving students in the Netherlands. *Elementary School Journal, 104*(3), 233–251.

Laugksch, R. C. (2000). Scientific literacy: A conceptual overview. *Science Education, 84*(1), 71–94.

Lavonen, J., & Laaksonen, S. (2009). Context of teaching and learning school science in Finland: Reflections on PISA 2006 results. *Journal of Research in Science Teaching, 46*(8), 922–944.

Marsh, H. W. (1990). The structure of academic self-concept: The Marsh/Shavelson model. *Journal of Educational Psychology, 82*(4), 623–636.

Mayer, R. E. (2004). Should there be a three-strikes rule against pure discovery learning? The case for guided methods of instruction. *American Psychologist, 59*(1), 14–19.

Matthews, M. R. (2008). *Philosophical and pedagogical problems with constructivism*. Sydney: University of New South Wales. Retrieved from http://per.physics.helsinki.fi/Tutkijakoulun_kesaseminaari_2008/Michael_R_Matthews_helsinki_2008.pdf.

Nentwig, P. M., Demuth, R., Parchmann, I., Grasel, C., & Ralle, B. (2007). Chemie im kontext: Situating learning in relevant contexts while systematically developing basic chemical concepts. *Journal of Chemical Education, 84*(9), 1439–1444.

Nieswandt, M. (2007). Student affect and conceptual understanding in learning chemistry. *Journal of Research in Science Teaching, 44*(7), 908–937.

Palmer, D. H. (2009). Student interest generated during an inquiry skills lesson. *Journal of Research in Science Teaching, 46*(2), 147–165.

Pecrun, R. (2009). Emotions in school. In K. R.Wentzel & A. Wigfield (Eds.), *Handbook of motivation at school* (pp. 575–604). New York, NY: Routledge.

Pintrich, P. R., & Schunk, D. H. (1996). *Motivation in education: Theory, research, and applications*. Englewood Cliffs, New Jersey: Prentice Hall.

Randler, C. (2009). Association between emotional variables and school achievement. *International Journal of Instruction, 2*(2), 3–10.

Reeve, J., Deci, E. L., & Ryan, R. M. (2004). Self-Determination theory: A dialectical framework for understanding sociocultural influences on student motivation. In D. M. McInerney & S. V. Etten (Eds.), *Research on sociocultural influences on motivation and learning* (Vol. 4, pp. 31–60). Greenwich: Information Age.

Rennie, L. J. (1990). Student participation and motivational orientation: What do students do in science? In K. Tobin, J. Butler, & B. J. Fraser (Eds.), *Windows into science classrooms: Problems associated with higher-level cognitive learning* (pp. 164–198). London: The Falmer Press.

Rheinberg, F., Vollmeyer, R., & Rollett, W. (2000). Motivation and action in self-regulated learning. In M. Boekaerts, P. R. Pintrich, & M. Zeidner (Eds.), *Handbook of self-regulation* (pp. 503–531). San Diego: Academic Press.

Rieck, D. F. (1998). Providing direction and motivation for students to review topics from previous chemistry classes. *Journal of Chemical Education, 75*(7), 850.

Romance, N. R., & Vitale, M. R. (1992). A Curriculum strategy that expands time for in-depth elementary science instruction by using science-based reading strategies—Effects of a year-long study in grade 4. *Journal of Research in Science Teaching, 29*(6), 545–554.

Romance, N. R., & Vitale, M. R. (2001). Implementing an in-depth expanded science model in elementary schools: Multi-year findings, research issues, and policy implications. *International Journal of Science Education, 23*(4), 373–404.

Rothbart, M. K., & Hwang, J. (2005). Temperament and the development of competence and motivation. In A. J. Elliot & C. S. Dweck (Eds.), *Handbook of competence and motivation* (pp. 167–184). New York, NY: The Guilford Press.

Rudd, J. A., Greenbowe, T. J., Hand, B. M., & Legg, M. J. (2001). Using the science writing heuristic to move toward an inquiry-based laboratory curriculum: An example from physical equilibrium. *Journal of Chemical Education, 78*(12), 1680–1686.

Ryan, R. M., & Deci, E. L. (2000). Intrinsic and extrinsic motivation: Classic definitions and new directions. *Contemporary Educational Psychology, 25*(1), 54–67.

Schiefele, U., & Rheinberg, F. (1997). Motivation and knowledge acquisition: Searching for mediating processes. *Advances in Motivation and Achievement, 10,* 251–301.

Seery, M. K. (2009). The role of prior knowledge and student aptitude in undergraduate performance in chemistry: A correlation-prediction study. *Chemistry Education Research and Practice, 10*(3), 227–232.

Schunk, D. H. (1998). An educational psychologist's perspective on cognitive neuroscience. *Educational Psychology Review, 10*(4), 297–342.

Schunk, D. H., & Pajers, F. (2009). Self-efficacy theory. In K. R.Wentzel & A. Wigfield (Eds.), *Handbook of motivation at school* (pp. 35–54). New York, NY: Routledge.

Shachar, H., & Fischer, S. (2004). Cooperative learning and the achievement of motivation and perceptions of students in 11th grade chemistry classes. *Learning and Instructions, 14*(1), 69–87.

Stipek, D. (1998). *Motivation to learn: From theory to practice*. Boston, MA: Allyn and Bacon.

Urdan, T., & Schoenfelder, E. (2006). Classroom effects on student motivation: Goal structures, social relationships, and competence beliefs. *Journal of School Psychology, 44*(5), 331–349.

Taasoobshirazi, G., & Glynn, S. M. (2009). College students solving chemistry problems: A theoretical model of expertise. *Journal of Research in Science Education, 46*(10), 1070–1089.

Tarhan, L., & Sesen, B. A. (2010). Investigation of the effectiveness of laboratory works related to "acids and basis" on learning achievements and attitudes toward laboratory. *Procedia-Social and Behavioral Sciences, 2*(2), 2631–2636.

Thiel, A., Eurich, C. W., & Schwegler, H. (2002). Stabilized dynamics in physiological and neural systems despite strongly delayed feedback. In J. R. Dorronsoro (Ed.), *Artificial Neural Networks (Proceedings International Conference on Artificial Neural Networks, ICANN 2002)* (pp. 15–20). Berlin; Heidelberg: Springer.

Thiele, R. B., & Treagust, D. F. (1994). An interpretive examination of high school chemistry teachers' analogical explanations. *Journal of Research in Science Teaching, 31*(3), 227–242.

Vrtačnik, M. (2011). Development of the basic science competencies through experimental work. *The Teachers' Guidelines for Implementing New Chemistry Curricula for High School, General and Inorganic Chemistry*, (pp. 19–23). Slovene Board of Education.

Vrtačnik, M. (2009). » Foam, foam « Teaching unit based on the design and construction laboratory approach, Project European Social Fund and Slovenian Ministry for Education and Sports, » Development of Science Competencies « .

Woodburn, J. H. (1977). Classroom mechanics—using applied chemistry to tackle motivation problems. *Journal of Chemical Education, 54*(12), 763.

Zangyuan, O. (2003). The application of adaptive learning environment on oxidation-reduction reactions. *International Journal of Instructional Media, 30*(1), 223–235.

Zimmerman, B. J., & Cleary, T. J. (2009). Motives to regulate Learning: A social cognitive account. In K. R.Wentzel & A. Wigfield (Eds.), *Handbook of motivation at school* (pp. 247–264). New York, NY: Routledge.

Chapter 12
Contexts as Learning Catalysts for Students and Teachers: Approaches and Exemplary Results from the Projects *Chemie im Kontext* and CHEMOL

Ilka Parchmann, Nina Dunker and Wiebke Endres

Introduction and Background

Active learning for students' demands successful stimuli and supporting structures. The teaching tradition in German classrooms often uses experiments as stimuli. However, the results of empirical studies show that experiments rather enhance "activities of hands" than "activities of minds" (Lunetta 1998; Euler 2002; von Aufschneiter and Riemeier 2005). Students carry out experiments as "cookbook-recipes," not as a scientific approach to gain new insights, following certain rules and processes as shown in Fig. 12.1.

Additionally, students do not connect experiments and basic concepts to phenomena in daily life and classroom teaching with just a high number of experiments does not automatically lead to successful learning processes either (Prenzel et al. 2007). We can therefore state that experiments on their own do not work as "catalysts" for active learning, initiating the development of applicable and sustainable chemical knowledge and competencies.

The projects *Chemie im Kontext* for secondary level and CHEMOL for primary level use contexts derived from the students' daily-life experiences or contexts connected to important societal issues to raise questions which can be investigated by groups, using different approaches and techniques. The active learning of the students is supported by scaffolding material and a variety of teaching and learning methods, involving different roles for teachers and students.

I. Parchmann
IPN, Kiel, Germany
e-mail: parchmann@ipn.uni-kiel.de

N. Dunker (✉)
University of Bremen, Bremen, Germany
e-mail: dunker@uni-bremen.de

W. Endres
University of Oldenburg, Oldenburg, Germany

I. Devetak and S. A. Glažar (eds.), *Learning with Understanding in the Chemistry Classroom*, DOI: 10.1007/978-94-007-4366-3_12,

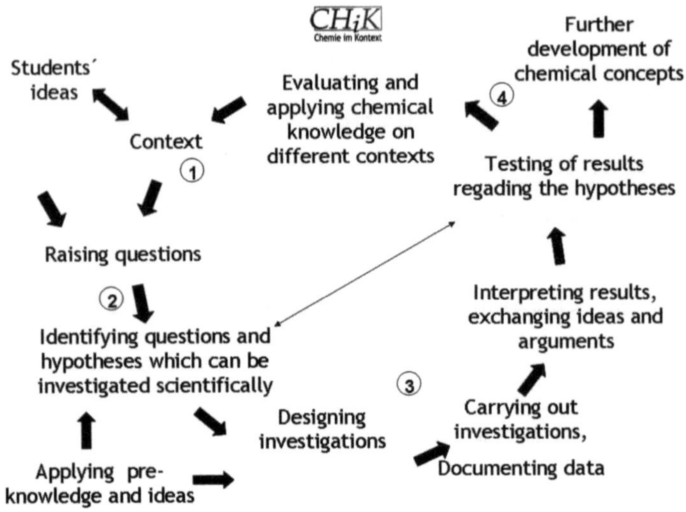

Fig. 12.1 Procedures of an experimental scientific research approach (The numbers refer to the four phases of *Chemie im Kontext*, see below.)

The implementation of both projects into school practice was supported by the involvement of teachers or teacher students already in the design process of the material (Parchmann et al. 2006; Steffensky and Parchmann 2007). This cooperation of perspectives from research and practice assured the ecological and theoretical validity of the conceptual framework and the material. Therefore, the term "active learning" can also be applied to the learning of the involved teachers and researchers, who learned from each other in a "symbiotic" way (Parchmann et al. 2006).

This chapter of the book will describe the structure of both projects, give examples of teaching and learning processes and discuss exemplary results from qualitative and quantitative research studies.

Chemie im Kontext

Active Learning for Students—Contexts as Learning Catalysts

The conceptual framework of *Chemie im Kontext* is based on three principles [for further information on Chemie im Kontext (CHiK) see Parchmann et al. 2006; Nentwig et al. 2007]:

1. **Context-based learning**: Learning environments are considered "in context," when learners acquire knowledge and competence on a need-to-know-basis in

dealing with a relevant issue, starting with their questions and ideas. Examples are: "Food design—why, how and where?;" "Carbon dioxide and climate change?;" "Materials by design;" "A mouth full of chemistry."

2. **Development of basic concepts**: To develop a basic knowledge foundation that can be applied to new contexts and situations, the main principles of chemistry must be derived and abstracted from the contexts. These principles are described as "basic concepts," they structure and summarize the factual knowledge (see the basic concepts of the National Standards).
3. **Variety of teaching and learning methods**: A variety of teaching and learning methods is one of the key elements for a successful chemistry education, (a) because it considers the diversity of interests, pre-knowledge, capabilities, and learning styles and (b) because it offers the students situations in which they can develop and apply competencies in all areas as demanded by the National Standards in Germany.

All teaching and learning units are structured by four phases: (1) phase of contact (aiming at the students' motivation and an activation of their pre-knowledge), (2) phase of curiosity and planning (aiming at the development of the students' questions and structuring the following learning process), (3) phase of development and presentation, and (4) phase of summary, deepening, exercise and abstraction, and transfer.

Opportunities for active learning of students are given in all four phases but with different meanings. In the first phase of contact, the students are expected to bring in their own ideas and questions into the discussion and further planning. To do this, the students have to connect the chosen topic with their pre-knowledge and daily-life experience, with is often a lack of competence already (see above). The second phase demands the students to decide on relevant questions which could be answered based on scientific inquiry (see Fig. 12.1). This competency is mentioned as one of the central goals in the definition of scientific literacy, according to the OECD-PISA-consortium: "Scientific literacy is the capacity: (1) to use scientific knowledge, (2) to identify questions, and (3) to draw evidence-based conclusions in order to understand and help make decisions about the natural world and the changes made to it through human activity." The students have to learn the characteristics of the specific "Nature of Science (NoS)" to differentiate between scientific questions, more detailed chemical questions, and others. Therefore, this approach also fosters the development of an understanding of the Nature of Science (Lederman 1992; McComas 2000). Of course, especially in the beginning of chemistry classes, the students will get exemplary questions which they could then use as analogies or "templates" for further units to choose and to define their own ideas and research questions.

The most student-oriented phase is the third one: In this phase of development, the students often work in groups. The teachers prepare learning environments that enable the students to carry out investigations according to their own interests, abilities, and time needed, for example by using the method of expert groups (Leerhoff et al. 2002). Another often applied method was the design of learning cycles or stations (Leerhoff et al. 2000): All students get a list of obligatory stations and additional stations which they have to work on. The sequence of stations can be free or given in advance. Usually, not all stations incorporate experiments to enable the teacher to observe the experimental stations in particular, while others do not need any specific observation and can be carried out by the students themselves.

The most open approach is the design of tasks which only describes a situation and for which the students have to develop their own methods of investigation, often experiments. The students enjoyed this creative work very much and the results were even better than the teachers had expected (Kandt 2008). Of course, this very open approach could not be used for every topic and in every situation, but the teachers were asked to incorporate it, if possible.

The last phase of summary and abstraction (see Fig. 12.1) is the most teacher-based one. Students often cannot decide themselves which aspects of a topic to be the most important ones, where to make connections to other contexts or how to deepen their understanding of basic concepts (such as the concept of matter and particles or energy). Therefore, as one result of the trials of CHiK, the teachers usually had to guide this phase and help the students to summary, to reflect and to "decontextualize" their knowledge.

Last but not least, the success of every teaching and learning unit does not only depend on the learning activities but also on the testing and reflecting of gained knowledge and competencies. The tests developed in *Chemie im Kontext* expected the students to be able to name and to apply basic concepts and factual knowledge as well as to create ideas for investigations, to translate daily-life situations (and language) into scientific questions (and language) and to evaluate decisions, as described in the definition of scientific literacy. The results overall were very satisfying: the students showed better or equal results in cognitive areas, appreciated the relevance of chemistry better and did not show the same dramatic loss of interest as reported in other chemistry classes (Demuth et al. 2008; Parchmann et al. 2006).

Active Learning for Teachers and Researchers: Learning Communities as "Catalysts"

CHiK was not developed as a complete curriculum, it was developed as a framework with exemplary units to enable teachers in different states and schools to adopt it to their syllabi and conditions (Parchmann et al. 2006). Hence, the implementation of CHiK was also part of the further development of teaching and learning units and material, based on the idea of "learning communities" (see Fig. 12.2). Such communities enabled a close cooperation between teachers in practice and university educators and researches, which assured the CHiK approach to consider the demands of research findings and school practise at the same time.

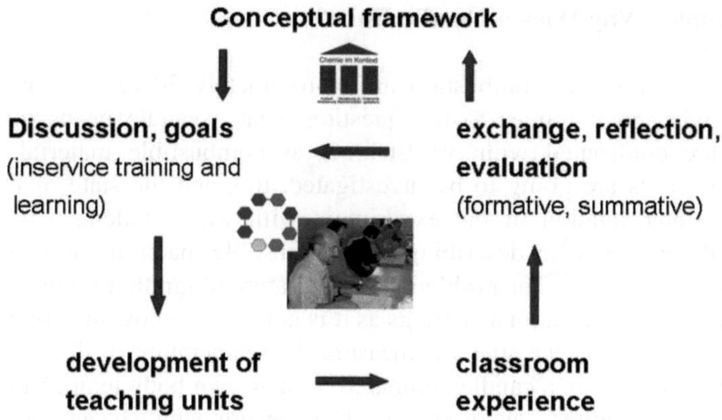

Fig. 12.2 Active learning of teachers and researchers in "symbiotic" learning communities

As such groups did not only carry out units and trial material but developed and designed their own ones, their learning can also be classified as active learning. The special situation of the close cooperation between different experts—teachers and researchers—led to an exchange of ideas, arguments, and expertise which does normally not happen in in-service training workshops, where the researchers are the "teachers" and the participants the "students." Looking at the results again, the teachers had not only enjoyed this work very much but also appreciated the worth of the learning communities to change and develop their teaching practise (Demuth et al. 2008; Parchmann et al. 2006).

CHEMOL

Active Learning for Students: Stories as Learning Catalysts

The CHEMOL-Project invites elementary school students to the chemistry laboratory for expanding their experimental abilities in learning environments focussed on science tasks. The teaching method that is mainly used is based on constructivistic ideas such as exploratory learning. The units, which are to acquire in CHEMOL, refer to four major topics: fire and combustion, water and solving, the gaseous state of matter and acids and bases. The learning setting is based upon students' preconceptions and is orientated toward their questions.

Active learning in CHEMOL means that students are confronted with tasks that they can survey because of the focusing upon one aspect, the possibility to develop own experiments and interpret the outcome together in a group. Additionally students expand their experimental abilities through real problem solving tasks.

One Example: Why Does a Candle Burn?

For the matter fire and combustion an approximately 20 experiments comprehensive environment replies to the question what a candle needs for burning. Students are confronted with wax/stearine as combustible material. Different burning materials are going to be investigated, too, and the state of matter also affects the interpretation of the experiment with wax. Students will be given several different tasks for describing the combustible material and classify them into different categories. The problem of temperature of ignition is very different to broach the issue in educational settings as it is not easy to show in experiments. To approach the problem, the students measure the temperature of different sources. They find out that even a candle compared to their own body temperature used as reference is much hotter—about 800 °C. The question why we can burn ourselves in a flame can be answered by this insight which can easily be achieved by elementary school students.

Inalienable and essential for students to gain the cognition of oxygen as needful for fire to burn is the conception of oxygen as invisible, gaseous matter. Therefore, students are brought to cognitive conflicts concerning first the missing air in experiments with fire covered with glasses. The task which brings the conflict lays in the question to save the fire of the candle just before it has been extinguished. The students learn to lift the glass a little bit so that air (oxygen) feeds the combustion and the flame is saved (see Fig. 12.3). By varying the variables in only one component, the students also learn a very essential idea of scientific experiences: varying only one variable at a time. Otherwise, the task will soon get too complex to be solved by elementary school students. Often different volumes of glasses are used to show the need of air or even oxygen as needful condition for combustible processes, but cognitive capacities of elementary school students are often overstrained coping with the variable of air and volume together.

In an investigation focussing on the subserving methods for elementary school students to gain insight of abstract ideas in science, the method of concept mapping is used for investigating effectiveness. The quantitative evaluation of used concepts shows impressively that special scientific terms such as oxygen, temperature, or combustible material are very important for effective learning processes (e.g., see Fig. 12.4). In the CHEMOL-project these terms are introduced additionally to the experimental work and inserted in established cognitive structures by using them in different situations and different contexts. The learning

Fig. 12.3 Saving flame by lifting glass

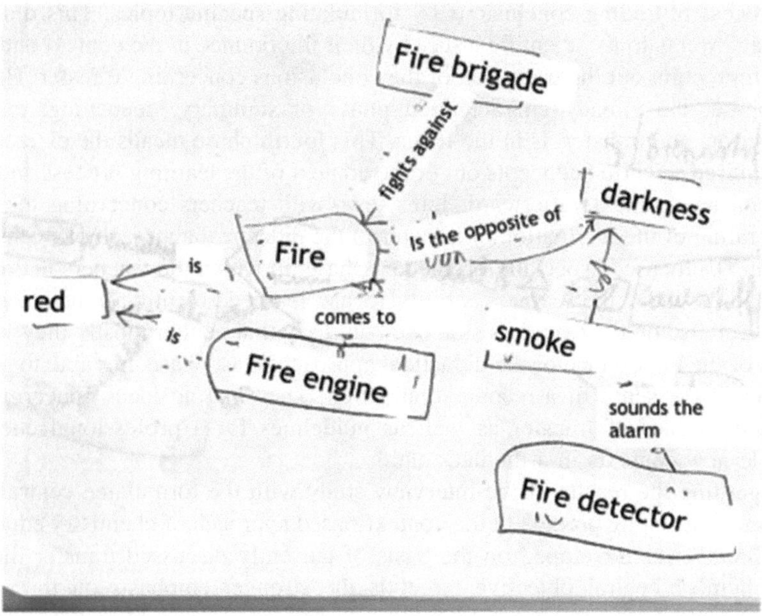

Fig. 12.4 Concept map

of these special terms builds up the cognitive structures for concepts of burning and combustion and works against rote learning by linking concepts. In an explicit way these linked concepts can be visualized by methods such as concept mapping. It can easily be used for diagnostic senses and promotes meaningful learning processes even for young learners.

For students in teacher training the "hands-on" activity in CHEMOL, where they structure educational topics for learners, and realize them with two or three children, is a very effective way to initialize "active learning" processes. The reflection of their own activity helps to overcome difficulties having regards to the educational structuration process as well as the teaching situation and fosters preparing them for their profession as a teacher (Steffensky and Parchmann 2007).

Outlook: "ProChem"—A Synthesis Between CHiK and CHEMOL with a Special Focus on the Problem of Transfer

The didactical approach of "Chemistry in context" puts the focus of the lessons in a thematic unit onto themes or problems of daily life, which are relevant and interesting for the learners. Such contexts motivate and structure what happens in the lessons in such a way that questions can be deduced and be answered during the process of finding conclusions by formulating specific topics. This didactical approach wants to fix scientific issues by their importance in the context and at the same time point out the transition of the conclusions concerning transfer. Here the last phase, the already characterized phase of summary, deepening, exercise, abstraction, and transfer, is in the focus. This fourth phase means the extraction of the acquired scientific concepts out of the context of the learning process and using them in new contexts afterward. Interviews with teachers concerning the qualitative rating of the facilitation of transfer in the didactical approach of "Chemistry in context" focused especially on the last phase, in which the teachers in common complained about the learners' ability to use learned scientific contents in new tasks. But also on the teacher's side one could say that neither mostly they weren't aware of the importance of the didactical approach's last phase, nor able to support transfer in the sense of a recontextualization. They had no ideas concerning the necessary basics of transfer as well as guidelines for a professional choice of suitable new contexts in a thematic unit.

Regarding the results of the interview study with the formulated central problems in the teaching process in the context-based approach of chemistry education, "ProChem" was developed on the basis of currently discussed transfer theories. "ProChem's" central objective target is the stronger emphasis on the idea of transfer in the learning process by using a variety of open problem situations in a specific scientific topic. Due to the contexts which demand near transfer as well as far transfer the scientific concept lying behind the thematic unit is steadily broadened and deepened. This is because of the analogies and regularities the problem situations point out so the learners are able to formulate generalizations. For half-a-year, the approach of "ProChem" for interfering scientific basic knowledge—in the sense of ideas and concepts—for elementary age has been in evaluation. The focus of this didactic direction of this approach is the embedding of particular chemistry contents in open problem situations, superordinated in a story, for which the arrangement of the experiments of the before described CHEMOL-project is important. The development of the scientific concept shows three phases which result directly out of the chosen contexts. Starting with a first phase to acquire the fundamental chemistry issues of a thematic unit, a second phase with a new context demands near transfer. Depending on the performance and previous knowledge of the learners, the first phase can be left out and a flexible start with the second one is possible. Of great importance is the third phase of each content, which demands far transfer and thus gives in insight into the learning process of the student and the state of knowledge. The exemplary teaching unit on the content of "air" described below concretely demonstrates the structure described above.

The heroes of the adventure story, which enable the students to get an emotional entrance into the learning process, want to salvage a treasure deep down at the bottom of the sea, but they are extremely afraid of water. This, being very motivating for kids in elementary school problem because of the immanent drama, results in the suggestion to use a diving bell. The simple experiment to explain the principal functionality of the diving bell contains a discovery that is astonishing for learners of this age: in the vessel used in this experiment isn't "nothing"—there is air in it. At this point they experience the material qualities of the invisible air. These properties are underlined as things developed in the story itself, where a hole in the diving bell leads to a lack of air. In a "rescue operation" air from one diving bell to the defect one is decanted. Now transfer is demanded in the directly following context, in which a supposedly genie in the bottle is presented to the students. The cold bottle, which is closed on the wet bottleneck with a five cent coin, is warmed up by the hands of the kids and the ghost in the bottle seems to speak through the hopping cent coin on the bottleneck. Near transfer is demanded in this case, since the learners have to recognize in analogy to the diving bell before, which is the existence of "air" in the bottle. The kids of this age in elementary school in our exemplary study groups had no problem to recognize the similarity to the context before and to transfer the knowledge and use it for explaining this new phenomenon. An enhancement represents the scientific approach of the expansion of air, when it is warmed up, which is responsible for the phenomenon of the "ghost in the bottle." The learners, personifying the air, which is typical for this age, realize the connection between the rise in temperature and the expansion of air very fast. This scientific issue is revived in a more complex context—the construction of a hot air balloon—at the end of the story. At this point of the teaching process, we observed a concrete reference to the experiment we described before, the "ghost in the bottle," in all our teaching experiments. The construction and the functionality of the balloon have caused no problems to the students in primary school, since they recognized the analogies and are able to formulate scientific laws.

To sum up, the teaching approach of "ProChem" uses the advantages of the approach of "chemistry in context" in case of the learning theory and the theory of motivation lying behind by emphasizing the ideas of transfer by using several contexts in the way of open problem situations which are staggered to support the ability of the learners to transfer their knowledge for solving relevant problems.

References

Demuth, R., Gräsel, C., Parchmann, I., & Ralle, B. (Eds.). (2008). *Chemie im Kontext – Von der Innovation zur nachhaltigen Verbreitung eines Unterrichtskonzepts [Chemie im Kontext – From an innovation to a sustainable dissemination of a teaching and learning approach]*. Münster, New York, München, Berlin: Waxmann.

Euler, M. (2002). Lernen durch Experimentieren [Learning with experiments]. In U. Ringelband, M. Prenzel, & M. Euler (Eds.), *Lernort Labor. Initiativen zur naturwissenschaftlichen Bildung zwischen Schule, Forschung und Wirtschaft. Bericht über einen Workshop IPN* (pp. 13–42). Kiel.

Kandt, W. (2008). Offenes Experimentieren im Anfangsunterricht. Band 5. In I. Parchmann, C. Hößle, M. Komorek, & K. Wloka (Eds.), *Studien zur Kontextorientierung im naturwissenschaftlichen Unterricht*. Tönning, Lübeck und Marburg: Der Andere Verlag.

Lederman, N. G. (1992). Students' and teachers' conceptions of the nature of science: A review of the research. *Journal of Research in Science Teaching, 29*(4), 331–359.

Leerhoff, G., & Eilks, I. (2002). Schülerinnen und Schüler erarbeiten sich den Atombau - Erfahrungen mit einem Gruppenpuzzle. *Praxis Schule 5–10, 13*(5), 49–56.

Leerhoff, G., Möllering, J., & Eilks, I. (2000). Lernzirkel zur Behandlung der Stoffeigenschaften. *Der Mathematische und Naturwissenschaftliche Unterricht, 53*(4), 231–234.

Lunetta, V. N. (1998). The school science laboratory: Historical perspectives and concepts for contemporary teaching. In B. J. Fraser & K. G. Tobin (Eds.), *International Handbook of Science Education* (pp. 249–262). Dordrecht: Kluwer.

McComas, W. F. (Ed.). (2000). *The nature of science in science education: Rationales and strategies (pp. 83–126).* The Netherlands: Kluwer Academic Publishers.

Nentwig, P., Parchmann, I., Gräsel, C., Ralle, B., & Demuth, R. (2007). Chemie im Kontext – A New Approach to Teaching Chemistry; Its Principles and First Evaluation Data. *Journal of Chemical Education (JChemEd), 84*(9), 1439–1444.

Parchmann, I., Gräsel, C., Baer, A., Nentwig, P., Demuth, R., & Ralle, B. (2006). Chemie im Kontext – A symbiotic implementation of a context-based teaching and learning approach. *International Journal of Science Education (IJSE), 28*(9), 1041–1062.

Prenzel, M., Artelt, C., Baumert, J., Blum, W., Hammann, M., Klieme, E., et al. (Eds.). (2007). *PISA 2006 – Die Ergebnisse der dritten internationalen Vergleichsstudie [PISA 2006 – Results of the third international comparative study].* Münster: Waxmann.

Steffensky, M., & Parchmann, I. (2007). The project CHEMOL: Science education for children – teacher education for students! *Chemistry Education: Research and Practise (CERP), 8*(2), 120–129.

von Aufschnaiter, C., & Riemeier, T. (2005). Experimente im naturwissenschaftlichen Unterricht [Experiments in science classroom education]. *Lernchancen, 47*(8), 6–10.

Chapter 13
How Does Level of Guidance Affect Understanding When Students Use a Dynamic Simulation of Liquid–Vapor Equilibrium?

Sevil Akaygun and Loretta L. Jones

Introduction

Learning chemistry involves understanding chemistry phenomena at three levels; macroscopic (the phenomena we can see, feel, and hear), symbolic (chemical formulas and equations), submicroscopic (the individual atoms and molecules), and the connections between them (Johnstone 1993). Because molecules are not visible and the concepts can be abstract, it is difficult for novices to visualize and make connections involving the submicroscopic level. Instructors desire to provide their students with appropriate guidance in learning these abstract concepts. But how much guidance is required? Too much guidance could even inhibit learning (Spencer 1999). This chapter discusses misunderstandings students have of molecular behavior in a simple system: liquid–vapor equilibrium. Approaches to helping students understand these concepts are introduced and the role of guidance discussed. Types of guidance strategies found to be effective are then outlined, followed by a research study in which some of these strategies were used. Finally, implications for instruction are presented.

Theoretical Background

Helping Students to Understand Physical Equilibrium

Understanding physical equilibrium at the submicroscopic level has been shown to pose problems for many learners (Haidar and Abraham 1991; Kelly and Jones 2007).

S. Akaygun (✉)
Bogazici University, Istanbul, Turkey
e-mail: sevil.akaygun@boun.edu.tr

L. L. Jones
University of Northern Colorado, Greeley, CO, USA

I. Devetak and S. A. Glažar (eds.), *Learning with Understanding in the Chemistry Classroom*, DOI: 10.1007/978-94-007-4366-3_13,

Even tertiary students do not easily make connections between observable physical change and submicroscopic explanations (Lekhavat and Jones 2009). Learners also have misconceptions about physical processes. For example, many precollege students believe that water splits into hydrogen and oxygen when it evaporates (Osborne and Cosgrove 1982). Gopal et al. (2004) found that the tertiary students they studied tended not to refer to the submicroscopic level when describing the processes of evaporation and condensation. They also exhibited weaker understanding of condensation than of evaporation and tended to believe that the level of an open container of water would remain constant during evaporation. Azizoğlu et al. (2006) found that students preparing to become teachers held many misconceptions about phase equilibria, even after 6 weeks of instruction. For example, many of the students believed that the vapor pressure of a liquid depends on the volume of its container and that the freezing point is independent of pressure. Canpolat et al. (2006) found additional misconceptions in their study of students preparing to become teachers. For example, the students tended to believe that vaporization does not begin until a liquid boils and that different liquids boiling at atmospheric pressure have different vapor pressures at their boiling points.

Computer animations and simulations have been shown to help students visualize submicroscopic phenomena and thus enhance the learning of chemistry (Ardac and Akaygun 2004; Burke et al. 1998; Gil and Paiva 2006; Kelly and Jones 2007; Sanger et al. 2000; Tezcan and Yilmaz 2003; Williamson and Abraham 1995; Xie and Tinker 2006). Molecular animations and simulations may also help students better understand the submicroscopic nature of physical equilibria. However, visualizations of molecules can be difficult for novices to interpret (Jones et al. 2005). Therefore, students may need additional guidance to benefit from the visualizations. Supplementary materials such as worksheets, assignments, questions, and exercises have been recommended to enhance learning from simulations and animations (Jong and Joolingen 1998; Robinson 2000). This study aimed to investigate the effect of level of guidance provided by worksheets used by students as they interact with a simulation of liquid–vapor equilibrium.

The Role of Guidance

The use of guidance provided during instruction has been investigated over the years (Ausubel 1964; Craig 1956; Mayer 2004). Some researchers have suggested that learners benefit most when the level of guidance provided is minimal, because learners construct most of the information by themselves (Bruner 1961; Steffe and Gale 1995). On the other hand, some have argued that direct instructional guidance on the concepts and procedures should increase learning (Mayer 2004; Sweller 2003). Positive effects of direct instructional guidance on learning have been supported by some controlled experimental studies (Moreno 2004; Tuovinen and Sweller 1999).

The reduced cognitive load experienced by learners has been cited as justification for providing guidance during instruction (Hmelo-Silver et al. 2007; Kirschner et al. 2006; Van Merriënboer et al. 2003). Cognitive load has been defined as the amount of mental activity required by working memory while performing a particular task (Sweller 1988). A difficult task, or one that requires recalling and combining a variety of content information, will have a higher cognitive load than a simpler task (Paas and van Merriënboer 1994; Sweller et al. 1998). Kalyuga et al. (2003) suggest that a learner's prior knowledge determines the cognitive load the individual will experience. The cognitive load of the learner when studying a particular content area then decreases as the expertise of the learner increases. For example, novice students may solve equilibrium problems by setting up tables of data in order to determine how to set up a quadratic equation, but an expert might simply set up the quadratic equation directly.

Kirschner et al. (2006) compared constructivist, discovery, problem-based, experiential and inquiry-based teaching. In their analysis the authors argue that unguided or minimally guided instructional approaches are less effective and less efficient than instructional approaches that provide extensive guidance. They claim that guided instruction helps learners engage in cognitive activities, produces expert-like skills, and provides minimum cognitive load. They also argue that minimally guided instruction may put too high a burden on working memory (the items kept in mind when solving a problem) and the accumulation of information in long-term memory.

Schmidt et al. (2007) did not agree with the manner in which Kirschner et al. (2006) equated problem-based learning (in which groups of learners are presented with a complex problem and must work out how to solve it) with minimally guided instruction. In their commentary, Schmidt et al. (2007) argued that problem-based learning also allows flexible adaptation of guidance and is compatible with the organization of learners' cognitive structures. Hmelo-Silver et al. (2007) also disagreed with Kirschner et al. (2006) and suggested that problem-based learning and inquiry learning are not minimally guided, rather highly scaffolded; therefore, cognitive load is reduced. In scaffolded instruction extensive guidance is provided at the start, but then is gradually withdrawn as learners develop competence (Reiser 2004).

One method of providing guidance is the use of written materials such as process worksheets and worked examples (Van Merriënboer 1997; Kirschner et al. 2006). According to Kirschner et al. (2006) such worksheets provide students with an outline of the phases they go through when solving the problem and also hints that they may need to complete each phase successfully. Worksheets have been used to help chemistry students to remedy their misconceptions and to attain better conceptual understanding of fundamental concepts such as chemical equilibrium (Costu and Unal 2004), phase changes (Costu et al. 2003), and acids and bases (Ozmen and Yildirim 2005), as well as to improve science process skills (Karsli and Sahin 2009). In this study, worksheets having different levels of guidance were provided along with a computer simulation in order to investigate the amount of guidance necessary for comprehension of liquid–vapor equilibrium.

Guidance Strategies

A variety of guidance strategies that emphasize different aspects of the learning process have been identified and incorporated into learning environments, including computer-based learning environments. Jackson et al. (1994) described three guidance strategies implemented in their dynamic computer modeling environment:

(1) *Grounding in experience and prior knowledge:* They believe that the learning environment should allow learners to create models based on their prior experiences and knowledge so that the models are meaningful for them.
(2) *Bridging representations:* They argue that analogies, examples, and multiple visuals should be used as a bridge to connect new representations to learners' current understanding.
(3) *Coupling actions, effects, and understanding:* They propose that the interactive learning environment provide a coupling between the learners' actions and mental representations, because learners test their mental models while they are interacting with the simulation.

The investigators concluded that the guidance strategies they used helped students run and revise the model artifacts in the simulation and their own mental models.

Another type of guidance system, "Knowledge Integration Environment (KIE)," is a framework used with an online platform of resources and software that is used to help students improve their understanding of science (Bell et al. 1995; Linn 1996). KIE Activities include guidance to support students as they integrate their ideas (Bell and Davis 2000). The guidance provided in the KIE learning environments includes four main principles or strategies:

(1) *Making science accessible:* Encouraging students to build on their scientific ideas as they develop more powerful scientific principles; and to revisit their scientific ideas regularly.
(2) *Making thinking visible:* Modeling students by illustrating how links and connections are made, scaffolding them to explain their ideas, and providing multiple visual representations from media.
(3) *Helping students learn from each other:* Encouraging students to listen and learn from their peers; and designing social activities to promote productive social interactions.
(4) *Promoting lifelong science learning:* Encouraging students to reflect on their scientific information and to continue to engage in knowledge integration. (Bell and Davis 2000; Linn 2000).

Bell and Davis (1996), Hannafin (1999), and Cagiltay (2006) identified guidance strategies for electronic learning environments and suggested the following four main types of strategies:

(1) *Conceptual Guidance:* Guiding the learners in what to consider by identifying key conceptual knowledge related to a problem or revealing conceptual organization.
(2) *Procedural Guidance:* Guiding students in what to do by emphasizing how to utilize available resources and tools.
(3) *Strategic Guidance:* Providing logistical support to accomplish the activity by helping students to identify and select needed information, evaluate available resources, and relate new to existing knowledge and experience.
(4) *Metacognitive (reflective) guidance:* Providing guidance in how to think during learning and reflect on the goal(s). Metacognitive guidance may emphasize specific ways to think about a task.

In another study of guidance provision in a computer-mediated learning environment, Ping and Swe (2004) described the guidance strategies used by teachers to engage students in computer-mediated lessons. In their study, Ping and Swe (2004) identified four categories of guidance:

(1) *Orienting activities* to direct student attention to key variables, concepts, and visual cues.
(2) *Peer interactions* to facilitate cognitive thinking and metacognitive skills.
(3) *Prompts* to promote knowledge integration.
(4) *Modeling* to guide students to generate questions and elaborate thinking.

The authors included question prompts as a guidance strategy, since these prompts were designed to promote connections between the new ideas and prior knowledge and experiences.

In this study the level of guidance in the worksheets that learners completed as they worked through a computer simulation was manipulated using the four types of strategies (*conceptual, procedural, strategic, and metacognitive* guidance) recommended by Bell and Davis (1996), Cagiltay (2006), and Hannafin (1999). Specifically, in one type of worksheet (A), extensive conceptual guidance was introduced by directing the students' attention to key concepts and variables; procedural guidance was provided by asking questions in a stepwise manner; metacognitive (reflective) guidance was provided by adopting the strategy of predict-observe-explain; and prompting questions were also included to promote knowledge interaction and reflection. In the second type of worksheet (B), none of the guidance strategies were used; instead, only an open-ended (unguided) three-part problem was provided. Worksheet B could be described as providing a problem-based learning environment.

These strategies were chosen for this study because each of these strategies focuses on a particular understanding the students may lack. In addition, even though these strategies were designed for online environments and response systems, they were easy to adopt and apply to the worksheets accompanying online instruction.

Purpose of the study

This study was part of a larger investigation of student mental models of physical equilibrium (Akaygun 2009). The research question examined was, "How does the level of guidance provided in worksheets that accompany a simulation of liquid–vapor equilibrium affect understanding of the dynamic nature of equilibrium?"

Method

Participants

Study participants were 191 first-semester general chemistry students at a medium-sized public research university in the western United States. Students were in 11 different laboratory sections taught by six different teaching assistants. Participants were randomly assigned to work with either a guided or open-ended worksheet while working on a computer simulation. At the end of the computer lesson, the novices completed the equilibrium post-test and a personal evaluation questionnaire (PEQ). After the implementation of the study, selected volunteers were interviewed as they worked through the simulation. Approval for the study was obtained from the university's institutional review board.

Instruments and Materials

Liquid–Vapor Equilibrium Simulation

A simulation of liquid–vapor equilibrium based on research data from the literature and from observations of student work was developed by the authors of this chapter (Akaygun and Jones 2007). The simulation was programmed in Adobe Flash by the CADRE design group in Sydney, Australia. A motion algorithm that simulates the Brownian motion of polar particles was used to calculate the separations, orientations, and interactions of water molecules in the liquid phase. In the gas phase the relative rates at which the molecules evaporate at two different temperatures were calculated and used to create a realistic simulation. One of the screens from the simulation is shown in Fig. 13.1.

As seen in Fig. 13.1, the simulation shows simultaneous macroscopic and submicroscopic views of water in an open and a closed flask that are placed side by side. The simulation allows students to observe the processes occurring in the liquid, at the surface, and in the vapor by clicking the corresponding regions in the flasks. The molecular view for the surface includes a counter displaying the number of evaporating and condensing molecules that was designed to help

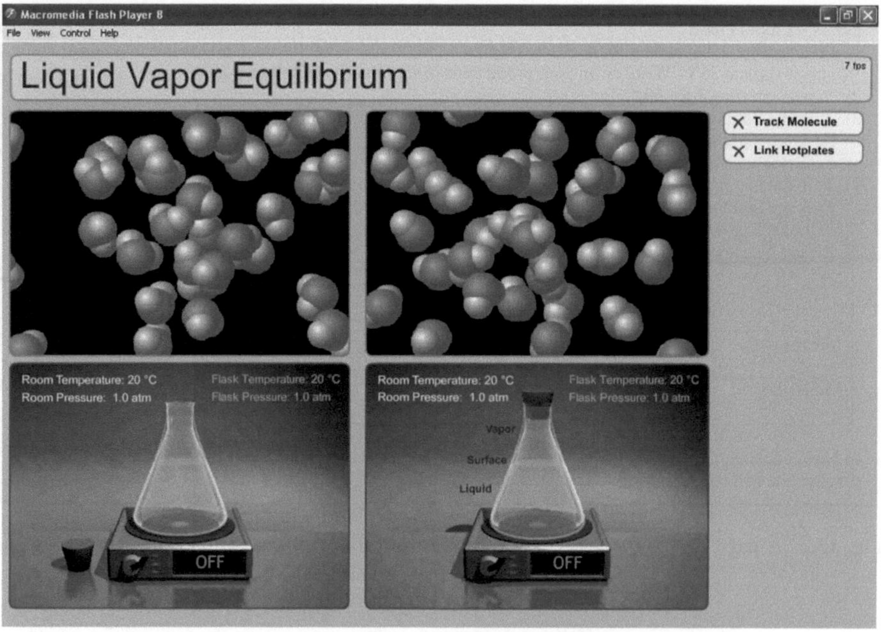

Fig. 13.1 A screen shot from the liquid–vapor equilibrium simulation, which shows simultaneous processes in the open flask (*left*) and the closed flask (*right*)

the students to visualize the dynamic nature of the equilibrium condition. The molecular view for the vapor displays another counter showing the number of molecules condensing on the wall of the flask, so that students may compare condensation rates in the open and closed flasks. In addition, the simulation was designed to show the processes at two different temperatures: 25 and 60 °C. The simulation is available online as the second item listed at http://artsci.drake.edu/honts/molviz/page2/page2.html.

Worksheets

Handouts containing instructions on navigating the simulation and a set of detailed questions to be answered as the students worked with the simulation. Worksheet A was designed to be more structured and to provide more guidance by using the strategies described in the previous section of this chapter (Fig. 13.2). Worksheet B was designed to provide less guidance. It contained the same set of instructions as Worksheet A but, instead of questions, had only a three-part open-ended problem to solve using the liquid–vapor equilibrium simulation (Fig. 13.3). Follow-up questions to be answered when finished with the simulation were the same for both worksheets (Fig. 13.4).

Part A – Liquid Phase

1) Predict what is happening at the macroscopic and molecular level for the liquid water in an *open* and a *closed* flask at 25°C. Write down your predictions in the table given below
2) Repeat question 1 for 60°C.
3) Turn the hot plate to 25°C for the *open flask*, click on the lower part of the flask so that the edges of the flask are highlighted with a yellow line. Describe what you observe at macroscopic and molecular levels in the table given below.
4) Turn the hot plate to 25°C for the *closed flask*, click on the lower part of the flask so that .the edges of the flask are highlighted with a yellow line. Describe what you observe at macroscopic and molecular levels in the table given below.
5) Repeat questions 3 &4 for 60°C

		25°C		60°C	
	Liquid Phase	Open Flask	Closed Flask	Open Flask	Closed Flask
Macroscopic	Prediction				
	Observation				
Molecular	Prediction				
	Observation				

6) Now, click on the "link hot plates" button and compare the flasks. Was there a difference in the following between the two temperatures; 25°C and 60°C?

Fig. 13.2 A page from Worksheet A shows the extensive guidance provided to students using this worksheet

Mission: You are given open and closed flasks at two different temperatures, 25°C and 60°C. Investigate these flasks at liquid, surface and vapor phase by considering the characteristics of the systems and

a) propose *one main difference* between the open flask system and closed flask system at the molecular level; i. e. behavior of the molecules in the two types of flask systems
b) label one (or both) of them according to this difference.
c) justify your reasoning for this main difference.

Fig. 13.3 The problem presented in Worksheet B shows the minimal guidance provided to students using this worksheet

3) Does the molecular structure of the water molecules change when the molecules move from the liquid to gas phase?
   a) No, they don't change.
   b) Yes, the molecules decompose into individual atoms
   c) Yes, the molecules combine to form new molecule

4) Do molecules expand as they move from liquid to gas phase?
   a) Yes
   b) No

5) How does the rate of evaporation compare to the rate of condensation in the *open* flask?
   a) The rate of evaporation is equal to the rate of condensation.
   b) The rate of evaporation is greater than the rate of condensation.
   c) The rate of evaporation is smaller than the rate of condensation.

Fig. 13.4 Sample follow-up questions used on both worksheets

Conceptual Pre-and Post-test on Liquid–Vapor Equilibrium (Pre-test)

True/false and multiple choice questions on liquid–vapor equilibrium (Fig. 13.5). The questions were designed to assess misconceptions identified in the literature and discovered in previous research (Akaygun and Jones 2007). The same questions were used on both tests; only the order of the items was changed.

Personal Evaluation Questionnaire (PEQ)

A questionnaire containing open-ended questions to evaluate the effectiveness of the study through personal comments.

Procedure

During the 10th week of the semester, at the beginning of their laboratory period, participant volunteers completed a demographic form and the Pre-test on Liquid–Vapor Equilibrium, which took approximately 10 min. Next, the participants were randomly assigned to work with either Worksheet A (guided) or Worksheet B (open-ended) as they completed the liquid–vapor computer simulation (35–45 min).

The study took place in a computer lab where students worked individually on desktop computers. Each participant was assigned a code, which was used throughout the study. No introductory material or lecture was provided; students had only their previous understandings on which to rely. Depending on the type of worksheet, students answered either guided (type A) or open-ended (type B) questions while they were working and answered follow-up questions at the end.

1) Circle True or False for each of the following explanations of what happens when liquid water evaporates to form a gas.

True / False: Water molecules expand in size.
True / False: Water molecules separate into H and O atoms.
True / False: Attractions between individual water molecules are broken.
True / False: Molecules move further apart.

2) Assume that water is being heated from 25°C to 90°C in a *closed* flask. Circle True or False for each of the following statements about this process.

True / False: Steam molecules get smaller in size since they get trapped
True / False: The high pressure in the closed flask keeps the water molecules from moving much.
True / False: When the temperature stabilizes at 90°C, the rate of evaporation equals the rate of condensation.
True / False: More boiling occurs in a closed flask than an open flask because the closed flask has more heat content.

Fig. 13.5 Sample items from the post-test on liquid-vapor equilibrium. Each question was designed to assess misconceptions that had been identified in students

The students were observed while they worked on the simulation, but the instructors did not interact with the students. When students completed the simulation, they were given the Post-test on Liquid–Vapor Equilibrium (about 10 min) and the Personal Evaluation Questionnaire (about 5 min).

Four interviews were held approximately one week following the implementation of the study. Two participants were randomly selected from the participants who worked with worksheet type A and two were selected from those who worked with worksheet type B. The 20–25 min interviews were designed in a think-aloud format, in which the students were asked to explain what they thought while working with the simulation (Bowen 1994). The interviews were audio and video-recorded.

Results and Discussion

Demographics

Of the participants, 39 % were male and 63 % were female. The ethnicity of the participants was as follows: 83 % white, 7 % Hispanic, 4 % black, 3 % Asian, and 3 % others. The participants were found to be in various stages of their studies: 58 % freshmen, 24 % sophomore, 17 % junior, and 2 % senior. The majority of the students stated that they were pursuing a medical career or planned to enter a natural science field such as biology, physics, or chemistry.

Conceptual Pre- and Post-test on Liquid–Vapor Equilibrium

The average scores on the conceptual pre- and post-test on liquid–vapor equilibrium were compared by a paired-sample *t* test. The average scores of both groups improved significantly ($p < 0.05$), as shown in the third and fourth entries in Table 13.1.

As can be seen in Table 13.1, no significant difference ($p > 0.05$) between the Pre-test scores of the groups who worked with worksheet type A or B was found, indicating that the two groups of students held equivalent levels of prior knowledge. In addition, no significant difference ($p > 0.05$) was found between the Post-test scores of the same two groups. On the other hand, a significant improvement between the Pre-test and the Post-test ($p = 0.000$) was found in the scores of the students who worked with either type of worksheet. This result implies that the use of the simulation had helped the two groups to reach the same level of conceptual understanding, regardless of whether the more or less guided worksheet had been used. Despite the fact that both groups showed a significant improvement in

Table 13.1 Average scores on the pre- and post-tests (Max. Score = 26)

	Type of worksheet	N	Mean	*T*	*df*	Sig. (two-tailed)
Pre-test	A	99	14.15	−1.025	189	0.307
	B	92	14.65			
Post-test	A	92	16.05	−1.624	189	0.106
	B	99	16.97			
Pre versus post-test	A	92	14.15 16.05	−5.101	91	0.000
Pre versus post-test	B	99	14.65 16.97	−6.387	98	0.000

understanding, the scores were still low (61.7 % for Group A and 65.3 % for Group B). Students may not have yet entirely mastered the concepts.

The responses to specific items in the Pre-test and Post-test of students who worked with each kind of worksheet were compared by a related-samples non-parametric sign test. The analysis showed that the responses of students to 13 of the 26 items improved significantly from the Pre-test to the Post-test ($p < 0.05$).

After completing the simulation students in both groups showed a better understanding of evaporation and were able to correct misconceptions such as, "Water molecules separate into H and O atoms during evaporation." and "Steam molecules get smaller in size." regardless of the type of worksheet being used.

A significant difference between the students who worked with worksheet type A or B was seen on only one item. Significantly more students who worked with worksheet type B (less guided) selected the correct answer on Item 2 in the Post-test, as shown in Table 13.2.

Item 2: Circle True or False for each of the following explanations of what happens when liquid water evaporates to form a gas: Water molecules expand in size. (Answer: False)

This difference might be due to the fact that students who worked with the less guided worksheet spent more time working with the simulation than students who

Table 13.2 Average scores on question 1, item 2 of the conceptual liquid-vapor equilibrium pre- and post-tests

Item 2	Type of worksheet	N	Correct answers in pre-test (%)	Correct answers in post-test (%)	*df*	Sig. (two-tailed)
Pre and Post-test	A	92	72	77	91	0.424
Pre and Post-test	B	99	71	88	98	0.004

Table 13.3 Comparison of scores of students who worked on different types of worksheets

	Type of worksheet	N	Mean	F ($df = 189$)	Sig. (two-tailed)
Worksheet score	A	92	4.06	189	0.000
	B	99	2.43		
Follow-up questions	A	92	5.50	189	0.496
	B	99	5.66		

used the more guided worksheet, who divided their time between viewing the simulation and answering the questions in the worksheet. Students who had more time to focus on the simulation might have noticed features in the simulation that were not mentioned in the worksheets.

Student responses to the questions on the worksheets were evaluated and scored out of a total of 5 points in each case. Next, the students in each group were compared by independent sample t-test with respect to worksheet score and score on the follow-up questions, which were the same for both types of worksheets. The results of the t-test analysis are shown in Table 13.3.

The only significant difference found between the scores of students who used different types of worksheets was found to be in the worksheet score itself ($p = 0.000$). This difference merely indicates that the questions on the more extensively guided worksheet were easier to answer than the more open-ended questions. The scores of the two groups on the follow-up questions, which were the same for each worksheet, were not significantly different.

The students were asked to rate the difficulty of the worksheet and the simulation based on their performance, the mental effort they spent, and the frustration they experienced. Students in both groups rated the difficulty of their worksheets as "average." Students who had worked with the more guided worksheet (type A) also rated the difficulty of the simulation as "average." On the other hand, students who had worked with the less guided worksheet (type B) rated the difficulty of the simulation as "less than average." This response is the reverse of what would have been expected on cognitive load considerations alone, because the lower guidance of the open-ended worksheet should have resulted in a higher cognitive load.

Personal Evaluation Questionnaire

The Personal Evaluation Questionnaire consisted of open-ended questions designed to assess student opinions about the helpfulness of the simulation, aspects of the simulation and worksheet they liked or disliked, their suggestions for the improvement of the study, and what part of the simulation they found to be the most challenging. The responses of the students were coded and a frequency analysis was performed.

The responses of students who worked with the two types of worksheets were compared by Chi square analysis. The results of the analysis are shown in

Table 13.4 Attitudes of students toward the helpfulness of the computer lesson

Group	N	Helpfulness	Number	*df*	Pearson Chi square	Sig (two-tailed)
Worksheet A (more guided)	92	Not helpful	21			
		Partially helpful	12			
		Helpful	59			
				2	11.888	0.003
Worksheet B (less guided)	99	Not helpful	6			
		Partially helpful	11			
		Helpful	82			

Table 13.5 Comparison of features of the computer lesson mentioned by students using different types of worksheets

Aspect	*df*	Pearson Chi square	Sig (two-tailed)
Helpfulness	2	11.888	0.003
Reasons for helpfulness	9	17.376	0.043
Aspects liked	11	25.035	0.009
Aspects disliked	15	27.276	0.027
Suggestions	4	17.077	0.002
Most challenging part	7	8.871	0.262

Table 13.4. As seen in Table 13.4, significantly more students who had worked with Worksheet B (less guided) said that the computer lesson was helpful than students who had worked with Worksheet A (more guided).

Significant differences in the attitudes toward the computer lesson were found between the groups of students when compared by Chi square analysis, as summarized in Table 13.5.

Reasons students gave for the helpfulness of the lessons, the aspects they liked or disliked, and the suggestions they made for improving the lesson differed significantly between the two groups. Table 13.6 lists the major categories of comments in which differences were seen between students in the two groups.

When suggesting how the simulation was helpful students in Group B, who worked with a less guided worksheet, were more likely to refer to conceptual understanding in their comments; they were more likely to focus on the visualization rather than on the questions and procedures. This finding implies that students who work with open-ended worksheets may focus more on the conceptual aspect of simulations, whereas students who work with more guided worksheets may focus more on the interactivity of simulations and on the worksheet questions.

Comments on the aspects liked suggest that students who worked with a more guided worksheet liked the visual, graphical, and the design aspect of the simulation more than students who worked with a less guided worksheet. On the other hand, students who worked with a less guided worksheet liked the conceptual

Table 13.6 Differences in comments made by students in the two groups

	Group A (92)	Group B (99)
How the simulation was helpful		
It made the molecular processes visible	15	29
It was hands-on or interactive	31	13
It helped in understanding the concepts	36	47
Aspects of the simulation liked		
Easy to understand	0	8
Helped in understanding the concepts	2	4
A specific feature, such as the molecule counter	3	9
Ability to compare the different phases	4	13
Its interactivity	36	30
The ability to visualize the concepts	38	30
Aspects of the simulation disliked		
A specific feature, such as not being able to zoom out	2	8
The graphics	3	6
The time length of the lesson	20	8
Suggestions were made for improvement of		
The simulation	10	35
The worksheet	22	13
The implementation	27	20

aspect and the specific features that emphasized the conceptual aspects more than students who worked with a more guided worksheet. It may be that as the amount of guidance provided by the worksheets decreased, students spent less time answering specific questions and more time exploring the simulation; they may have focused more on the conceptual aspects of the simulation and have valued them more than the other group.

Significantly different aspects were disliked by the two groups of students. The finding that more students in Group A disliked the amount of time required for the lesson might be related to the observation that students in Group A spent more time answering their questions and less time exploring their own interests than the students in Group B. Students in Group B were more likely to indicate disliking a specific feature of the simulation or the graphics of the simulation. This finding may be related to the observation that because students in Group B spent less time answering worksheet questions than students in Group A, therefore, they may have paid more attention to the features of the simulation.

The students in the two groups also made significantly different suggestions for improving the lesson. Once again the students in Group B appeared to be more focused on the simulation itself, while students in Group A were more focused on the worksheet questions and on the implementation of the study. Students who used less guided worksheets may have spent more time with the simulation and may have paid more attention to the specific features of the simulation, focusing more on conceptual understanding than students who had worked with the more guided worksheet. On the other hand, students in Group A may have spent more

Table 13.7 Some student quotes showing attitudes toward the computer lesson

Aspect	Student quotes
Helpfulness	Group A: "No, not enough time"
	"Yes, it gives you a visual representation of what is going on a molecular level"
	Group B: "Yes, because it was nice to visualize the information"
	"Yes, it made the topic easier to understand"
Aspects liked	Group A: "Animation was cool"
	"Visual"
	Group B: "The pictures of molecules helps get involved"
	"Being able to control different aspects"
Aspects disliked	Group A: "Too many questions and repetitive charts"
	"Time consuming"
	Group B: "The numbers of the temperatures didn't match"
	"Top three buttons were rather slow/unresponsive"
Suggestions	Group A: "More things to click"
	"Making less questions to answer"
	Group B: "The molecules should have been smaller so a larger area could be seen"
	"I thought overall the lab was set up very well providing all the necessary info. One suggestion is maybe providing molecular speeds of the molecules"

time with their worksheets and thus were more focused on the guiding questions than on the simulation itself. Some student quotes, which show their attitudes toward the computer lesson, are given in Table 13.7.

Classroom Observations and Interviews

Students in each section were observed as they worked with the simulation. Because students were randomly assigned to the two groups, it was difficult to measure exactly how much time students in the two groups spent on the lesson. However, observers noted that the students who took the longest time tended to be in Group A.

Findings from the interviews conducted support the findings from the classroom observations and the Personal Evaluation Questionnaire. The two students who had used the open-ended Worksheet B both indicated that they had liked the worksheet. One mentioned that it helped her make her own decisions, but still provided the basic guidance needed; the second student mentioned that she found it easy to use. Both of these students found the simulation easy to use, but the first student added that she needed to think and figure out why things were happening as she used the simulation. The two students who had used the more guided Worksheet A also indicated that they had found the worksheet helpful. However, one mentioned having difficulty making the connection between the questions and the molecular motion in the simulation. The other student indicated that, although the worksheet questions were easy to answer, she found completing them to be a long and frustrating process.

Conclusions

Students in both groups showed significant learning gains after working with the simulation, as seen in Table 13.1. When performance on specific items on the Pre- and Post-test was examined, in 13 of the 26 items students exhibited significantly better understanding of evaporation and condensation. For instance the majority of students in both groups overcame the misconception of "separation of water molecules into H and O atoms" when evaporating.

The only difference in achievement between the students who worked with the two worksheets was seen for the misconception that water molecules expand in size during evaporation, a misconception addressed only in a follow-up question on both worksheets. Only students who worked with the less guided worksheet (Group B) did significantly better on this item on the Post-test (Table 13.2), suggesting that students using the open-ended worksheet may have paid more attention to aspects of the simulation not mentioned on the worksheet, while students in Group A may have been more focused on the worksheet questions.

Overall, as described in Tables 13.1 and 13.3, no significant difference in average scores on the Post-test or follow-up questions were found for students who had used the two types of worksheets. This finding might indicate that even the minimal guidance of open-ended Worksheet B was sufficient to help students learn the concepts needed to answer the questions on the Post-test and the follow-up questions at the end of each worksheet (Costu and Unal 2004). On the other hand, the effects of the different levels of guidance provided by the worksheets might not have been revealed by the assessments used in the study. The different types of worksheets might have had an effect on other aspects of learning and it would be worthwhile to investigate other possible effects of varying the amount of guidance provided to students.

In this study students were able to improve their understanding of liquid–vapor equilibrium after viewing a simulation accompanied by worksheets having two different levels of guidance. These findings suggest that when students learn other chemistry concepts with simulations and animations the accompanying worksheets can be either highly guided or open-ended. Students in this study had positive attitudes toward the computer lesson, regardless of the level of guidance. However, students using the less guided worksheets had more positive attitudes toward their worksheets than did students using the more guided worksheets. In addition, the responses to the Personal Evaluation Questionnaire suggest that the students using the open-ended worksheet were more focused on the concepts that they were learning. The fact that students who used the open-ended worksheet found the computer lesson to be more helpful than students who used the highly guided worksheet suggests that worksheets used with computer lessons should have a minimum of guidance. Students might have enjoyed discovering the simulation through their interaction with the computer instead of being directed (and perhaps distracted) by the questions on the worksheet. In addition, most of the students who worked with the highly guided worksheet stated that the worksheet was time-consuming or lengthy; hence, they made suggestions for the modification of the worksheet or the implementation rather than the simulation itself.

Student attitudes toward the computer lesson varied significantly for students using worksheets with different levels of guidance. Significantly more students who worked with the less guided worksheet thought the simulation was helpful (Table 13.4). In addition, their reasons for finding the simulation helpful were also significantly different. Students who worked with the less guided worksheet were more likely to report that the simulation helped them conceptually understand the processes, whereas students who worked with a more guided worksheet were more likely to report that the simulation was helpful due to being hands-on (Table 13.6). Similarly, the aspects liked, disliked, and suggestions for improvement were significantly different between the groups in that students who had worked with the less guided worksheet wrote comments that focused more on the chemistry concepts, whereas students who worked with the more guided worksheet focused more on the graphical-visual aspects of the simulation. It may be that as the students spent more time and effort exploring the simulation in an open-ended fashion, they were paying more attention to the chemistry concepts than students who mostly focused on answering the larger number of questions in the more guided worksheet.

Because no significant difference was found between the level of guidance and the Post-test scores, minimal guidance in an open-ended format may be sufficient guidance for students using computer simulations of molecular behavior. No evidence was found that strategies recommended for reducing the cognitive load of instruction (Bell and Davis 1996; Cagiltay 2006; Hannafin 1999; Kirschner et al. 2006) were helpful in this case. In fact, differences in attitude between students using more guided and less guided worksheets suggest that students using the less guided worksheet focused more on the conceptual basis of the computer lesson. Further investigation of the type of questions and answers might reveal whether any other variables might have been affected by the difference in guidance.

Implications for Teaching

The findings of this study suggest that it may be preferable to use either minimal guidance in simulation worksheets or to provide scaffolding in which students move from more guided to less guided questions in the same lesson. When the level of guidance is high learning can become a tedious experience and student attention can be distracted from conceptual understanding as they struggle with answering a large number of questions.

1. Run the simulation by connecting the hot plates so that both open and closed flasks are simultaneously at the same temperature.
2. Observe and compare the following:
a) the motion of the molecules in the *liquid* phase in *open* and *closed* flasks.

b) the motion of the molecules at the *surface,* in *open* and *closed* flasks.

c) the motion of the molecules in the *gas* phase, in *open* and *closed* flasks.

d) the number of *evaporating* & *condensing* the molecules from the surface, in *open* and *closed* flasks.

e) the *rate of evaporation* & *condensation* in *open* and *closed* flasks.

Fig. 13.6 A page from a sample worksheet with a transitional level of guidance that could be provided to students using the liquid–vapor simulation

In this study students may have been able to learn the content primarily from the simulation, with a need only for minimal guidance. For more difficult topics and when simulations are not available, an ideal situation might be a true scaffolded approach, with the first session closely guided, the second transitional, and the third open-ended. In other words, in such an approach a single worksheet may contain three sections: Highly guided, transitional, and lightly guided. The highly guided questions in the beginning of the worksheet may be designed by applying conceptual and procedural guidance (Bell and Davis 1996; Cagiltay 2006), in which detailed directions, tables/charts, and concrete cases are given. In the transitional section the level of guidance/scaffolding could be gradually decreased by providing supports such as cues, hints, and coaching comments. Finally, minimally guided questions may be presented so that students can organize their cognitive structures by using understandings gained from the guidance provided earlier (Schmidt et al. 2007). An example of worksheet questions for the application in this chapter that uses an intermediate level of guidance is shown in Fig. 13.6.

Acknowledgments The research study described in this chapter was funded in part by the National Science Foundation under Grant No. REC 0440103. The authors of this chapter would like to acknowledge fruitful consultations with Barbara Tversky, Columbia University, and the helpful feedback and support provided by faculty and students of the Chemical Education Research Group at the University of Northern Colorado.

References

Akaygun, S. (2009). The effect of computer visualizations on students' mental models of dynamic nature of physical equilibrium (Doctoral dissertation, 2009). University of Northern Colorado.

Akaygun, S., & Jones, L. L. (2013). Research—based design and development of a simulation of liquid–vapor equilibrium. *Chemistry Education Research and Practice*, *14*(3), 324–344.

Ardac, D., & Akaygun, S. (2004). Effectiveness of multimedia-based instruction that emphasizes molecular representations on students' understanding of chemical change. *Journal of Research in Science Teaching, 40*(4), 317–337.

Ausubel, D. P. (1964). Some psychological and educational limitations of learning by discovery. *The Arithmetic Teacher, 11*(5), 290–302.

Azizoğlu, N., Alkan, M., & Geban, Ö. (2006). Undergraduate pre-service teachers' understandings and misconceptions of phase equilibrium. *Journal of Chemical Education, 83*(6), 947–953.

Bell, P., Davis, E. A., & Linn, M. C. (1995). The knowledge integration environment: Theory and design. In *Proceedings of the Computer Supported Collaborative Learning Conference* (CSCL 1995: Bloomington, IN) (pp. 14–21). Mahwah, NJ: Lawrence Erlbaum Associates.

Bell, P., & Davis, E. A. (1996, April). *Designing an activity in the knowledge integration environment.* Paper presented at the annual meeting of the American Educational Research Association (AERA), New York, NY.

Bell, P., & Davis, E. A. (2000). Designing mildred: Scaffolding students' reflection and argumentation using a cognitive software guide. In B. Fishman & S. O'Connor-Divelbiss (Eds.), *Fourth international conference of the learning sciences* (pp. 142–149). Mahwah, NJ: Erlbaum.

Bowen, C. W. (1994). Think-aloud methods in chemistry education. *Journal of Chemical Education, 71*(3), 184–190.

Bruner, J. S. (1961). The art of discovery. *Harvard Educational Review, 31*(1), 21–32.

Burke, K. A., Greenbowe, T. J., & Windschitl, M. A. (1998). Developing and using conceptual computer animations for chemistry instruction. *Journal of Chemical Education, 75*(12), 1658–1661.

Cagiltay, K. (2006). Scaffolding strategies in electronic performance support systems: Types and challenges. *Innovations in Education and Teaching International, 43*(1), 93–103.

Canpolat, N., Pinarbasi, T., & Sözbilir, M. (2006). Prospective teachers' misconceptions of vaporization and vapor pressure. *Journal of Chemical Education, 83*(8), 1237–1242.

Costu, B., Karatas, F. O., & Ayas, A. (2003). Kavram ogretiminde calisma yapraklarinin kullanilmasi (Using worksheets in teaching concepts). *Pamukkale Universitesi Egitim Fakultesi Dergisi, 2*(14), 33–48.

Costu, B., & Unal, S. (2004). The use of worksheets in teaching Le-Chatelier's Principle. *Yüzüncü Yıl Üniveristesi, Online Eğitim Fakültesi Dergisi, 1*(1), 1–22. Retrieved November 05, 2009 from http://efdergi.yyu.edu.tr/makaleler/cilt_I/bayram_suat.doc.

Craig, R. (1956). Directed versus independent discovery of established relations. *Journal of Educational Psychology, 47*(4), 223–235.

Gil, V. M. S., & Paiva, J. C. M. (2006). Using computer simulations to teach salt solubility. The role of entropy in solubility equilibrium. *Journal of Chemical Education, 83*(1), 170.

Gopal, H., Kleinsmidt, K., Case, J., & Musonge, P. (2004). An investigation of tertiary students' understanding of evaporation, condensation and vapour pressure. *International Journal of Science Education, 26*(13), 1597–1620.

Haidar, A. H., & Abraham, M. R. (1991). A comparison of applied and theoretical knowledge of concepts based on the particulate nature of matter. *Journal of Research in Science Teaching, 28*(10), 919–938.

Hannafin, M. J. (1999). Learning in open-ended environments: Tools and technologies for the next millennium. Retrieved September 15, 2010, from http://it.coe.uga.edu/itforum/paper34/paper34.html.

Hmelo-Silver, C. E., Duncan, R. G., & Chinn, C. A. (2007). Scaffolding and achievement in problem-based and inquiry learning: A response to Kirschner, Sweller, and Clark (2006). *Educational Psychologist, 42*(2), 99–107.

Jackson, S. L., Stratford, S. J., Krajcik, J., & Soloway, E. (1994). Making dynamic modeling accessible to precollege science students. *Interactive learning environments, 4*(3), 233–257.

Johnstone, A. H. (1993). The development of chemistry teaching: A changing response to changing demand. *Journal of Chemical Education, 70*(9), 701–704.

Jones, L. L., Jordan, K. D., & Stillings, N. A. (2005). Molecular visualization in chemistry education: The role of multidisciplinary collaboration. *Chemistry Education Research and Practice, 6*(3), 136–149.

Jong, D. T., & Joolingen, V. (1998). Scientific discovery learning with computer simulations of conceptual domains. *Review of Educational Research, 68*(2), 179–201.

Karsli, F., & Sahin, C. (2009). Developing worksheet based on science process skills: Factors affecting solubility. *Asia-Pacific Forum on Science Learning and Teaching, 10*(1), 4–16.

Kalyuga, S., Ayres, P., Chandler, P., & Sweller, J. (2003). The expertise reversal effect. *Educational Psychologist, 38*(1), 23–31.

Kelly, R. M., & Jones, L. L. (2007). Exploring how different features of animations of sodium chloride dissolving affect students' explanations. *Journal of Science Education and Technology, 16*(5), 413–429.

Kirschner, P. A., Sweller, J., & Clark, R. E. (2006). Why minimal guidance during instruction does not work: An analysis of the failure of constructivist, discovery, problem-based, experiential, and inquiry-based teaching. *Educational Psychologist, 41*(2), 75–86.

Lekhavat, P., & Jones, L. L. (2009). The effect of adjunct questions emphasizing the particulate nature of matter on students' understanding of chemical concepts in multimedia lessons. *Educacion Quimica, 20*(3), 351–359.

Linn, M. C. (1996). Key to the information highway. *Communications of the ACM, 39*(4), 34–35.

Linn, M. C. (2000). Designing the knowledge integration environment. *International Journal of Science Education, 22*(8), 781–796.

Mayer, R. E. (2004). Should there be a three-strikes rule against pure discovery learning? The case for guided methods of instruction. *American Psychologist, 59*(1), 14–19.

Moreno, R. (2004). Decreasing cognitive load in novice students: Effects of explanatory versus corrective feedback in discovery-based multimedia. *Instructional Science, 32*(1–2), 99–113.

Osborne, R. J., & Cosgrove, M. M. (1982). Children's conceptions of the changes of state of water. *Journal of Research in Science Teaching, 20*(9), 825–838.

Ozmen, H., & Yildirim, N. (2005). Effect of work sheets on student's success: Acids and bases sample. *Journal of Turkish Science Education, 2*(2), 64–67.

Paas, F., & van Merriënboer, J. J. G. (1994). Instructional control of cognitive load in the training of complex cognitive tasks. *Educational Psychology Review, 6*(4), 51–71.

Ping, L. C., & Swe, K. M. (2004). Engaging junior college students in computer-mediated lessons using scaffolding strategies. *Journal of Educational Media, 29*(2), 97–112.

Reiser, B. J. (2004). Scaffolding complex learning: The mechanisms of structuring and problematizing student work. *The Journal of the Learning Sciences, 13*(3), 273–304.

Robinson, W. R. (2000). A view of the science education research literature: Scientific discovery learning with computer simulations. *Journal of Chemical Education, 77*(1), 17.

Sanger, M. J., Phelps, A. J., & Fienhold, J. (2000). Using a computer animation to improve students' conceptual understanding of a can-crushing demonstration. *Journal of Chemical Education, 77*(11), 1517–1520.

Schmidt, H. G., Loyens, S. M. M., van Gog, T., & Paas, F. (2007). Problem-based learning is compatible with human cognitive architecture: Commentary on Kirschner, Sweller, and Clark (2006). *Educational Psychologist, 42*(2), 91–97.

Spencer, J. N. (1999). New directions in teaching chemistry: A philosophical and pedagogical basis. *Journal of Chemical Education, 76*(4), 566–569.

Steffe, L., & Gale, J. (Eds.). (1995). *Constructivism in education*. Hillsdale, NJ: Lawrence Erlbaum Associates Inc.

Sweller, J. (1988). Cognitive load during problem solving: Effects on learning. *Cognitive Science, 12*(1), 257–285.

Sweller, J. (2003). Evolution of human cognitive architecture. In B. Ross (Ed.), *The psychology of learning and motivation* (Vol. 43, pp. 215–266). San Diego, CA: Academic.

Sweller, J., van Merriënboer, J. J. G., & Paas, F. (1998). Cognitive architecture and instructional design. *Educational Psychology Review, 10*(3), 251–296.

Tezcan, H., & Yilmaz, U. (2003). Kimya ogretinde kavramsal bilgisayar animasyonlari ile geleneksel anlatim yontemin basariya etkileri (The effects of conceptual computer animations and traditional lecturing method used in teaching chemistry on achievement). *Pamukkale Universitesi Egitim Fakultesi Dergisi, 2*(14), 18–32.

Tuovinen, J. E., & Sweller, J. (1999). A comparison of cognitive load associated with discovery learning and worked examples. *Journal of Educational Psychology, 91*(2), 334–341.

Van Merriënboer, J. J. G. (1997). *Training complex cognitive skills.* Englewood Cliffs, NJ: Educational Technology Publications.

Van Merriënboer, J. J. G., Kirschner, P. A., & Kester, L. (2003). Taking a load off a learner's mind: Instructional design for complex learning. *Educational Psychologist, 38*(1), 5–13.

Williamson, V. M., & Abraham, M. R. (1995). The effects of computer animation on the particulate mental models of college chemistry students. *Journal of Research in Science Teaching, 32*(5), 521–534.

Xie, Q., & Tinker, R. (2006). Molecular dynamics simulations of chemical reactions for use in education. *Journal of Chemical Education, 83*(1), 77.

Chapter 14
Evaluation of the Predict-Observe-Explain Instructional Strategy to Enhance Students' Understanding of Redox Reactions

David F. Treagust, Zuzi Mthembu and A. L. Chandrasegaran

Introduction

It is generally agreed by researchers that the most important thing that learners bring to their classes are their preinstructional conceptions (Ausubel 1968; Driver and Oldham 1986). A wide range of empirical studies summarised in Duit (2009) provide ample evidence that learners hold preinstructional conceptions in many fields that differ substantially from scientifically accepted conceptions. Learners' conceptions have been described as an individual's idiosyncratic mental representations, whereas scientifically acceptable concepts are firmly defined or widely accepted (Duit and Treagust 1995). Traditional instructional strategies generally do not recognise learners' preconceptions and often fail to take into account the meaning of specific words that have different meanings when used by teachers and students (Treagust and Chittleborough 2001). When developing effective instructional strategies, the preinstructional conceptions that learners bring with them to the learning situation have to be considered. The Predict-Observe-Explain (POE) strategy is one instructional strategy that takes learners' preinstructional conceptions into account.

The POE Instructional Strategy

Originally designed as the Demonstrate-Observe-Explain (DOE) strategy by Champagne, Klopfer and Anderson (1979) to probe thinking of first-year physics students, the strategy was redesigned as the POE by Gunstone and White (1981). The POE strategy requires learners to first predict and record the outcome of an

D. F. Treagust (✉) · Z. Mthembu · A. L. Chandrasegaran
Science and Mathematics Education Centre, Curtin University,
6845 Perth Western, Australia
e-mail: D.Treagust@curtin.edu.au

I. Devetak and S. A. Glažar (eds.), *Learning with Understanding in the Chemistry Classroom*, DOI: 10.1007/978-94-007-4366-3_14,

event giving reasons to support their prediction. The event to illustrate a concept that is being introduced by the teacher is then observed by the students who record their observations. If students find that there is a contradiction between their observations and their original predictions, they must provide an explanation. The POE is used in conjunction with a teacher's demonstration or student's practical activity or with a video excerpt as a means of encouraging discussion and debate among students. The use of practical-based activities in POEs makes science learning more meaningful for students. An important aspect of the POE is that learners are required to write down their predictions with reasons, before sharing their ideas with members of the group. Any changes in their ideas can then be monitored so that mistakes are not repeated. Fensham and Kass (1988) refer to inconsistent or discrepant events when observations of a phenomenon are not consistent with students' expectations, resulting in cognitive conflict which in turn can be a useful motivation for learning. According to Ballantyne and Baine (1995), instructional strategies that are effective in enhancing understanding of concepts are those that induce a state of cognitive conflict within the learner. Such strategies assist learners to articulate and explore ideas and theories that they hold about a concept. Provoking cognitive conflict can be expected to lead to effective learning and also to the development of relevant thinking skills.

The instructional approach of the POE dispels the absorbtionist view of learning with the teacher as a transmitter of knowledge (Prawat 1989). Teachers who use this strategy are able to provide students with indirect instructional intervention as the goal is to facilitate learners to construct their own knowledge. Despite learners often experiencing difficulty with reconciling any discrepancy between their prediction and observation, with encouragement from teachers they should feel free to express their views. The students' explanations reveal their understanding of particular concepts (Gunstone and White 1981).

Several research studies that used POEs to probe understanding of science concepts among secondary school students have been documented in the research literature (Liew and Treagust 1995; Tao and Gunstone 1997; White and Gunstone 1992). Gunstone and White (1981) undertook an investigation to determine secondary students' understanding of gravity and related principles of mechanics. They deduced that despite having acquired extensive factual knowledge learners were unable to relate their knowledge to the everyday world. In their study of solubility and electricity concepts, Liew and Treagust (1995) found that POEs were effective in eliciting useful information on learners' conceptual understanding of the epistemology and ontology of the concepts that were investigated.

Students' Understanding of Redox Reactions

Chemistry is generally viewed as a difficult subject by students because the subject involves learning 'abstract and formal explanations of invisible interactions at molecular levels' (Carr 1984, p. 97). To engage in chemical reasoning, the learner may need to shift between four representational systems (the macroscopic, submicroscopic, symbolic and algebraic), and students often experience difficulty shifting between these systems (Nakhleh and Krajcik 1994; Treagust and Chandrasegaran 2009). Redox reactions, in particular, are regarded by students as one of the most difficult topics because, in addition to moving between the four systems, students have to contend with four models of redox reactions, namely the oxygen model, the hydrogen model, the electron model and the oxidation numbers model (Harrison and Treagust 1998).

Learners' difficulties in understanding redox reactions are both conceptual and procedural. The conceptual difficulties include the interdependence of the oxidation and reduction processes, the concept of relative strengths of oxidising and reducing reagents and the process of transfer of electrons resulting from students' general inability to understand the concept of oxidation numbers (De Jong, Acampo, and Verdonk 1995; Garnett and Treagust 1992). One procedural difficulty is the classification of reactions as redox reactions because students who preferred to use the criterion of electron transfer instead of change in oxidation numbers did not recognise equations without charges and electrons as redox reactions (Ringnes 1995). In other instances, students believed that they could identify redox reactions based on changes in the charges of polyatomic species in an equation (Garnett and Treagust 1992). In a study with first-year university students, Niaz (2002) observed that students developed only procedural knowledge about electrochemical reactions as redox reactions merely by memorising the formulas. Another procedural difficulty that has been identified in research studies involved identifying reactants as oxidising or reducing agents because imprecise terminology and the linguistics of some statements used by teachers resulted in confusion among students (De Jong et al. 1995). For example, in several instances, teachers referred to a substance when they meant particles and vice versa. In other instances, teachers were vague about the oxidising and reducing agents involved when they referred to, for example, copper as an oxidant without specifying whether the reference was to the copper atom or copper ion. The mixing of context-specific meanings, especially the phenomenological meaning with the particulate meaning, has been a source of difficulty for students when identifying reactants as oxidants or reductants (De Jong and Treagust 2002). While this mixing of meanings is not unusual for experts (experienced teachers) because they are able to move from one representation to another with ease, this can be confusing for novice learners.

There is concern that the four different models that are used in schools when introducing redox reactions may not be very successful in explaining the concept of redox reactions (Ringnes 1995) because of the linguistic reasoning of the definitions referred to earlier, when instead a historical development of the concepts would have been more appropriate. Furthermore, there is incompatibility of the various models when applied to the same type of reaction. For example, the hydrogen and oxygen models cannot explain why the reaction between sodium and chlorine to produce sodium chloride is a redox reaction. Also, the electron model has its limitations because electron transfer is not reflected in all redox reactions.

The unique language of chemistry that uses words from everyday speech but with different meanings frequently gives rise to learning difficulties (Treagust and Chittleborough 2001). When terms are used in a scientific context, it is assumed by teachers and textbook authors that the scientific meaning will be understood by students (Garnett, Garnett and Hackling 1995; Nakhleh and Krajcik 1994; Schmidt 1997). When the historical development of redox reactions is not used, confusion arises with the use of everyday terms. For example, 'reduce means to gain electrons, contrasting with decrease in everyday language and oxidation need not comprise reactions with oxygen' (Ringnes 1995, p. 77). In addition, the 'ox' in 'redox' has misled students into thinking that oxygen must be involved in all redox reactions (Schmidt 1997).

Rationale and Purpose of the Study

The focus of this study was to facilitate students' improved understanding of redox reactions through the professional enrichment of teachers in South African secondary schools in the Durban area by augmenting their knowledge, skills, behaviour and attitudes using inservice education programmes with follow-up support in the classroom setting. Support is necessary when teachers integrate new skills within their existing instructional repertoires and are faced with overcoming initial uncertainties created by the required change (Thijs and van den Akker 1997). Among the skills that teachers needed to acquire, one pertained to promoting learner-centred instructional strategies that engage students in academically challenging experiences that are characterised by inquiry and open-ended questions to promote deep understanding of science concepts (Anderson 1997). Effective use of the POE as a learner-centred instructional strategy provides opportunities for greater teacher–learner and learner–learner interactions that in turn promote skills like negotiating, compromising, reasoning and making decisions.

The study was designed to answer the following research questions:

(1) How proficient were students in performing the POE activities on redox reactions?
(2) What was the effect of the use of POEs involving redox reactions on students' achievement in the test on redox concepts?
(3) How did students perceive the usefulness of the POE activities in facilitating their understanding about redox reactions?
(4) What was the effect on the use of POEs involving redox reactions on students' attitudes towards learning science?

Methods and Procedures

Participants

Following a teacher inservice programme about the use of the POE instructional strategy to enhance students' understanding of redox reaction concepts, its efficacy was evaluated in an intensive case study involving 66 students from two Grade 11 classes of one of the teachers who participated in the inservice programme. The school was an urban comprehensive school in the greater Durban area of South Africa and had achieved satisfactory matriculation results ranging from 90 to 95 % in the years 1995–2000. The teacher was interested in trialling the new POE strategy because she realised the potential in achieving the aims of the POE, while also realising that using the strategy effectively would contribute to her own professional growth.

Research Design

The second author conducted the research in an urban comprehensive school in the greater Durban area of South Africa using a case study method that was implemented in two phases. During this period, the researcher collected both quantitative data using test instruments as well as qualitative data from observations of classroom lessons and interviews with students. Phase 1 of the study (involving two lessons) focused on classroom instruction to introduce the classification of chemical reactions into various reaction types with emphasis on enabling students to identify redox reactions. Redox reactions were identified using two methods: (1) electron transfer and (2) oxidation numbers. The balanced chemical equations for redox reactions were derived using half reactions. Prior to instruction, quantitative data were collected in pre-tests by administering a modified version of the Test of Science-Related Attitudes (TOSRA) questionnaire (Fraser 1981) and a test on redox reactions consisting of 25 multiple-choice items. Phase 2 of the study involved implementation of eight POE activities over a period of 8 weeks. Following this intervention programme, the TOSRA questionnaire and the redox reactions test previously used were administered to the students as post-tests.

Students' attitudes to science learning questionnaire: The modified Test of Science-Related Attitudes (TOSRA) questionnaire consisting of 46 items in 4 scales was administered before and after completion of the redox reactions topic in order to assess students' attitudes to science learning.

Redox reactions test: The test consisted of 25 multiple-choice items based on the South African Grades 11 and 12 chemistry syllabuses and was administered as a pre-test and as a post-test. Examples of several items are provided in Fig. 14.1.

POE activities: Eight practical activities involving redox reactions were designed for use in lessons conducted over an 8-week period. The complete set of

Item 3: Which of the following pairs will react with each other?

A $Mg^{2+}(aq)$ and $Cu(s)$ C $Cu^{2+}(aq)$ and $Ag(s)$

B $Ag^{+}(aq)$ and $Pb(s)$ D $Mg^{2+}(aq)$ and $Zn(s)$

Item 6: Which of the reactions below involves a transfer of electrons?

A $ZnSO_4(aq) + BaCl_2(aq) \longrightarrow ZnCl_2(aq) + BaSO_4(s)$

B $2H_2(g) + O_2(g) \longrightarrow 2H_2O(l)$

C $NaOH(aq) + HCl(l) \longrightarrow NaCl(aq) + H_2O(l)$

D $HCl(l) + H_2O(l) \longrightarrow H_3O^{+}(aq) + Cl^{-}(aq)$

Item 11: The oxidation half-reaction for the reaction

$Hg(s) + 2AgNO_3(aq) \longrightarrow Hg(NO_3)_2(aq) + 2\,Ag(s)$

is:

A $Hg(s) \longrightarrow Hg^{2+}(aq) + 2e^{-}$

B $Ag(s) \longrightarrow Ag^{+}(aq) + e^{-}$

C $Hg^{2+}(aq) + 2e^{-} \longrightarrow Hg(s)$

D $Ag^{+}(aq) + e^{-} \longrightarrow Ag(s)$

Item 17: Which item is oxidised in the following reaction?

$Zn(s) + 2H^{+}(aq) \longrightarrow H_2(g) + Zn^{2+}(aq)$

A $Zn(s)$ C $H_2(g)$

B $H^{+}(aq)$ D $Zn^{2+}(g)$

Item 23: H_2S is bubbled through an acidified solution of $FeCl_3$ and the following reaction occurs:

$2FeCl_3(aq) + H_2S(g) + HCl(l) \longrightarrow 2FeCl_2(aq) + S(s) + 3HCl(l)$

The ions oxidised and reduced are respectively…

	Ions oxidised	Ions reduced
A	$Fe^{3+}(aq)$	$S^{2-}(aq)$
B	$S^{2-}(aq)$	$Fe^{2+}(aq)$
C	$Fe^{2+}(aq)$	$S^{2-}(aq)$
D	$S^{2-}(aq)$	$Fe^{3+}(aq)$

Fig. 14.1 Examples of test items on redox reactions

POE activities is available from the first author. These activities involved: (1) adding acetic acid to steel wool, (2) dipping a zinc strip in aqueous copper (II) sulphate, (3) dipping a copper strip in aqueous zinc sulphate, (4) immersing a coil of copper wire in aqueous silver nitrate, (5) dipping strips of magnesium, zinc, iron, lead and copper in various aqueous solutions of metal ions, (6) bubbling chlorine gas into aqueous potassium iodide, (7) burning candles of varying lengths under a sealed bell jar of air and (8) inverting a test tube containing sulphur dioxide over another test tube containing hydrogen sulphide. An example of the instructions for carrying out one of these activities (Activity 2) is provided in Fig. 14.2 while the expected changes that occur are shown in Fig. 14.3.

Observations: Lessons were observed in each class for a total of 24 h with the second author as the researcher playing the role of "observer as participant" (Merriam 1988, p. 93) moving to and fro along the continuum as a participant and as an observer (Bogdan and Biklein 1992).

Interviews: A total of 36 students, based on their performance in class (two students from the top, middle and bottom third of each class), were selected to be interviewed. Responses to 15 questions were solicited in structured interviews. The purpose of the interviews was to evaluate students' opinions on the usefulness

Activity 2: immersing a zinc strip in aqueous copper(II) sulfate
Instructions to students:
1. You will investigate the redox reaction that occurs when a zinc strip is dipped in beaker containing some aqueous copper(II) sulfate.
2. Collect the materials and solution required for this activity.
3. Predict whether a chemical reaction will take place. Write a brief explanation or reason for your prediction.
4. Share your prediction with members of your group and come to an agreement of what you would expect to happen.
5. Perform the experiment. What changes can you observe? Record all changes that occur. Were your observations similar to your earlier predictions?
6. Write down your explanations for all changes that you observed in terms of the redox reaction that had occurred. Compare your observations with your prediction. Are these in agreement? If not, discuss with members of your group to reconcile any differences.

Fig. 14.2 An example of a POE activity used in the study

Activity 2

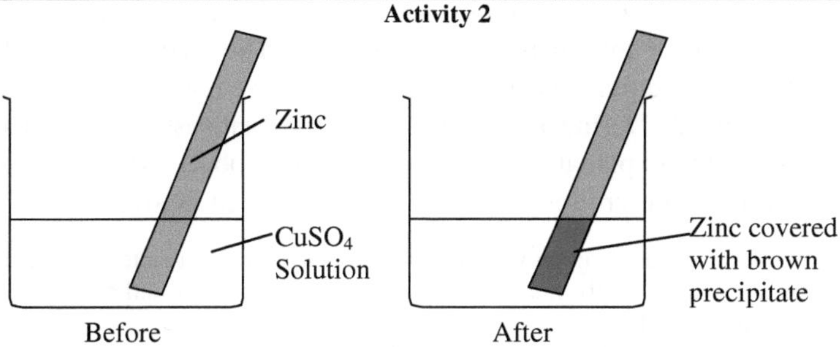

Zinc displaces copper from aqueous solution as zinc is more reactive than copper. The blue copper(II) sulfate solution fades and becomes colourless due to the formation of aqueous zinc sulfate, and a reddish-brown deposit of copper is produced. Zinc reduces Cu^{2+} ions to copper and is itself oxidised to Zn^{2+} ions.

$$Zn(s) + Cu^{2+}(aq) \rightarrow Zn^{2+}(aq) + Cu(s)$$

Fig. 14.3 Illustration of the changes in the POE *Activity 2*

of POEs in facilitating their learning (Questions 1–5 and 13–15) as well as to probe their understanding of basic redox concepts (Questions 6–12). The 15 questions that were used in the structured interviews are listed in Table 14.2. The responses that were solicited were organised along the following themes: (1) satisfaction, (2) enjoyment, (3) sharing, (4) conflict resolution, (5) acquisition of conceptual knowledge and (6) knowledge gained. The discussion on students' knowledge gained about redox reactions focused on students' responses based on (1) understanding charge and oxidation number, (2) identifying redox reactions, oxidising and reducing agents, (3) balancing redox equations and (4) why half reactions are used. The interviews were conducted on completion of the topic on redox reactions and lasted about 20 min each.

Results and Discussion

The POE Activities

In each class, students performed the POE activities in five groups. The summary of students' predictions, observations and explanations for each of the eight POE activities shown in Table 14.1, provides a response to research question 1: How proficient were students in performing the POE activities on redox reactions? A discussion of four of the Activities (1, 2, 7 and 8) provides more details of the experimental procedure and students' responses.
Activity 1—Reaction between steel wool and dilute acetic acid

The general prediction was that the balloon would become inflated and that the steel wool would rust. When the experiment was performed, students observed that the steel wool did rust but the balloon actually deflated, causing cognitive conflict. However, students were unable to provide satisfactory explanations for what was observed. No scientific explanations relating to the redox reaction that had occurred were given for the phenomenon observed. The teacher herself failed to relate the chemical reaction that had occurred to redox reactions that had previously been introduced. During the whole class discussion students expected the teacher to provide the explanations for what had been observed. The response of students from one of the groups to Activity 1 is provided below.

> *Prediction:* The size of the balloon would increase. When a metal reacted with an acid a gas would be liberated … (and) … enter … the balloon; steel wool would rust … (as it) … would react with the acid.
> *Observation:* The balloon deflated; steel wool changed to brown; could not see any reaction between steel wool and acetic acid.
> *Explanation:* Reaction was slow.

From the authors' perspective, the data suggest that these students' prior knowledge and beliefs, and hence their expectation of the outcomes, influenced their observations. Several students, who predicted that a metal oxide would be produced by the reaction between steel wool and the acid, saw an agreement between their prediction and observation, although the oxide was formed by the reaction between steel wool and oxygen in the flask in the presence of water. The belief that oxygen from the acid would react with a metal showed lack of understanding of the chemical reaction that had occurred.
Activity 2—Reaction between zinc and aqueous copper (II) sulphate

Although the five groups predicted a colour change, none predicted all the changes that were expected. Various groups were more specific about the actual colour change from blue to colourless, the formation of a 'precipitate' of copper, and a decrease in size of the zinc strip. All groups observed the formation of a 'precipitate' which they described as either copper or as a brown layer that had formed on the zinc strip. Other observations included a decrease in size of the zinc strip, with one group suggesting that zinc had been 'eaten up and became rough'.

Table 14.1 A summary of examples of students' responses to the eight POE activities

Predictions	Observations	Explanations
Activity 1: Addition of acetic acid to steel wool in a conical flask fitted with a balloon over the neck of the flask		
The balloon would be inflated with air from the flask	The balloon deflated	A metal oxide is produced when the metal reacts with the acid
The steel wool would rust	The steel wool turned brown	
Activity 2: Placing a zinc strip in a beaker of aqueous copper (II) sulphate		
The solution would change from blue to colourless because Cu^{2+} ions are removed from solution	The blue solution turned colourless	The solution changed colour due to the removal of Cu^{2+} ions from solution
The zinc strip would decrease in size	The zinc strip decreased in size	Copper was produced because the zinc dissolved
A precipitate of copper would be produced	A brown layer formed on the zinc strip	Zinc is more reactive than copper
Activity 3: Placing a copper strip in a beaker of aqueous zinc sulphate		
The solution would change colour	No reaction occurred	Zinc cannot donate electrons
Activity 4: Immersing a coil of copper wire in aqueous silver nitrate		
The solution would turn blue because copper goes into solution	Shiny crystals form on the copper wire	Copper displaces silver
Activity 5: Placing strips of (a) magnesium, (b) zinc, (c) iron, (d) lead, and (e) copper in separate beakers containing aqueous solutions of $ZnSO_4$, $FeSO_4$, $Pb(NO_3)_2$ and $CuSO_4$		
(a) with magnesium	Magnesium went into solution and metal precipitates were produced in each case	Magnesium displaced the other metals from solution because it is more reactive
A precipitate would form with aqueous solutions containing Zn^{2+}, Fe^{2+}, Pb^{2+} and Cu^{2+} ions because magnesium is more reactive	A reddish-brown precipitate was produced with Cu^{2+} ions and the blue solution changed colour	
	A 'colourless' precipitate was produced in the other solutions	
	A precipitate was produced in each of the solutions	

(continued)

Table 14.1 (continued)

Predictions	Observations	Explanations
(b) with zinc A precipitate would form with aqueous solutions containing Fe^{2+}, Pb^{2+} and Cu^{2+} ions The colour would change with aqueous solutions of Cu^{2+} ions A precipitate would be produced with aqueous solutions of Pb^{2+} and Cu^{2+} ions	Grey and brown precipitates of lead and copper, respectively, were produced No precipitate was produced in aqueous solution of Zn^{2+} ions	Zinc is more reactive than iron, lead and copper
(c) with iron The colour would change with aqueous solutions of Cu^{2+} ions A precipitate would be produced with aqueous solutions of Pb^{2+} and Cu^{2+} ions	Grey and brown precipitates of lead and copper, respectively, were produced No precipitate was produced in aqueous solution of Zn^{2+} ions	Iron is more reactive than copper and lead
(d) with lead A precipitate would form with aqueous solution containing Cu^{2+} ions and the colour of the solution would change	A brown precipitate was produced and the blue colour of the Cu^{2+} solution turned colourless	Lead has displaced copper
(e) with copper There would be no reaction with all solutions	No reaction was observed in all cases	Copper is the least reactive metal
Activity 6: Bubbling chlorine gas into a test tube of aqueous potassium iodide containing a layer of xylene		
The xylene layer would change from clear to purple as iodine is produced	The xylene layer turned from clear to purple	The purple colour of the xylene layer indicates the presence of iodine. Chlorine is more reactive than iodine
Activity 7: Burning three candles of varying lengths under a sealed bell jar of air		
The shortest candle would be extinguished first because carbon dioxide is heavier and sinks to the bottom	The longest candle was extinguished first, followed by the next shorter candle then the shortest	More oxygen settled at the bottom
Activity 8: Inverting a test tube containing sulphur dioxide over another test tube containing hydrogen sulphide		
There would be a reaction and a colour change	Yellow precipitate mixed with drops of water was produced	There was a redox reaction. Hydrogen sulphide was a reducing agent

The groups were generally not consistent in verifying their predictions against their observations. For example, several groups predicted a change in colour but did not report it as an observation. Explanations included removal of copper ions from solution and that zinc was more reactive than copper. Once again, there was no explanation in terms of the redox reaction that had occurred. To reinforce concepts associated with redox reactions, students were required to write ionic equations for this and other similar reactions and to identify the oxidising and reducing agents. The response of students from one of the groups to Activity 2 is provided below.

> *Prediction:* Colour would change as copper ions were removed from solution; decrease in the size of zinc; a precipitate of copper would form.
> *Observation:* ... solution changed to colourless; decrease in size of zinc; brownish layer formed on zinc.
> *Explanation:* The change in ... colour was due to removal of copper ions from solution; the copper precipitate formed because zinc dissolved. Zinc was more reactive than copper.

Activity 7—Burning of candles under a bell jar

Most of the groups predicted that the candles would eventually be extinguished because of lack of oxygen or that the oxygen was being used up or that the candles would burn for some time before being extinguished because no more air was allowed under the bell jar. Three of the groups specified the order in which the candles would be extinguished: two groups stated that the long candle would be extinguished first and the shortest last as oxygen is heavy and sinks to the bottom while the third group stated that the shortest candle would be extinguished first as carbon dioxide is heavier and would sink to the bottom. All the groups observed that the long candle was extinguished first followed by the middle candle, then the shortest one. The observations were explained in various ways: (1) carbon dioxide rose up because it was warm, hence extinguishing the longest candle first, (2) oxygen sank to the bottom because it was warm, hence extinguishing the shortest candle first, (3) no more oxygen was available to support the burning candles and (4) the burning candles gave out heat.

The group that predicted that the shortest candle would go out first because carbon dioxide was heavier and sank to the bottom changed their minds after observing the experiment and suggested that the carbon dioxide produced by the burning candles rose up, resulting in a conflict between their prediction and observation. To some students, the term 'air' meant oxygen since they predicted that the candles would go out because no more air was allowed under the bell jar.

When making their observations, students focused on the order in which the candles were extinguished to the exclusion of other changes that occurred. They did not report other possible changes, indicating that students were selective in their observations. In this activity the students used their prior knowledge that oxygen supports combustion. Though the formation of carbon dioxide is alluded to in the observations, no group predicted that it would be produced as the product of combustion in a redox reaction. The response of students from one of the groups to Activity 7 is provided below.

> *Prediction:* The candles would (be extinguished) because of lack of oxygen; the long one would go out first, the shortest (one) the last.
> *Observation:* The long one went out first followed by the middle one then the shortest was last; vapour formed on the walls of the jar.
> *Explanation:* Carbon dioxide was warm, (became) light (and) rose up.

Activity 8—Reaction between hydrogen sulphide and sulphur dioxide

Three groups predicted a change in colour with one of these groups also mentioning the liberation of hydrogen. Only one group predicted the formation of a solid deposit. The fifth group predicted the formation of liquid as a product without specifying the nature of the liquid. All groups reported the same observation that a yellow 'precipitate' and droplets of a liquid (most probably water) were produced. The groups confirmed that a redox reaction had occurred. A few students indicated confusion by predicting the production of hydrogen gas and that sulphur changed into a liquid. The response of students from one of the groups to Activity 8 is provided below.

> *Prediction:* Sulphur dioxide would … form a precipitate.
> *Observation:* A yellow precipitate was formed.
> *Explanation:* This was a redox reaction.

In response to research Question 1, data show that there was initial reluctance among students to write down their predictions prior to performing the activities. This tendency was partly due to their unfamiliarity with expressing themselves as the usual practice of writing during instruction involved merely copying down notes from the blackboard. Students were also selective in reporting their observations, noting only the observations that they had initially predicted when in actual fact the activity involved a number of visible changes. In addition, except in the case of Activity 8, there was no reference in students' explanations to the species that were reduced and oxidised or to the fact that a redox reaction had actually occurred; no ion-electron or redox equations were provided for any of the activities.

Students' Performance on Redox Reactions Test

The knowledge and understanding gained by the students about redox reaction concepts was evaluated by administering a 25-item multiple-choice test on redox reactions, providing a response to research question 2: What was the effect of the use of POEs involving redox reactions on students' achievement in the test on redox concepts? The test was administered as a pre-test a week before teaching redox reactions and again as a post-test at the conclusion of instruction using the POEs. As there were no available standardised tests on redox reactions, the test was constructed using examination-related questions as well as questions found at the end of the chapter on redox reactions in one of the prescribed textbooks. Examples of five items were provided earlier in Fig. 14.1. The questions involved recall of knowledge (7 items), application of knowledge (15 items) and providing explanations (3 items). The main purpose of the test was to use students' achievement to evaluate the efficacy of the POEs in classroom instruction. A paired samples *t* test analysis was performed to compare the pre-test and post-test means of students' scores using a SPSS (version 16) software programme. There was a significant increase in students' mean scores from the pre-test ($M = 6.19$, $SD = 2.03$) to the post-test [$M = 11.09$, $SD = 3.53$, $t\ (43) = 14.01$, $p < 0.01$]. Although there was a significant increase in mean scores, the post-test mean score is far from satisfactory. However, this change in students' performance must be viewed from the perspective of the prevailing instructional practice that was highly teacher-centred with limited opportunities for students to be involved in laboratory or investigative activities. Hence, it may be concluded that the POE activities had to some extent enhanced students' knowledge and understanding of redox reaction concepts.

Student Interviews

In order to assess students' perceptions of the POE strategy, structured interviews were conducted with 36 students using 14 questions listed in Table 14.2 and as explained in the methods section. The interviews provided response to research Question 3: How did students perceive the usefulness of the POE activities in facilitating their understanding about redox reactions? Students' responses are discussed along six themes as follows:

Satisfaction (Question 1): The students used expressions like 'right' and 'good' when asked how they perceived the use of POEs. Most students indicated that the new teaching-learning strategy was better than the strategies previously used to teach science. These are examples of their perceptions about the use the POEs:

> It is right because if you predict something you have to think.
> We expect teachers to tell us everything.
> I think it okay; you get to know how people think.

The students echoed the view that introducing the strategy at Grade 11 still gave them a chance to apply it at Grade 12 level. One student, however, felt that using POEs was a waste of time, with the comment: "This is very long and tiresome. It should be started at Grade 6. This is a waste of time. We want to sit and be taught". This student was comfortable with the traditional transmission method. She seemed to think it was more useful when the teacher spoke while the students listened.

Enjoyment (Question 5): Most students found POEs enjoyable because the teaching technique placed emphasis on engaging learners in hands-on activities. For several students, doing practical work was a novel experience. A selection of several of the comments highlighting this issue is given below:

Table 14.2 Students structured interviews protocol

Number	Question
1	How do you feel about the use of the POE strategy in the classroom?
2	How did you feel when you had to share your prediction with others?
3	When your prediction did not agree with your observation were you able to resolve conflict?
4	Does the use of POE give you a better understanding of redox reactions?
5	Do you think now you enjoy science better than before?
6	When we talk of charged particles what do you think of?
7	What does it mean when we say that something is positively charged or negatively charged?
8	Which of the following equations represent redox reactions? Explain your answer. (a) $2\ Mg + O_2 \rightarrow 2\ MgO$ (b) $Mg + 2\ HCl \rightarrow MgCl_2 + H_2$ (c) $H^+ + OH^- \rightarrow H_2O$ (d) $CO_3{}^{2-} + 2H^+ \rightarrow H_2O + CO_2$
9	In one of the reactions above identify the oxidising and the reducing agents.
10	What can you say about oxidation number?
11	Balance the following equations. What sort of reactions do they represent? $HBr + H_2SO_4 \rightarrow SO_2 + Br_2 + H_2O$ $H^+ + Zn + NO^{3^-} \rightarrow Zn^{2+} + NO + H_2O$
12	Why do we use half reactions?
13	What do you think you have gained by using POEs?
14	What other comments do you wish to make about POEs?

Enjoyed it better by engaging in hands-on activities, you remember things better.
I enjoyed it because we did not do much practicals before but when you see things it is easy to understand.

The students were relieved from the fear and embarrassment of making inappropriate statements when the teacher did not confront them for giving wrong answers. Students noted that there was a change from the traditional way of teaching in which the teacher gave all the answers to the students resulting in students learning facts without sufficient understanding. They believed that the non-judgemental approach adopted by the teacher when using POEs helped them to actively participate in class discussions.

Sharing (Question 2): Some students thought that sharing their predictions was a good idea as each student contributed his or her views for discussion and evaluation, hence enabling the group to decide which prediction was the most suitable. They appreciated the fact that they were able to put forward their ideas and develop these as a group. This view is supported by the following comments:

It is better, for learners to bring their views and promote a debate until a consensus had been reached. Sharing enables one to rectify his or her faults.
Sharing is better if you made a mistake you are able to rectify it.

Other students expressed the feeling that sharing gave them a chance to listen to views of their peers as reflected by the following statement: "It was a good idea to know what the other learners thought".

However, several other students did not feel happy about sharing because they felt that there were students who did not do any work but simply copied from others, or who lacked confidence in their work and therefore would not like to reveal what they thought to other students. The following statements support this view.

> Sharing is not right because the other people do not bring forth any predictions and expect to copy from you.
> I do not feel happy because I don't want others to see my work.

Conflict resolution (Question 3): Several students claimed that they were able to resolve any conflict that existed between their predictions and observations by working in consultation with others in their group. This view is echoed by the comments:

> You look at your work and ask others where you might have gone wrong.
> Yes after observing I could see where I went wrong.

There were also students who were not very certain about resolving conflict or felt discouraged when conflict between their predictions and observations could not be resolved. These students commented that:

> Sometimes you get it and other times you don't.
> I get discouraged when it does not match.

Acquisition of conceptual knowledge about redox reactions (Questions 4 and 14): There was overwhelming agreement by students that the use of POEs led to an improvement in their knowledge of redox reactions (despite the fact that students seldom explained the redox reactions that had occurred in the POEs in terms of the redox concepts that were involved). They felt that this instructional strategy was different from what normally happened in the class. Several of the many comments which support this view are:

> Most of the time you cram information but now you do (an) experiment and understand it better when you do it.
> Previously, you wrote an equation not knowing where it came from but now because you actually do the experiment and take note of the reactants and changes that occur.
> If you predict, observe and explain you understand it better than being told.

Knowledge gained about redox reactions (Questions 6–12): Several examples of students' responses are discussed in this section. In Question 6, students were asked what they understood when we refer to charged particles. Most students when asked about charged particles in this question were able to provide satisfactory explanations. Three students linked charged particles to cations and anions: they indicated that a cation is positively charged and an anion is negatively charged: "A charged particle could be either an anion (negative) or cation (positive)". Two students explained charged particles in terms of loss or gain of electrons with one specifically indicating that it is an atom that will lose or gain electrons: "An atom that has lost or gained electrons". Only two out of 12 students could not explain what charged particles were.

In Question 7, students were asked what was meant when something was positively or negatively charged. In response, most students demonstrated good understanding of what a positively or negatively charged entity was. Three of the students clearly associated a positively charged entity with a loss of electrons and a negatively charged entity with a gain of electrons, typified by "A positively charged body loses electrons and a negatively charged body has gained electrons". Six students merely described a positively charged entity as the one that had lost electrons without detailed explanation, demonstrating lack of experience in providing scientific explanations. Two students explained a positively charged entity as one that had excess protons.

In Question 8, students were required to identify redox reactions from a list of given equations and in Question 9 they were asked to state which substances were oxidising and reducing agents from the given list. Five students could correctly identify Eqs. (a) and (b) as redox reactions by using oxidation numbers. Four of these students correctly identified magnesium as a reducing agent and oxygen and hydrogen ions as oxidising agents. The fifth student, in addition to Eqs. (a) and (b), identified Eq. (c) as a redox reaction but could not work out oxidation numbers for this equation. Two students experienced difficulty in assigning oxidation numbers and hence could not identify the redox reactions. One student chose Eq. (d) as an example of a redox reaction and stated that H^+ had undergone a change. Two others made no attempt to answer the questions. Another student chose Eq. (c) as a redox reaction but could not specify the reducing and oxidising agents. One student assigned oxidation numbers to Eq. (c) as follows:

$$\overset{0}{H^+} + \overset{0}{OH^-} \quad \overset{+1\,-2}{H_2O}$$

This student assigned a value of zero for oxidation numbers of the ions H^+ and OH^-, suggesting that the ions were in their elemental state; she could not differentiate an element in its elemental state from an element in a compound. Students generally regarded the oxidation state of an element to be the same as the charge of a monatomic ion of that element. Also, the concept of individual and total oxidation numbers of polyatomic species was not evident to them.

In Question 10, the students were asked to explain what they understood about oxidation numbers. Students had a problem explaining what oxidation numbers were. One student explained it in operational terms, that is, if the oxidation number increases the substance is oxidised and if the oxidation number decreases the substance is reduced. Another stated that oxidation number is for identifying redox reactions. Six students did not provide an answer whereas two stated that they did not know.

Most students were able to balance the redox equations in Question 11. In Question 12 the students were asked why half reactions are used. Only two students were able to provide satisfactory explanations, such as:

To identify whether the reaction is a redox reaction.
To identify reducing and oxidising agents.

The interdependence of reduction and oxidation was generally a problem for students. Though students claimed to have gained knowledge and understanding in the redox reactions test, they still had problems understanding the concepts related to redox reactions.

With reference to research Question 3, students were generally positive about the use of POEs, stating that the activities were enjoyable and challenged them to think. Several students were engaged in hands-on activities for the first time and had gained better experience as a result of handling chemicals and seeing for themselves the colour changes and precipitates that were produced. Students were appreciative of the fact that they were being made responsible for their own learning, and appreciated the fact that their views were considered. However, the section of the interviews that required students to provide written responses to questions about redox reactions indicated that several students were not explicit about the correct use of terms like *oxidise, reduce, oxidising agent* and *reducing agent*. Students were also generally unable to distinguish between the oxidation number of a particular element and the sum of the oxidation numbers of a group of atoms.

Students' Attitudes Towards Science

A modified version of the TOSRA questionnaire consisting of 46 items was administered to students before and after implementation of the POEs in order to assess changes, if any, in students' attitudes towards learning about redox reactions using this strategy. Due to absenteeism and other reasons, responses from 61 of the total of 66 students were used in the analysis. The administration of the TOSRA prior to and after implementation of the POE activities provided a response to research Question 4: What was the effect on the use of POEs involving redox reactions on students' attitudes towards learning science? The Cronbach's alpha reliability coefficients for the four dimensions of the modified TOSRA that was administered as a pre-test and a post-test ranged from 0.63 to 0.83 and are deemed satisfactory for the purpose of this study (Nunal and Bernstein 1994). The means and standard deviations for each of the four scales and results of the paired samples t-test comparisons of the pre-test and post-test results are provided in Table 14.3. The results indicate statistically significant improvement in students' attitudes as a result of the use of POEs in instruction on redox reactions in only the Adoption of Scientific Attitudes scale.

Table 14.3 Scales means and standard deviations of students' attitudes toward science measured by the modified Test of Science-Related Attitudes (TOSRA) ($N = 61$)

Scales	Number of items	Pre-test mean (sd)	Post-test mean (sd)	*t*-value
Attitudes to scientific inquiry (I)	11	3.92 (0.50)	3.85 (0.58)	0.99
Adoption of scientific attitudes (A)	11	3.76 (0.58)	3.92 (0.49)	2.53*
Enjoyment of science lessons (E)	11	4.12 (0.48)	4.17 (0.51)	0.83
Leisure interest in science (L)	13	4.01 (0.63)	4.06 (0.53)	0.86

* $p < 0.05$

Conclusions, Limitations and Implications

This study has revealed several outcomes relating to the use of POEs in science instruction on redox reactions in these two South African classes. The information obtained suggests that POEs are effective in capturing a range of possible observations and predictions made by students. The POEs were also effective in obtaining quality information on students' existing knowledge. There were, for example, instances when students were able to successfully draw on their prior knowledge when confronted with the problems in the activities that they were engaged in as in the burning of candles under a bell jar (Activity 7): several students were aware that oxygen was essential for combustion. However, in the reaction between steel wool and acetic acid (Activity 1), several students inappropriately used their prior knowledge because they were convinced that hydrogen gas was produced in the reaction.

With reference to research Question 1 (How proficient were students in performing the POE activities on redox reactions?), it was observed that several students did not record their predictions and explanations, or if they did, it was often very brief. Students were not used to writing down their own opinions as the most common form of writing in the classroom involved copying notes written on the chalkboard by the teacher. Also, students were generally selective in reporting their observations. Where a POE involved a number of observations, students did not report all observations but only reported the phenomenon that captured their interest or only the ones that they had predicted. In Activity 7, for example, most students only reported the order in which the candles were extinguished and omitted other observations like the warming of the bell jar and liquid condensation on the inner surface of the jar. Students were not able to see some of the changes in a chemical reaction unless their attention was drawn to the expected changes.

Through the use of POEs students were able to understand what was meant by a chemical reaction with respect to reactants and products. As a result, students learned how to write chemical equations as demonstrated in the following comment:

> Previously I wrote a chemical equation not knowing where it came from, but now I actually do the experiment and take note of the reactants and changes that occur.

The students also learned how to write half reactions and work out which substance transfers an electron (is oxidised) and which substance receives an electron (is reduced). Students had to label the half reactions as oxidation and reduction half reactions so that they could work out which substance was the reducing agent or oxidising agent. One student expressed the following comment "I can identify half reactions and oxidising as well as reducing agents". Further, these POEs created awareness that the chemical reactions could be classified into various types.

With reference to research Question 2 (What was the effect of the use of POEs involving redox reactions on students' achievement in the test on redox concepts?), despite practically no reference in their written explanations to the redox reactions that had occurred in the eight POEs that were used, students' improved understanding about redox reactions was evident from the significant difference between the performance of students in the redox reactions test prior to using POEs and after using POEs. It is likely that the specific reference by the teacher to the redox reactions that were involved after the POE activities (except in Activity 1) could have facilitated improved understanding about redox reactions.

Students were generally positive about the use of POEs with regard to research Question 3: How did students perceive the usefulness of the POE activities in facilitating their understanding about redox reactions? Data solicited through student interviews revealed responses that were affective as well as conceptual in nature. Students described POEs as being enjoyable, and helped them improve their understanding of concepts because the POEs challenged them to think. Some learners were engaged in hands-on activities for the first time and stated that they gained more experience when they manipulated the equipment and chemicals because they were actually able to see the colour changes as well as the formation of deposits in the reactions concerned. Students were more appreciative of the fact that they were being made to be responsible for their learning as there was a shift away from learners getting all the information from the teacher. Student–student interactions promoted respect for each other's views when they worked in groups. Students became more accountable for their actions and were happy that their ideas were considered. They were encouraged to say what they thought without fear of being ostracised. One of them remarked, "No one has ever asked me what I thought before"; as a result her self-confidence was boosted. Through the use of POEs, students learned to share ideas through their predictions and observations. While some students welcomed the idea of sharing their views there were those who felt uneasy about this. They feared being laughed at. This behaviour was common amongst students who possessed weak content knowledge and hence, were afraid of 'losing face' amongst their peers. The students, however, recognised the shift from traditional teaching (that mainly involved knowledge transmission by the teacher) to greater student-centred learning, resulting in their improved ability, for example, to identify and balance redox reactions, suggesting that students did understand what they were learning about.

However, although students appeared positive about what they knew when they were interviewed, their positive indications did not translate into improved understanding about redox reactions as indicated by their responses to the pen-and-paper section of the interview. For example, several students did not fully understand the meaning of certain terms and were, therefore, unable to use and explain the terms correctly. These included terms such as 'reduced' when used with reduction as against 'oxidised' when used with oxidation, as well as other terms like 'oxidising agent' and 'reducing agent'. Students were also unable to differentiate between the oxidation number of a single entity like an ion or an atom and the sum of oxidation numbers of the molecule of a compound, and experienced difficulty assigning oxidation numbers to the elements in a polyatomic ion. Ringnes (1995) believed that the way redox is taught in secondary schools does not promote understanding, the chemical terms used are in conflict with everyday usage and a historical development of the concepts is seldom given. Anderson (1997) stressed that careful choice of language is important in the teaching of chemistry as students' cognitive frameworks are not sufficiently developed, and hence may misrepresent the words used by teachers and textbook authors giving rise to alternative conceptions.

The TOSRA that was administered before and after the implementation of the POE activities indicated limited changes in students' attitudes towards learning about redox reactions in response to research Question 4: What was the effect on the use of POEs involving redox reactions on students' attitudes towards learning science? Comparison of the pre- and post-test scales means for the four dimensions of the TOSRA indicated statistically significant improvement in students' attitudes as a result of the use of POEs in instruction on redox reactions only in the Adoption of Scientific Attitudes.

One of the main limitations of the study was the teacher's unfamiliarity with POEs and hence, it would take time for her to be proficient in using the new instructional strategy. The limited availability of equipment and chemicals could further stifle teachers' enthusiasm to continue using the strategy with their classes as teachers do not have sufficient time and expertise to design appropriate learning materials of their own. While the language of instruction was English, this was not the students' first language and they may have been unwilling to express their ideas from the POEs when they were unsure of the English language terms. The use of the TOSRA posed another limitation because the language may not have been very familiar to the students. In addition, the students in an African cultural setting may not readily understand the sociocultural factors incorporated in the TOSRA that were developed in a different cultural setting.

Several implications for classroom instruction have emerged from this study. First, personnel involved in the planning and implementation of inservice programmes for teachers can be agents of change in classroom practice among teachers by promoting constructivist teaching and learning strategies like the POE. Continued support should be provided to teachers, both through regular inservice programmes, classroom visits and by making available essential resources to meet the needs of teachers. Teachers themselves should be more proactive by engaging in peer observations during implementation of POE instructional strategies. Peer coaching is a powerful strategy for staff professional enrichment. Advocates of peer coaching recognise the role that peer coaching can play in assisting teachers to incorporate newly acquired skills into their own teaching repertoires.

In conclusion, use of POEs in science instruction involving redox reactions was generally seen to benefit both the teacher and the students. Students learned more readily and benefited from group support and enjoyed the discussion with peers which encouraged knowledge construction. A more relaxed classroom environment would probably make learners free to express their opinions and listen to the ideas of peers. Sharing of ideas contributed to the development of social skills like listening, empathy and respect for other people's views. When using POEs, students receive immediate feedback. The teacher is able to evaluate his/her lessons by monitoring group participation; the teacher is also able to diagnose students' alternative conceptions as well as monitor their progress. POEs give the teacher the opportunity to devise ways of dealing with the students' conceptual problems and help them lower the status of these conceptions in favour of scientifically acceptable concepts.

References

Anderson, O. R. (1997). A neurocognitive perspective on current learning theory and science instructional strategies. *Science Education,81*(1), 67–89.

Ausubel, D. P. (1968). *Educational psychology*. New York: Holt, Rinehart and Winston.

Ballantyne, R., & Bain, J. (1995). Enhancing environmental conceptions: An evaluation of cognitive conflict and structured controversy learning units. *Studies in Higher Education,20*(3), 293–303.

Bogdan, R. C., & Biklein, S. K. (1992). *Qualitative research for education: An introduction to theory and methods*. Boston: Allyn & Bacon.

Carr, M. (1984). Model confusion in chemistry. *Research in Science Education,14*(1), 97–103.

Champagne, A. B., Klopfer, L. E., & Anderson, J. H. (1979). *Factors influencing the learning of classical mechanics*. Pittsburgh: University of Pittsburgh.

De Jong, O., Acampo, J., & Verdonk, A. (1995). Problems in teaching the topic of redox reactions. Actions and conceptions of chemistry teachers. *Journal of Research in Science Teaching,32*(10), 1097–1110.

De Jong, O., & Treagust, D. F. (2002). The teaching and learning of electrochemistry. In J. K. Gilbert, O. De Jong, R. Justi, D. F. Treagust, & J. H. Van Driel (Eds.), *Chemical education: Towards research-based practice* (pp. 317–335). The Netherlands: Kluwer Academic Publishers.

Driver, R., & Oldham, V. (1986). A constructivist approach to curriculum development in science. *Studies in Science Education,13*(1), 105–122.

Duit, R. (2009). *Students' and teachers' conceptions and science education*. Retrieved August 13, 2009 from http://www.ipn.uni-kiel.de/aktuell/stcse/stcse.html.

Duit, R., & Treagust, D. F. (1995). Students' conceptions and constructivist teaching approaches. In B. J. Fraser & H. J. Walberg (Eds.), *Improving science education* (pp. 46–49). Chicago: The National society for the Study of Education.

Fensham, P. J., & Kass, H. (1988). Inconsistent or discrepant events in science instruction. *Studies in Science Education,15*(1), 1–16.

Fraser, B. J. (1981). *Test of Science-Related Attitudes*. Melbourne: The Australian Council for Educational Research Limited.

Garnett, P. J., Garnett, P. J., & Hackling, M. W. (1995). Students' alternative conception in chemistry: A review of research and implications for teaching and learning. *Studies in Science Education,25*(1), 69–96.

Garnett, P. J., & Treagust, D. F. (1992). Conceptual difficulties experienced by senior high school students of electrochemistry: Electric circuits and oxidation and reduction equations. *Journal of Research in Science Teaching,29*(2), 121–124.

Gunstone, R. F., & White, R. T. (1981). Understanding of gravity. *Science Education,65*(3), 291–299.

Harrison, A. G., & Treagust, D. F. (1998). Modeling in science lessons: Are there better ways to learn with models? *School Science and Mathematics,98*(8), 420–429.

Liew, C. W., & Treagust, D. F. (1995). A predict-observe-explain teaching sequence for learning about students' understanding of heat and expansion of liquids. *Australian Science Teachers Journal,41*(1), 68–71.

Merriam, S. B. (1988). *Qualitative research and case study applications in education*. San Francisco: Jossey-Bass Publishers.

Nakhleh, M. B., & Krajcik, J. S. (1994). Influence of levels of information as presented by different technologies on students' understanding of acid, base and pH concepts. *Journal of Research in Science Teaching,31*(10), 1077–1096.

Niaz, M. (2002). Facilitating conceptual change in students' understanding of electrochemistry. *International Journal of Science Education,24*(4), 425–439.

Nunally, J. C., & Bernstein, I. H. (1994). *Psychometric theory* (3rd ed.). New York: McGraw-Hill.

Prawat, R. S. (1989). Teaching for understanding: Three key attributes. *Teaching and Teacher Education,5*(4), 315–328.

Ringnes, V. (1995). Oxidation-reduction learning difficulties and choice of redox models. *School Science Review,77*(279), 477–478.

Schmidt, H. J. (1997). Students' misconceptions—looking for a pattern. *Science Education,81*(2), 123–135.

Tao, P., & Gunstone, R. F. (1997). *Conceptual change in science through collaborative learning at the computer*. Oak Brook: Paper Presented at the National Association for Research in Science Teaching.

Thijs, A., & van den Akker, J. (1997). *Peer coaching as a promising component of science teachers' inservice training*. Paper presented at the South African Association for Research in Mathematics and Science Education, University of Witwatersrand: Johannesburg.

Treagust, D. F., & Chandrasegaran, A. L. (2009). The efficacy of an alternative instructional program designed to enhance secondary students' competence in the triplet relationship. In J. K. Gilbert & D. F. Treagust (Eds.), *Models and modelling in science education: Multiple representations in chemical education* (pp. 151–168). The Netherlands: Springer.

Treagust, D. F., & Chittleborough, G. (2001). Chemistry: A matter of understanding representations. In J. Brophy (Ed.), *Subject-specific instructional methods and activities* (Vol. 8, pp. 239–267). Oxford: Elsevier Science Ltd.

White, R. T., & Gunstone, R. F. (1992). *Probing understanding*. London: Farmer Press.

Chapter 15
Application of Case Study and Role-Playing in Forensic Chemistry and Analytical Chemistry Education: Students', Graduates' and Teachers' Points of View

Iwona Maciejowska, Renata Wietecha-Posłuszny, Michał Woźniakiewicz and Paweł Kościelniak

Introduction

Active participation of students in their education represents a paradigm shift in education over the past 50 years. Many different teaching and learning methods and techniques have been proposed, among which are: case study, role-playing, brainstorming, projects etc. Some of these methods have a long history. These methods include didactic courses, where students are taught to assume certain characters playing specific roles. Such methods are applied mainly in vocational education, the teaching of arts, and educational activities and are based on a simulation of reality. As early as the seventeenth century, a kind of lyceum was established, to which pupils from local schools were sent for training. By the nineteenth century, these lycea had become a part of most trade schools, such as Commercielles Laboratoires, bureau modele, practica de operationes de commercio, banco modello etc. (Hopf 1973). In these schools there were classes where the set-up and organisational system of various enterprises were simulated (Nowacki 1999). Since that time, these methods have been used in the education of: managers,[1] politicians etc., and finally scientists, including chemists (Walters 1991). Collections of interesting cases have even been built up, which are shared with a

[1] E.g. in the Yale School of Management (USA).

I. Maciejowska (✉)
Department of Chemical Education, Jagiellonian University, Krakow, Poland
e-mail: maciejow@chemia.uj.edu.pl

R. Wietecha-Posłuszny · M. Woźniakiewicz · P. Kościelniak
Laboratory for Forensic Chemistry, Department of Analytical Chemistry, Jagiellonian University, Kraków, Poland

P. Kościelniak
Institute of Forensic Research, Kraków, Poland

I. Devetak and S. A. Glažar (eds.), *Learning with Understanding in the Chemistry Classroom*, DOI: 10.1007/978-94-007-4366-3_15,

wider community of teachers and lecturers, who can use them during their natural science classes.[2] In the UK, the University of Glasgow (Johnstone et al. 1978; Johnstone and Percival 1978; Reid 1976, 1980b), University of Hull, University of Plymouth (Belt et al. 2002a–e), and the Higher Education Academy, Physical Sciences Centre[3] have been leaders in this field.

Among the methods in this field, the following routines can be mentioned: simulation, drama (role-playing method), and staging. Simulation and drama constitute a reflection of a situation that is potentially real (training of social skills), whereas staging may be of a more abstract nature ("just imagine you are an oxygen particle"). Simulation models real-life phenomena and is usually more structured than role-playing. These definitions differ somewhat depending on the language-culture area where they are used. For the practising teacher (lecturer), however, the terminology is not so important. Jackson and Walters (2000) have written: "The role-play construct, especially the laboratory component, was based largely on Professional experiences of industrial and academic analytical chemists. This is where differences between role-playing and simulation bear reemphasis."

Typically, the introductory text of the role-playing method is a case description, which requires a versatile analysis. The roles are then acted out in a small group of people. The student's behaviour in a given role can be precisely defined in instructions given by the lecturer or it can be based on general assumptions presented earlier, or left completely to the student's imagination. The preparation of roles can take place at the beginning of a given class, e.g. by working with a text. However, by giving students more time (e.g. by introducing the topic during previous classes), they have access to various information sources and this allows them to develop the role better. As far as large training groups are concerned, students who do not play any role can be observers, but they also have to have a concrete task assigned to carry out, e.g. to complete observation sheets which would allow them to prepare themselves well for a final discussion and summary.[4] Decisions made when working with the role-playing method have to be finally analysed from the point of view of their optimality. In role-playing, participants adopt characters or parts that have personalities, motivations and backgrounds that are different from their own. This allows the students to explore management roles, specialist roles etc. Sometimes they even dress in clothes that are typical of the characters they are playing (Kimbrough et al. 1995; Klich et al. 2005; Maciejowska 2004).

A variety of role-playing methods is common in chemistry classes in Polish schools. One of these is the debate, e.g. judgement over acids (Stobiński 1974),

---

[2] The National Center for Case Study Teaching in Science. Case Collection, retrieved June 29, 2009, from http://ublib.buffalo.edu/libraries/projects/cases/ubcase.htm.

[3] HEA,C/PBL Case Studies. Retrieved September 01, 2010, from http://www.heacademy.ac.uk/physsci/home/networking/sig/CPBL.

[4] Maciejowska, I. (2009). Metoda ról/inscenizacja/drama/symulacja. Retrieved December 22, 2009, from http://chemia.zamkor.pl/images/materialy/metoda_rol.pdf.

alcohol (Babczonek-Wróbel 1999), saccharides (Borowska and Panfil 2001) or radioactivity (Maciejowska and Odrowąż 2008). During the discussion, organised in the form of a court trial, students playing the role of defence counsels and prosecutors present the positive and/or negative effects of using a given substance or phenomenon in human life. In such a case, a great number of students can be involved in the class as witnesses, for whom role descriptions are prepared by their colleagues-advocates. The role-playing method is also applied at universities—from the very first year. In an exercise titled "Amsyn" (Johnstone et al. 1981), students form three groups: management, local authority, and trade unions. The goal of the exercise is to find a compromising solution to the problem of a river polluted by dyestuff intermediates produced by a local factory.

In the late 1970s and early 1980s, Johnstone noticed that there is an unmet need to teach students so-called 'transferable skills' (Johnstone et al. 1978, 1981; Johnstone and Sharp 1979) and also that it is difficult to obtain a consensus as to the list of such skills that are necessary for chemists (Johnstone and Percival 1978). Since that time, much has changed. Learning outcomes for chemistry graduates, worked out under the 'Tuning Educational Structures in Europe' project have been presented in the "Budapest chemistry descriptors". For example:

First cycle degree graduates will[5]: (1) have the ability to gather and interpret relevant scientific data and make judgments that include reflection on relevant scientific and ethical issues; (2) have the ability to communicate information, ideas, problems and solutions to informed audiences; (3) have developed those learning skills that are necessary for them to undertake further study with a sufficient degree of autonomy.

Second cycle degree graduates will: (1) have the ability to apply their knowledge and understanding, and problem-solving abilities, in new or unfamiliar environments within broader (or multidisciplinary) contexts related to chemical sciences; (2) have the ability to integrate knowledge and handle complexity, and formulate judgements with incomplete or limited information, but that include reflecting on ethical responsibilities linked to the application of their knowledge and judgements; (3) have the ability to communicate their conclusions, and the knowledge and rationale underpinning these, to specialist and non-specialist audiences clearly and unambiguously.

The training methods described in earlier paragraphs develop not only core chemistry competencies, but also generic skills such as: decision-making, communication, collaboration. They help to educate students in how chemistry is applied in real life, by professionals working in teams. Students learn responsibility in carrying out their duties (Deavor 1994) because they have to think about consequences of chemical decisions on the social and economic welfare of the community (Johnstone and Sharp 1979). The results of research indicate the short-term effects as well as the long-term impact on graduates (Jackson and Walters 2000).

---

[5] Tuning Educational Structures in Europe. Retrieved September 01, 2010, from http://tuning.unideusto.org/tuningeu/images/stories/template/Template_Chemistry.pdf.

Walters (1991) has proved that the role-playing procedure has a positive impact on students' decisions concerning taking-up jobs in vocations connected with chemistry. Role-playing can be very useful in the study of complex systems (Aubusson et al. 1997). Topics of role-playing are often related to current and controversial issues, such as: environmental (Kimbrough et al. 1995; Smythe and Higgins 2008; Maciejowska 2004; Karaś and Maciejowska 2007) or forensic ones (Murcia et al. 1990). Such issues mobilise discussion participants to search for a compromise solution, and in this way to learn negotiation skills, listening to the arguments of the other party, accepting different views, opinions and solutions, etc. Walters (1991) noticed that practice in planning and implementing a team-based problem-solving strategy or developing a working relationship within a group aided them in establishing their future career. Reid (1980a) stated that noncognitive outcomes of simulation-type techniques are the most important ones, because other teaching techniques are less successful in achieving them.

Forensic Chemistry Case

The laboratory class presented below has been compulsory for forensic chemistry students since 2006. The class is conducted at the Faculty of Chemistry of the Jagiellonian University within the course "Chemical, criminal and toxicological investigations". The classes comprise two teaching/learning units, 7.5 lesson hours each (1.5 h—seminar and 6 h—workshop). Classes are for fourth year students of chemistry, who are members of the forensic chemistry specialisation. Each group consists of six students. The forensic chemistry panel comprises 12 students only, hence during the school year the classes are conducted with two groups.

The role-playing class is based on a real story that was highly publicised in the Polish media.[6] In June 2003, Krakow citizens were shocked by news of two offenders who had attempted to rob an armoured van. After a short shoot-out with a guard, the robbers were forced to retreat. A few hours later, officers spotted two suspects getting into a car that was under observation and tried to arrest them. Just before being clapped in handcuffs, the robbers ate a small amount of white powder. They were transported to a police station, but they got sick and then died despite medical treatment. The autopsy revealed that they were poisoned with arsenic.

This story became the basis for a case study for forensic chemistry students.[7] Some details had to be altered to make it appropriate for the classroom. Thus, instead of shooting and violence, a white-collar crime was proposed:

[6] A statement by the spokesperson of the public prosecutor's office in Kraków (Poland). Retrieved December 22, 2009, from http://www.krakow.po.gov.pl/rzecznik-prasowy,50,8,2004-marzec.html.

[7] Presented at the conference, "Variety in Chemistry Education", Plymouth (UK), 5–7 September, 2004; Michał Woźniakiewicz, Renata Wietecha-Posłuszny, Paweł Kościelaniak, Iwona Maciejowska, "Bank robbery??—example of forensic chemistry training".

Fig. 15.1 Discussion on theoretical topics and brainstorming

counterfeiting. In this case, the robbers forged a document allowing them to take about 400,000 PLN (100,000 Euro) from a bank.

The aim of the classes is to give students some practical 'experience' of issues related to the forensic expert's job: the way of analysis of a dossier, autopsy results, prosecutor's questions, material evidence inspection, appropriate preparation of exhibits for analysis, and using appropriate analysis techniques as well as drawing up a valid report from the inspection of examined evidences and their chemical analysis and formulating examination conclusions for the purposes of expert witnessing in court.

The seminar part is dedicated to introducing the "rules of the game" and presenting the theoretical basis. Each student plays the role of a forensic expert: a toxicologist or a questioned document examiner. Some also play the role of group manager. The prosecutor (lecturer) asks teams of 'forensic experts' to examine documents and evidence: a reporting officer's narrative, an autopsy report, autopsy samples and residues of white powder.

Students/experts, after having become acquainted with the case documentation, consider various variants of analysis of presented evidence materials for a given case. The most important task is to describe each piece of evidence and discuss the preferred methods of analysis (from the list: atomic fluorescence spectrometry, liquid chromatography, capillary electrophoresis etc.), taking into account the time, costs and type of evidence and available equipment—Fig. 15.1.

Within the practical part of the classes, students/experts, when beginning the analysis of the mentioned materials, confer and make decisions by themselves as to what exhibits are to be analysed and what techniques they are going to use to respond reliably and comprehensively to the questions asked by the prosecutor: what was the cause of death, what kind of substance was in the white powder, was the document they used authentic etc.?

Students who play the role of court experts-toxicologists receive rich autopsy material in the form of blood samples, biopsy segments of different internal

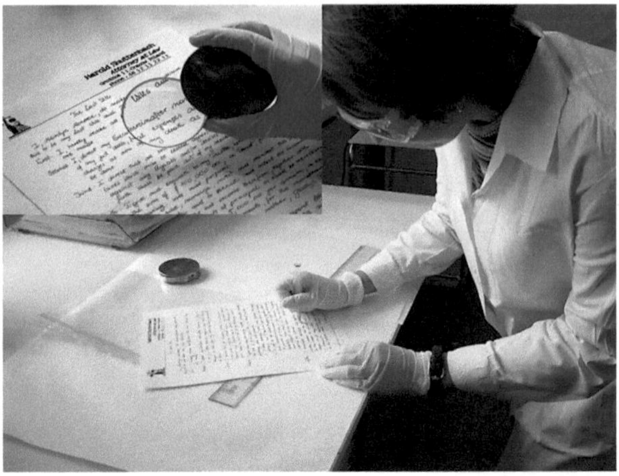

Fig. 15.2 Primary questioned document examination and taking ink samples

organs, and hair samples of individuals taking part in the event. Students start the analysis by preparing the biological material samples and extracting the xenobiotics from the matrix, using the method selected in the discussion. At the next stage, they carry out an analysis using the method selected earlier, e.g. atomic fluorescence spectrometry used in inorganic analysis. The culmination of this group of experts' (students') work is a laboratory note and an expert-toxicologist opinion. The laboratory note in this case contains a description of all actions carried out in the toxicological analysis, including applied procedures, analytic method parameters and obtained results.

The work of the crime detection experts' group proceeds similarly. Students receive a document that has allegedly been 'cooked up'; in this case, it is an authorisation note used by the suspects to withdraw money from a bank. The criminalistic experts (students) first of all carry out an initial scrutiny of the document with the naked eye and with the use of magnifying glass (Fig. 15.2). Then they use an optical microscope and a microscope used for examination of objects in infrared. When examining the document, they search for potential differences in the examined coating materials (e.g. their thickness, intensity of lines, colour) and select fragments of lines, from which samples for chemical analysis will be taken.

At this stage students perfect their practical skills connected with the work of court experts, who deal mainly with micro-quantities of samples and have to take into consideration the risk of contamination of samples, etc. They become acquainted with good laboratory practice principles (GLP). The next stage is electrophoretic separation with DAD detection and making a lab note, as well as interpreting results. At the end of the exercise the students—altogether as one group—write a so-called court (forensic) expert opinion.

The expert opinion consists of a detailed description of the investigation of all received material evidences, the method of examination and analysis, the reasons for the used analytical procedures, an interpretation of the obtained results and answers to all questions asked by the prosecutor. The expert opinion is a part of the final report and is judged by the person who conducts the classes.

Analytical Chemistry Case

A role-play teaching module was also introduced in the Analytical chemistry course for chemistry students (V semester). This is a compulsory course which consists of 30 h of lectures and ten laboratory classes, each 6 h long. Every year during laboratory classes about 130 students, divided into 19 groups (randomly changing after each class) learn how to apply ten of the most important instrumental techniques to identify and/or determine particular compounds. This particular exercise is aimed at showing students how important the optimisation of GC separation is.

The very first step is to organise the student group in the manner of a laboratory team, as all companies do, where everyone has their job and the final product is a combination of results obtained by all members of the team. Typically, a group of seven persons consists of two technicians, two analysts, two IT experts and a team leader, similarly to the roles proposed by Walters: Manager, Chemist, Hardware, Software (Walters 1991). The technicians' major task is to operate the equipment, including initiating the GC system, changing its parameters and running the measurements. To keep the situation real, each group is given original manuals (in Polish) issued by a GC system manufacturer (they were also scanned and available to students on the Internet before classes), so students have to face a situation that is quite common when they are hired by a laboratory just after graduation. This part can be as long as is interesting and the assistance of the rest of the group is welcome. Meanwhile, the analysts are given their task, with detailed instructions on how to process chromatograms. They also have to collaborate with an IT subteam, working with them on constructing an excel spreadsheet to calculate data and draw graphs.

The team leader watches the whole analysis and ensures that there is good communication between all team members. It is his job to supervise IT experts so that they build a spreadsheet that is readable by everyone and contains all the information necessary to write a laboratory report.

Method

The aim of the study was to check how students perceive the newly introduced didactic method, whether—in the opinion of students and lecturers—classes conducted with this method have an impact on teaching results and the selection of further career path, and also to investigate the potential possibilities of further development of this method and extending it to other domains and its optimisation.

The authors were particularly interested in the opinions of doctoral students, who in their career play both roles: students participating in classes (in the course of master's studies), and then academics (when they undergo pedagogical training at doctoral studies).

For many reasons, including ones related to the limited number of potential participants and the necessity of obtaining an extensive overview, the quantitative method of examination was selected.

In 2009, a poll was conducted amongst all ten current students of the forensic chemistry specialisation. The questionnaire was composed of 14 questions, including nine open and five closed ones (five-point Likert scale) with a possibility of adding commentary. A similar questionnaire was filled in by Tempus Program participants from Macedonia, after they have completed their classes. They were persons aged 26–39 years old, who deal professionally with the forensic sciences and related issues: police and public prosecutor's office staff, forensic doctors, employees of the Ministry of the Interior and Administration, and the Center of Public Health. Six interviews were filmed with graduates of the forensic chemistry specialisation, who have completed a master's degree within the last 4 years, including four doctoral students still studying this specialisation, who conduct didactic classes for students (90 h/year). The interview contained questions from the student's questionnaire extended by one related to pedagogic experiences: "Do you use the role-playing method in classes you teach at school/at university? Please substantiate the answer".

Finally, two extended interviews were carried out with persons who planned the classes mentioned above (the authors) and conducted them from the very beginning. The interviews were transcribed after carefully listening and watching tapes. After checking all the obtained statements, categories and similarities could be established.

At the end, an attempt was made to compare the answers of so many different respondents on the subject of the same methodological innovation.

Results

What Did Students Say About Role-Playing?

Responding to the closed questions, all students stated that classes were interesting ("strongly agree", or "agree somewhat"), mobilised independent and creative thinking ("agree somewhat") and supported their desire to start working as forensic expert. The students' attitude to the exercise was a bit more uniformly positive than previously reported by Johnstone et al. (1981). Students agreed with the statement that they would eagerly recommend these classes to other students.

In their responses to the open questions or commentaries, students also judged the lecturers' work. They liked "class leaders' commitment", "well prepared materials", "atmosphere" and the classes themselves. They liked: drawing up opinions and conclusions, interesting topic and props, independent work within small groups, the analysis of the case as a whole, creativity and the possibility of presentation on toxicology (poisoning with arsenic).

What is especially worth emphasising is the fact that students noted the motivational impact on educational output. In the commentaries, one of them wrote that "the case description and material evidence make it easier to understand the necessity of carrying out a concrete analysis and getting acquainted with a certain amount of knowledge", as well as that classes are certainly conducive to creativity, leading, among other things, to the proposal contained in the commentary that "an interesting element would be the defense of the decisions made by the expert".

What Did the Professionals (Tempus Project Participants) Say?

Professionals from Macedonia, who are mostly not involved vocationally in education processes, particularly liked the work in groups, and then the case description, the knowledge range of persons who conducted classes related to the given case and carrying out experiments.

All of them agreed with the statement that they would "very willingly" or "willingly" (fifty–fifty) recommend this type of class at the university in their country.

What Did Alumni/Doctoral Students Say?

Interviews with doctoral students showed that:

Similarly to what was described by Walters (1991), a real (life) context was essential:

> [Because] there was the whole history to that, associated facts and data, it was easier for us to find out and understand the given task (Alumnus N° 2 = A2).

> Classes were interesting because we were dealing with circumstances which really occur at the scene of the event, i.e. there was a bottle or glass of wine that can really be found at the felony scene. The advantage of these classes was that we could deal with a situation which, even if it is not real, perfectly imitates a real situation (A6).

The level of emotional commitment of students was very high, resulting in accuracy and diligence in performance of the task:

> We didn't think of it as a play. Because it was a real event (at least I felt so), we tried to carry out all our tasks so that everything was done well. We identified ourselves (I think

> so) with the role of such experts and were very careful and cautious not to spoil or damage anything (A2).

The way of conducting classes, in the opinion of doctoral students, led to educational effects, including the development of HOCS (Higher Order Cognitive Skills). Johnstone had already described the "cognitive gain" achieved despite the informal nature of discussions (Johnstone et al. 1981).

> By the fact the classes were so interesting, we could better remember what was going on in these classes, and generally we were more involved in this process, and we retained more knowledge and data at that time in our heads (A3).

"Every opportunity to express one's opinion causes one to have to go back in one's mind to already acquired knowledge, to what one has already learned it the course of studies, (…) check what one already knows, at which stage of skills one is at of those analytical chemists who would find their feet in their work position" (A6) and that is why the doctoral students now apply the method of role-playing in classes on similar subject matters, since they believe that it gives good results.

Playing a role has affected or can influence the decision to choose a profession. It draws one's attention to unknown facets of a professional job, e.g. responsibility.

> I think I am not quite sure if I would like to be this forensic expert because it is still a great responsibility, and I do not know if I would be up to it (A2).

The complexity of the classes is an essential factor. "I remember very clearly capillary electrophoresis classes where we investigated a series of documents, and there was a real thrill following the thread to the bitter end. It was such a complete experience, a complete Sherlock Holmes play" (A4).

They would willingly recommend this method to other persons who conduct classes, but perceive it rather in the context of their study years.

> I was conducting classes with students in computer lab, it encompassed learning of the Office Package. It would be very hard to introduce this method there (A2).

> The role-playing method in other university classes…?—but how would it be applied in practice? Because here, in forensic chemistry, we identify ourselves with forensic experts, whereas in other laboratories …? (A5).

What Did the Authors and Lecturers Who Introduced the Method Say?

The role-playing method was applied because of a low rating in evaluation surveys of particular students' classes. It was a remedy to improve educational quality, among other things, by increasing students' motivation to work.

It was important (for the authors of the exercise) to achieve the same or at least similar education effects in chemistry (subject related learning outcomes) as before.

> At classes on instrumental analysis in the third year (juniors) of studies, two parameters were optimised and the class was extremely boring. One operates 20 year old apparatus, it was like an "old worn out car". Nothing needs to be done when operating this apparatus, you just inject a sample from a vial to the appliance, keep seated, and look at it, and the apparatus does nothing but hum. In principle, you cannot do too much with these classes since this measurement consists in just that: you make the injection and do nothing, just observe (Teacher N° 1 = T1).

> Those analyses were assessed by students as the worst ones among other classes (T2).

The authors of the exercise described the gradual introduction of modifications:

> First, an expert's opinion appeared, students were to feel like forensic experts, but only in the final part of drawing up the report (project"X files"[8]). They were given exemplary opinions and on the basis of these judgements, they drew up their own expert opinions instead of a customary lab report. Later on an idea arose that classes conducted in our organisation. This way the "Bank robbery" project[9] Arose—described in this study (T1).

Lecturers were trying to raise educational standards by making classes more attractive:

> Because these are university classes, the report on the investigation carried out also has to be assessed, since not all analytic parameters are contained in the opinion itself; this would be incomprehensible to the court; the answers are: it was, it was as much as, it has poisoned, it did not poison, the signature has been forged, the signature has not been forged, whether in all probability, whether there is a certain probability, something has been added to the document with another tool, at a different time (T1).

From the Statements of the Authors of the Exercise it Transpires that Even They Sometimes Did Not Make Use of the All Possibilities Provided by the Role-Playing Method

> Many times we experienced 'apparatus failure'. We have selected a technique which is interesting but extremely erratic. This is discouraging for students (T1).

In response to the question concerning whether they can learn to use something such as an appliance, he said "no, they just think that our equipment is trashy and poor, that they do not approach this in the way that something is going on, that is

---

[8] Presented at the conference: XLV Annual Meeting of Polish Chemical Society, Kraków (Poland), 9-13 September 2002; Renata Wietecha, Joanna Mania, Michał Woźniakiewicz, "X-files – meeting in a panel of Forensic chemistry".

[9] See Footnote 7.

normal in laboratory work, that no planning of an investigation for exactly 4 h is possible …" (T1).

This attests to how difficult the process of changing the attitudes of educational staff of universities is.

In this method, the way the roles are assigned to students is extremely important, as are the duties connected with these roles. Decisions made by persons who conduct classes and apply strategies/procedures bring about various effects. This also became the subject of separate studies conducted at the Department of Chemistry, UJ (there is insufficient space in this article to describe them).

> Not everybody can be such a leader, some people are just completely not suited. At the very beginning I was checking if I can select the leader by myself, and if he/she is capable of subordinating the group. It does not work like that. It is the group that has to indicate the leader (…) (T1).

> An independent selection of the leader affected the whole work organisation within the group very positively and mobilised each and every "student-*expert*" better (T2).

The question "Does everybody have to develop his/her leadership skills?" was answered as follows:

> not everybody has to attain leadership skills, not everybody has to know it (T1).

> I personally believe that every student should set themselves the task of managing a group of people (having been trained in this). But I realise that not every individual is suited to such a role and I believe that such a person should finally decide by themselves if they feel good in the role of a manager or prefers to be an advisor, executor, etc. (T2).

In the classes discussed here, there was no role rotation within the group. Opinions given by persons conducting classes are not always consistent with the current trends in education and can differ significantly from each other. This once again confirms the conclusion known from other research that prior pedagogical knowledge and attitudes should be seriously taken into account during the training of teaching staff.

The Authors of Classes are Very Cautious When Assessing the Impact of Their Ideas on the Selection of Profession or Workplace; However, They Perceive the Usefulness of Presenting the Real Image of Professions Connected with Chemistry

> I think it is a bit too little, what we are able to show them, some individual talks held with the Institute of Forensic Research (IFR) employees have a great impact. They have lots of events, lots of cases, and can talk about how it was when they appeared before the court and referred their investigation results. I am not under the illusion that my classes would encourage somebody to become a forensic expert, but I hope that my classes can make

> somebody look at the forensic expert as not quite the same individual that can be seen in movies. Motion pictures create a very false image... Forensic experts do not conduct hearings of witnesses, they do not participate in inquiries (T1).

The influence of a prior meeting of students with persons doing various jobs had been previously observed, acknowledged and used in educational practice (Maciejowska 2007).

Classes conducted with the role-playing method and students' involvement in the work connected with this method increase the satisfaction of lecturers with their work: "The person who is conducting the classes feels better if the reception is so lively" (T2).

Preparation of a good description of the event circumstances requires time and constitutes a *conditio sine qua non* for success:

> They ask many questions, true, the questions concern the case. If the leading person thinks up such problems, he/she has to be very well prepared (T2).

About "Drug profiling" classes—"It is common knowledge that this subject is very interesting for every young individual as something forbidden, until it turns out that the substances are not drugs at all (...) when it turned out that the procedure is based, as far as I remember, on the determination of the differences at pH in various buffers—the glamour vanished. The entire class became for the student just another practical that had to be passed—then write a report, and that's it" (A6).

Conclusions

As has been suggested by others—lecturers, scientists and representatives of industry (Johnstone and Percival 1978; Byrne and Johnstone 1987; in the chapter on generic skills—*CEFIC* webpage), the authors assert that space can and should be made to develop generic skills within the framework of chemistry courses. Such generic skills can be taught with the help of interactive learning, e.g. the role-play method. This will help universities to educate open-minded scientist and citizens.

The topics selected for use with the above mentioned methods must be relevant to study curricula and everyday life in society.

When comparing the opinions of students, graduates, doctoral students, professionals and the authors of classes, it can be stated that authors are most careful about judging the impact of their classes on the further vocational careers of graduates. But because work in students' classes under university laboratory conditions does not provide a real-life picture of a chemist's job, each and every initiative which allows us to build bridges between the academic environment and the external world is important.

All agree that classes conducted with the role-playing method are interesting and motivating for both participants and lecturers. It is worth extending them to other subjects taught at the Faculty of Chemistry. The issue of assignment and/or role rotation requires some additional development.

Doctoral students and professionals emphasise the role of the real context of classes and the necessity of good preparation of materials as well as persons conducting classes, in this respect. Otherwise, the (simulated) situation is unrealistic and the probability of strong commitment on the part of students decreases.

The application of the method had a positive influence on the lecturers' and doctoral students' commitment to conducting classes. However, both doctoral students and the authors of classes showed limited confidence in the introduced novelty. This is nothing new and has already been described many times in various contexts (Johnstone and Sharp 1979; Jonasson 2008). Making this attitude more flexible can only be achieved with long-term activities/training.

By introducing modifications in teaching methods gradually, persons other than the authors of the ideas have more time to become accustomed to the changes. This decreases the risk of resistance on the part of both the usually rather conservative academic environment and students. In this way, one can progress from a few laboratory classes or a seminar to a course that is a few dozen hours long, and even to—and this already exists in the Faculty of Chemistry of the Jagiellonian University—a multi-hour interdisciplinary course covering legal proceedings comprehensively. "We intend to create a legal proceedings simulation with law students playing the role of defence counsels, prosecutors and judges on the one hand, and chemistry students playing forensic experts on the other" (T1).

References

Aubusson, P., Fogwill, S., Barr, R., & Perkovic, L. (1997). What happens when students do simulation-role-play in science. *Research in Science Education, 27*(4), 565–579.

Babczonek-Wróbel, D. (1999). Lekcje chemii w formie dramy. *Chemia w szkole, 5*, 282–283.

Belt, S., Overton, T., & Summerfield, S. (2002a). *The Titan Project. A problem solving case study in industrial and analytical chemistry*. Hull: University of Hull.

Belt, S., Overton, T., & Summerfield, S. (2002b). *Tales of the riverbank. A problem solving case study in analytical chemistry and environmental science*. Hull: University of Hull.

Belt, S., Overton, T., & Summerfield, S. (2002c). *The Pale Horse. A problem solving case study in analytical chemistry and forensic science*. Hull: University of Hull.

Belt, S., Overton, T., & Summerfield, S. (2002d). *New Drugs for Old. A problem solving case study in pharmaceutical and analytical chemistry*. Hull: University of Hull.

Belt, S., Overton, T., & Summerfield, S. (2002e). *A Dip in the Dribble. A problem solving case study in analytical,environmental and industrial chemistry. Hull:* University of Hull.

Borowska, B., & Panfil, V. (2001). *Metody aktywizujące w edukacji biologicznej, chemicznej i ekologicznej*. Bydgoszcz: Wydawnictwo TEKST.

Byrne, M. S., & Johnstone, A. H. (1987). Critical thinking and science education. *Studies in Higher Educarion, 12*(3), 325–339.

Deavor, J. P. (1994). Role-playing in the quantitative analysis lab. *Journal of Chemical Education, 71*(11), 980–982.

Hopf, B. (1973). *Bürosimulation in Rahmen kaufmannischen Grundbildung*. Hannover: Schroedel.

Jackson, P. T., & Walters, J. P. (2000). Role-playing in analytical chemistry: The alumni speak. *Journal of Chemical Education, 77*(8), 1019–1024.

Johnstone, A. H., & Percival, F. (1978). More tools for the job. *Chemistry in Britain, 14*, 507.

Johnstone, A. H., & Reid, N. (1981). Towards a model for attitude change. *International (European) Journal Science Education, 3*(2), 205–212.

Johnstone, A. H., & Sharp, D. W. A. (1979). Some innovations in university chemistry teaching. *Studies in Higher Education, 4*(1), 47–54.

Johnstone, A. H., Percival, F., & Reid, N. (1978). Simulations and games in the teaching of chemistry. *Perspectives on Academic Gaming & Simulation, 1&2*, 92–97.

Johnstone, A. H., Percival, F., & Reid, N. (1981). Is knowledge enough? *Studies in Higher Education, 6*(1), 77–84.

Jonasson, J. T. (2008). *Inventing Tomorrow's University. Who is to Take the Lead? An Essay of the Magna Charta Observatory*. Bolonia: University Press.

Karaś, A., & Maciejowska, I. (2007). Na rok przed 2008–10 lat protokołu z Kioto – inscenizacja (wirtualnej) konferencji zorganizowanej przez ONZ. *Nauczanie Przedmiotów Przyrodniczych, 24*, 21–24.

Kimbrough, D. R., Dyckes, D. F., & Mlady, G. W. (1995). Teaching science and public policy through role playing. *Journal of Chemical Education, 72*(4), 295–296.

Klich, J., Maciejowska, I., & Przybylski, W. (2005). Chemistry for students of humanities. Students as co-authors of lessons. *Annals of the Polish Chemical Society, II*, 495–498.

Maciejowska, I. (2004). Jak uczniów klas humanistycznych można zainteresować chemią. *Chemia w Szkole, 3*, 23–27.

Maciejowska, I. (2007). Field trips as a way of working with a class with an extended chemistry and biology syllabus. *Annals of Polish Chemical Society*, CD.

Maciejowska, I., & Odrowąż, E. (2008). Students – be active! outlines of chemistry lessons at secondary school. *Natural Science Education – Gamtamokslinis Ugdymas, 22*(2), 30–37.

Murcia, N. S., Lundquist, E. G., Russo, S. O., & Peters, D. G. (1990). Quincy meets perry mason: An experience in chemistry and law. *Journal of Chemical Education, 67*(7), 608–611.

Nowacki, T. W. (1999). *Aktywizujące metody w kształceniu*. Warszawa: CODN Pracownia Edukacji Zawodowej.

Reid, N. (1976). Simulations, games and case studies. *Education in Chemistry, 13*(3), 82–83.

Reid, N. (1980a). Simulation techniques in secondary education. *Affective Outcomes, Simulation & Games, 11*(1), 107–120.

Reid, N. (1980b). Understanding chemical industry - teaching materials. *Education in Chemistry, 17*(3), 78–80.

Smythe, A. M., & Higgins, D. A. (2008). (Role) playing politics in an environmental chemistry lecture course. *Journal of Chemical Education, 84*(2), 241–244.

Stobiński, J. (1974). *Kółko chemiczne w szkole podstawowej-* Warszawa: WSiP.

Walters, J. P. (1991). Role-playing analytical chemistry laboratories. *Analytical Chemistry*, 977A–985A, 1077A–1087A, 1179A–1191A.

Section III
Curriculum Reform and Teachers

The last section of this book deals with the chemistry curriculum and its changes with the connections with chemistry teacher education. Chemistry curriculums have changed over the decades from traditionally oriented chemistry teaching empathising symbolic and mathematical components of chemical concepts to more context-based enquiry learning oriented teaching supported by the different applications of the informational-communicational technology. We all emphasise that it is important to develop students' scientific/chemical literacy for students when adults are able to use their science knowledge in different in-life situations. For these reasons to implement curriculum innovations teachers should be adequately educated. This means that the teacher should in pre-service/university-level education develop their sense of permanent in-service education, so that they can instantly and effectively apply innovations that appear in the curriculum into their teaching. It is also important that teachers are aware of their possibilities to upgrade their teaching with outside school activities for students. Chemistry presented in museums, industry, agriculture, medicine, science centres, forensic TV shows… can influence students' interest to learn chemistry on a formal level. Teachers should for that matter use the informal ways of showing the importance of chemistry for human society to their advantage. Without proper teacher pre- and in-service education this can be neglected. Many books discuss science/chemistry curriculum reforms and teacher education aspects, but chapters under Section III of this book show some views of authors trying to illustrate novelties and specific aspects of different countries.

Coll, Dahsah, Chairam and Jansoon state in Chap. 16 that Thailand, like many countries worldwide, has engaged in major reforms to the science curriculum. A key focus of these reforms has been a shift towards a learner-centred science curriculum. In this chapter the authors report on a number of studies to show how a learner-centred science curriculum in Thailand places major importance on shifting the mindset of Thai students from a rather less active learning role in a strongly teacher-dominated classroom to a role in which they are active learners of chemistry.

In Chap. 17 Dori, Barak and Carmi present active learning in computerised chemical education environments. It is important to emphasise that informational-communicational technology (ICT) can play an important role in chemical education as a vehicle for learning chemistry actively. The authors present the results of studies that involved the integration of innovative learning environments as part of the chemistry curriculum. Taking into account curriculum reforms dealing also with the applications of the ICT the authors developed such learning environments that should promote more active chemistry learning from the perspective that students actively process information in order to learn in a meaningful way. Studies that were designed to study chemistry students' and teachers' learning outcomes in two technology-enhanced environments that enable active learning: Case-based Computerized Laboratories (CCL) and Computerized Molecular Modelling (CMM). The learning strategies included scientific inquiry, case study analysis, hands-on laboratory activities and project-based learning. Findings indicated that the integration of CLL and/or CMM enhanced students' conceptual understanding and their ability to mentally traverse across the four levels of chemistry understanding—macroscopic, microscopic, symbol. The three studies presented in this chapter along with the tools for evaluating the effect of active learning on chemistry students and teachers will enable teachers, educators and researchers, to investigate students' higher order thinking skills both qualitatively and quantitatively, teachers' professional development, and the process of implementation of a new learning unit in the classroom.

Chapter 18 by Ferk Savec and Wissiak Grm is the concluding chapter of this book. A well-educated teacher is an important factor during a student's chemistry learning, and for that reason teachers' education is presented at the end. It shows the importance of the pre-service chemistry teachers using student-centred learning during their teacher education pre-service programme. During practical pedagogical training, pre-service chemistry teachers—students of the third and fourth years at the Faculty of Education, University of Ljubljana must use the knowledge gained through a number of theoretical and pedagogical subjects in the framework of their tertiary education. This means that they must also use student-centred learning methods, which facilitate the learning of chemistry with understanding. The authors researched students' aspects about student-centred learning methods and their ability to use them effectively during their practical pedagogical training. Their results suggest that students recognise many advantages of using student-centred learning methods in the chemistry classroom but, due to their limited experience, they are unable to use them as effectively as they would like.

Chapter 16
Fostering Active Chemistry Learning in Thailand: Toward a Learner-Centered Student Experiences

Richard K. Coll, Chanyah Dahsah, Sanoe Chairam and Ninna Jansoon

Introduction

Teaching and learning in Thailand, as in many nations worldwide, is supposed to be learner-centered in nature (as a result of substantial curriculum reforms). But how learner-centered are Thai classrooms, and what pedagogies do Thai teachers use that foster active learning in Thai chemistry classes?

When Thailand reformed its curriculum (including the science curriculum) in the 1990s, the Institute for the Promotion of Teaching Science and Technology (IPST) was charged with leading the implementation of the reforms, and much research has been conducted since. In this work, we attempt to answer the two questions posed above, by reporting on three studies conducted by Thai researchers. We begin with an overview of Thai curriculum reforms, and then detail the chemistry topics involved. We provide a brief overview of the literature, and describe what has been reported as difficult about teaching and learning of the topics. Next describe the Thai context, as it relates to the teaching of the topic, and describe the leaner-centered pedagogies developed in the Thai-based research. We conclude by describing the findings and consider what they have to tell us about learner-centered education used to foster active chemistry learning in Thailand. The three chemistry topics we report on are *stoichiometry*, *kinetics,* and *dilution chemistry*.

R. K. Coll (✉)
University of Waikato, Hamilton, New Zealand
e-mail: r.coll@waikato.ac.nz

C. Dahsah
Srinakharinwirot University, Bangkok, Thailand

S. Chairam
Ubon Ratchathani University, Ubon Ratchathani, Thailand

N. Jansoon
Thaksin University, Songkhla, Thailand

I. Devetak and S. A. Glažar (eds.), *Learning with Understanding in the Chemistry Classroom*, DOI: 10.1007/978-94-007-4366-3_16,

Curriculum Reform in Thailand: A Driving Force for the Implementation of a Learner-Centered Science Curriculum

Thailand's educational reform movement arose from the so-called Asian economic crisis of the late 1990s, which highlighted an urgent need to develop Thai peoples' ability to keep up with the rapid change characterized by mass globalization. The 1999 National Educational Act was developed under the provision of the 1997 Thai Constitution, which emphasized the critical importance of change to the education system. The key issues of education reform identified in the 1999 National Education Act are: expanding basic education from 9 to 12 years of schooling, and extending compulsory education from 6 to 9 years of schooling; providing education to meet learners' basic learning needs, upgrading their skills, and encouraging their self-development; implementing internal and external quality assurance systems in schools and education institutions; reforming administration and management of education to encourage full participation of local educational authorities and local community; encouraging private sector participation in educational provision; reforming pedagogy by emphasizing learner-centered activities and establishing lifelong learning; reforming the curriculum, allowing for the contribution and participation of stakeholders, to meet new challenges and demands of different groups of learners with an emphasis on mathematics, science, and technology in parallel with the promotion of pride in national identity and cultural heritage; and reforming resource allocation at the national level on the basis of equity and encouraging local educational authorities and communities to mobilize their resources for education (International Bureau of Education 2004). As of May 2004, basic education was extended from 12 to 14 years including 2 years pre-primary schooling (Ministry of Education [MoE] 2004).

Thai School Structure and Curriculum

In the Thai education system, the school structure of the basic education as at 2008 is divided into five levels: 2 years of pre-primary (K1–K2); 3 years of lower primary (G1–G3); 3 years of upper primary (G4–G6); 3 years of lower secondary (G7–G9); and, 3 years of upper secondary schooling (G10–G12). Schooling at Grades 1 to 9 is now compulsory (Office of the Education Council [OEC] 2006). There are two curriculum documents that detail the curriculum for basic education: the 2003 Pre-primary Curriculum, and the 2001 Basic Education Curriculum (Office of the Education Council [OEC] 2006). The 2001 Basic Education Curriculum specifies learning is to occur over eight subjects: *Thai Language*; *Mathematics*; *Science*; *Social Science, Religion and Culture*; *Health and Physical Education*; *Art*; *Career and Technology*; and *Foreign Languages*. English is a part of the core curriculum for foreign languages, and is required at all levels

(MoE 2009). The 2003 curriculum for pre-primary education focuses on preparing children in terms of their physical, intellectual, emotional/mental and social readiness (Office of the Education Council [OEC] 2006). In a revision in 2008, the curriculum was called the "core curriculum 2551" or B.E. 2551 (Ministry of Education (MoE) 2009). The content and standards remain the same, with lower grades combined into G1–G9, that is, one grade level.

These national curricula act as a guide for schools, who are expected to construct their own school–based curriculum. The school-based curriculum is described as a 70:30 model, in which 70 % of the content is derived from the national curriculum, and 30 % from school contexts. Schools are expected to cooperate with individuals, families, community organization, local administration organizations, private person and organizations, professional bodies, religious institutions, enterprises, and other social institutions in order to strengthen their communities by encouraging learning in the communities themselves (National Education Act 1999, Section 29).

Learning Process

Learning reform lay at the heart of the Thai education reforms as detailed in the National Educational Act. The reform of the learning process, as indicated in Section 22 of the 1999 National Education Act and 2002 Amendments, is based on the principle that all learners are capable of learning and self-development, and learners are regarded as being the central focus of schooling. The teaching-learning process thus aims to enable learners to develop at their own pace, and to maximize their potential (Office of the National Education Commission [ONEC] 2003). The Act proposes that the learning process should address the following: provide substance and arrange activities in line with the learners' interests and aptitudes, and bearing in mind individual differences; provide training in thinking processes, management, how to face various situations, and application of knowledge for solving problems; organize activities for learners to draw from authentic experience to enable learners to think critically and apply this to their real lives; achieve in all subjects, a balanced integration of subject matter, integrity, values, and desirable attributes; enable instructors to create the instructional media and facilities for learners to learn, and enable them to benefit from research as part of the learning process; and enable individuals to learn at all times and in all places. This is taken in Thailand to mean that the learning process should be learner-centered in nature.

Teaching Stoichiometry in Thailand

Stoichiometry is a key topic in chemistry learning, as it includes the concepts that are essential for understanding both the macroscopic and microscopic conceptualizations of chemical reactions as well as solving many types of chemical problems. It is used to determine how much of each reactant is needed to produce a given quantity of product in a reaction, to calculate formulae of compounds from decomposition of products, to find volumes of gases under given conditions, to label concentrations of solutions, and much more (Chang 2003). Kolb (1978, p. 728) states that "there is probably no concept in the entire first year chemistry course more important for students to understand than the amount of substance (Mole) and one of main reasons the amount of substance (Mole) concept is so essential in the study of chemistry is stoichiometry".

Many science education studies suggest that stoichiometry is an important fundamental concept for the understanding of more complex chemical concepts (e.g., Camacho and Good 1989; Gabel and Bunce 1994; Schmidt 1991) and some authors report that students have misunderstandings of some important concepts of chemistry such as chemical equilibrium (Bergquist and Heikkinen 1990; Huddle and Pillay 1996; Kousathana and Tsaparlis 2002), and acid–base reactions (Carr 1984; Mettes et al. 1980) because they lack an understanding of stoichiometry.

Students' Difficulties in Learning Stoichiometry

Research suggests that stoichiometry is a difficult topic for many high-school students (Cain 1986; Dominic 1996). Three reasons have been suggested for this: students do not understand stoichiometry concepts; they often lack numerical problem solving skills or mathematical reasoning (BouJaoude and Barakat 2000; Schmidt 1994; Schmidt and Jigneus 2003); and they cannot transfer between the macroscopic, microscopic, and symbolic levels of representation (e.g., Gabel et al. 1987; Dori and Hameiri 1996, 1998; Robinson 2003).

Common student's alternative conceptions for concepts of stoichiometry reported in the literature are for the amount of substance (Mole) concept, balancing equations, limiting reagent, and concentrations (e.g., Camacho and Good 1989; Dominic 1996; Furio et al. 2002; Gabel and Sherwood 1984; Krishnan and Howe 1994). Whilst some research that investigated the relationship between problem solving and conceptual knowledge, suggests that students can produce the correct answer in chemistry problems without understanding the chemistry concepts (Gabel and Bunce 1994; Lythcott 1990; Nakhleh 1993; Nakhleh and Mitchell 1993; Sawrey 1990), other work suggests that students who hold misunderstandings of the concepts fail to solve stoichiometry problems (BouJaoude and Barakat 2000). Niaz (1995) provides some insights to this apparent contradiction. It seems that students can use formulae to help them solve simple problems, but they are less successful in solving more complex problems, that require conceptual understanding.

Similar to what is reported in the international science education literature, many Thai students also hold alternative conceptions for stoichiometry and have difficulty in solving numerical problems for stoichiometry (Dahsah and Coll 2007, 2008). The main alternative conceptions reported are: (1) one mole was the same as one molecule; (2) one mole of all substances contained 22.4 dm^3 at STP (they did not consider the different phases of substances); (3) the solution that contained the greatest amount of solute was the most concentrated solution (they did not consider the volume of solution); (4) the number of molecules ratio was the same as the mass ratio; and (5) the limiting reagent was the reactant present in the least amount (Dahsah and Coll 2008). For relations between conceptual knowledge and problem-solving skills, the studies suggested that students' problem solving skills of these Thai students depend on their conceptual understanding. Those students who did not understand the related-concepts in the questions could not solve numerical problems. Likewise, students who held alternative conceptions could not provide the full correct answers (Dahsah and Coll 2007).

Teaching Stoichiometry in Thai High Schools

The key concepts for stoichiometry topic taught in Thai high schools are detailed in the IPST textbook (Institution for Promoting Science and Technology [IPST] 2003a). The study of stoichiometry begins with the study of atomic mass, molecular mass, the amount of substance (Mole), chemical equations, concentrations, colligative properties, and quantity relationships of a chemical reaction (conservation of mass, Avogadro's law, limiting reagent, percent yield). These topics are taught to high-school students for a total of about 35 h in the second semester at Grade 10 or Grade 11 (Institution for Promoting Science and Technology [IPST] 2003b) depending on the school

Development of Stoichiometry Learning Units to Foster Active Learning of Stoichiometry

Stoichiometry Learning Units (SLUs) were developed by the second author in order to enhance Thai high-school students' understanding and problem-solving skills for stoichiometry, the content of which came from the Thai science curriculum (Institute for the Promotion of Teaching Science and Technology [IPST] 2003a). The learning process used in the SLUs is based on the guiding principles stipulated in Thailand National Educational Act B.E. 2542 (1999) (Office of National Education Commission [ONEC] 2003), in that it is learner centered, constructivist-based learning, and involves a conceptual change approach.

There were six essential features used as a guideline in development of SLUs: (1) All students should be encouraged to develop their competence in science to meet their interests and aptitudes and to achieve their potential; (2) Prior knowledge is important for students in the learning of new knowledge, bearing in mind individual differences; (3) Social interaction can facilitate successful science learning, hence, communicating of ideas and group work should be a feature in the learning process;

(4) Teachers should act as facilitators who encourage students to fully develop their potential, and students should be actively involved in 'hands-on' and 'minds-on' learning activities; hence, the students' role is that of an "active learner" and teachers' role is that of a "facilitator" of learning; (5) Productive science learning can be promoted by multiple, active and challenging learning activities with a variety of instructional materials. The activities can be used to create cognitive conflicts in students who hold conception different from the scientific ones, provided in a supportive environment to help students understand science concepts and ways of representing the concept using multiple modes of representations; and (6) Learning outcomes should be assessed using a variety of methods, such as observation on behavior, learning procedures, activities participation, students' journal, project work or portfolios, reports, as well as concept tests.

The learning process used in each unit was developed based on the guidelines above, and in particular the conceptual change teaching approach proposed by Stephans (1994). According to this model there are five steps to learning: express ideas, share ideas, challenge ideas, accommodate ideas, and apply ideas: (1) *Express Ideas*: students are 'activated' via a series of activities in which they formulate an outcome or prediction about a concept, to show their existing ideas; (2) *Share Ideas*: students are again activated to discuss and share their stated prediction or outcome—first with a peer—before sharing this with the whole class; (3) *Challenge Ideas*: students are activated through the activities or the experiments to test their predictions or observations and to determine the validity of their predictions; (4) *Accommodate Ideas*: students are activated to 'accommodate' the concept by resolving the conflicts between their existing ideas and their observations and/or by relating their ideas to an appropriate context, and (5) *Apply Ideas*: students are activated to extend and apply the concept they have learned to solve meaningful problems, and to use it in other situations.

The SLUs consisted of 16 learning units taught across 35 h. Details of the development of the SLUs are provided in Dahsah (2007) and Dahsah et al. (2009) and a sample is provided as an appendix.

Implementation of the Stoichiometry Learning Units

The SLUs were implemented by three volunteer teachers (all females) very experienced in teaching chemistry at high-school level (average of 27 years experience). The teachers were from different schools: one in Bangkok, the nation's capital city, and two in Nontaburi province, a suburban area close to Bangkok. The three schools educate from Grades 7 to 12. These were large schools, with school rolls of more than 3,000, about 40–50 students in a given class, and about 10 classes in each grade. Each teacher implemented the SLUs in her Grade 10 chemistry classroom over the second semester of the academic year. There were 50, 48, and 45 students in each intact class (143 students in total).

The SLUs were evaluated by using stoichiometry concept and problem-solving questionnaires, (see Dahsah and Coll 2007, 2008), classroom observations, semi-structured and informal interview, students' worksheet, and students' journal (see appendix for examples). The outcomes were reported in terms of students' understanding and problem-solving ability for stoichiometry, students' and teachers' opinion for the used of SLUs.

After learning by SLUs, more students held sound understanding in all concepts of stoichiometry. In particular, more than 70 % of the students held sound understanding for the concepts of: molecular mass, number of entities in one mole, molar unit, conservation of mass, and limiting reagent, and especially for the concepts of conservation of mass for which 86 % of the students held sound understanding. More than 60 % of the students held sound understanding for the concepts of concentration, and molar mass. However, some concepts less than 30 % of the students did not hold sound understanding. These are atomic mass, boiling point elevation, and chemical equation.

The problem-solving ability of the students is not as good as might be hoped. Whilst students were better at working with quantity relationships in chemical reaction, limiting reagent, and percent yield, they were not much better problem-solving. In particular, the results suggested that students' problem-solving ability seemed to be heavily related to their conceptual understanding. Students, who appeared to hold a sound understanding of the underlying stoichiometric concepts related to the question, were subsequently able to solve numerical problems and could give the correct answer. Some students who held partial understanding with alternative conceptions, or straight alternative conceptions, were able to construct an appropriate answer to numerical problems, but were unable to give comprehensive answers.

The results from observations, interviews, and students' journal suggest that most of the students actively participated in learning activities, and few students did not. The students enjoyed the activities especially doing the experiment and discussing their ideas both in group and in a whole-class setting. As shown in the students' journal (e.g., "SLUs are good units because students could learn through group activity, study and do experiments by themselves, try out their thinking, and make a conclusion. These could help students to understand the concepts more than learning by memorizing").

Using the SLUs, the constructivist-derived and conceptual change model-based pedagogies sought were very different from the norm in Thai classrooms and this is reflected in the teachers' concerns about the new approach. Despite much rhetoric and many IPST publications and workshops about learner-cantered education, few teachers actually understand the meaning of this term (Dahsah and Coll 2008). In addition, the learning in SLUs required more time for the students to work in constructing their own knowledge, compared with lectures by the teacher, especially for low achievement students. Thus, in the beginning, the teachers faced some problems in time management. As time went by the teachers became more accustomed to, and more enthusiastic about, these new pedagogies. The teachers were happy with the teaching in SLUs and they felt that the SLUs did help their students understand stoichiometry concepts, especially when doing experiments and group discussion.

As well as enhancing student's learning, the teachers also felt that they learned new teaching techniques through the use of the SLUs; such as the use of analogy, using everyday life examples to describe abstract concepts (e.g., using three kinds of beans to teach the concept of average atomic mass), and exploration techniques (e.g., demonstrations, questioning, card games) that helped them explore students' prior-knowledge. In particular, the teachers felt they had learned more about student-centered learning through the use of the SLUs. On a positive note, all of the teachers felt that the teaching using SLUs was effective and said they intended to use the SLUs in their teaching next year, and as a guide to improving their teaching for other topics.

Conclusions

The results in Thailand are similar to other reported work (see BouJaoude and Barakat 2000; Chiu 2001; Coşto 2007; Lin et al. 1996; Tinger and Good 1990). In addition, the results suggested that if the students achieved the first two criteria of problem-solving skills which are: understand the question, and select the appropriate information or concepts to use in solving the questions; then they always obtained the correct answer. This suggests the mathematical skills alone did not pose a problem for these students because all students who held a sound understanding for related concepts could get the correct answer. The learning process in the SLUs was based on a constructivist-based teaching and conceptual change approach, and in this students' prior knowledge and social interactions are important. The step of learning in the SLU included five steps: express ideas, share ideas; challenge ideas, accommodate ideas, and apply ideas. The hands-on and minds-on activities (i.e., demonstration, experiment, analogy—for example the bean analogy for teaching average atomic mass by asking students explore the average mass of black bean, soy bean, and green bean. Each type of bean represents different isotope of element and isotope demonstration as shown in the lesson plan—appendix) were used to allow students to express their ideas, to foster their conceptual conflict and to encourage students to accommodate new ideas. Questioning by teachers also used to activate student thinking in group and class discussions which aimed at enhancing student's learning and helps students see and resolve their conflicts. The 'apply' stage helped students become fruitful in the concepts, meaning that they could better apply their knowledge to solve other related problems. That is to say, all five steps are important to help students confront their alternative conceptions and reconstruct their conceptions in a scientific way. In particular, successful learning might not occur if some steps are omitted. Students' learning through these steps enhanced students understanding of the concepts at the macroscopic, microscopic, and symbolic level. The findings suggest that challenging activities such as experiments, demonstrations, and analogies work well when used to create cognitive conflict in students. Also, group discussion and teacher questioning are effective in helping students accommodate and reconstruct their ideas in a scientific way. Effective questions guide students to think step by step and give students time to think.

However, learning by SLUs required more time for the students to work in constructing their own knowledge, compared with lectures, especially for low achievement students. In addition, learning by SLUs needed active students, because the activities need students to do things and subsequently discuss their ideas both in group and in a whole-class setting to construct their own concepts. The teachers, after some initial concerns, were happy teaching using the SLUs and felt that the SLUs helped their students understand stoichiometry concepts better—as well as enhancing student problem—solving abilities. The teachers felt that they had learned new teaching techniques through the use of the SLUs, and learned more about learner-centered teaching. In addition, the teachers found that conceptual change teaching helped them understand their students better in terms of prior knowledge, and alternative conceptions.

Teaching Chemical Kinetics: Development of a Problem-Based Learning Approach

Chemical kinetics refers to the time dependence of chemical reactions. Interestingly, almost all 'everyday processes' involve kinetics in some way. For example, 'acid rain' is a problem in many large cities worldwide. Acid rain is caused by emission of sulfur dioxide and similar species from motor vehicle exhausts, and in recent years this has become a serious environmental problem in many large cities. Breathing acidic fumes can cause health problems, but probably the most visible evidence for acid rain is in the damage it does to buildings. It can severely damage buildings in a city, and it is obviously of interest to scientists and citizens alike to know how quickly such destructive reactions might take: this is the sort of information we get from the study of kinetics.

Like stoichiomtery, kinetics is often difficult for many students to comprehend. Despite its ubiquitous nature, Justi (2003) comment that there is a paucity of research about chemical kinetics teaching and learning at both the secondary and higher educational levels. In general, the teaching and learning of physical chemistry including chemical kinetics is teacher-dominated in approach at both the secondary school and tertiary levels. Many science teachers typically emphasize the qualitative aspects to aid understanding of the influence of variables such as temperature, concentration, and surface area on the rate of a chemical reaction. There are several reports in the literature describing experiments or equipment we might use to help the learning of kinetics. For example, Parkash and Kumar (1999) reported on experiments about chemical kinetics involving gaseous carbon dioxide formed from the reaction between ethanoic acid and sodium hydrogen carbonate at different time intervals. Choi and Wong (2004) investigated student understanding of experiments demonstrating first-order kinetics involving the application of a datalogger (a computer interfaced to one or more sensors). However, such experiments are expensive to set up, so are not always suitable for educational contexts for which there is limited access to sophisticated electronic instruments.

Much learning of kinetics involves the use of experiments or other laboratory/ practical activities. The laboratory has been given a central and distinctive role in science teaching including chemistry, and it is also claimed that rich learning benefits accrue when using practical experiments as part of pedagogy (Lazarowitz and Tamir 1994) especially for topics like kinetics. Laboratory activities are seen as a means of allowing students to pursue learning, having a variety of multi-sensory experiences, engaging them with experiences of concepts, and at the same time developing their practical science abilities and skills. There are many benefits of laboratory making connection between the laboratory work and lecture. Students can learn the basic practical skills of laboratory work. Practical work can be used as a way of demonstrations or engagement to students' concepts deal with in the lectures. If the laboratory activities are suitably challenging, then students are also provided with opportunities to do science as real scientists do (Hegarty-Hazel 1990; Wellington 1998; Woolnough 1991).

Despite their claimed potential, science practical activities including chemistry at both the school and higher educational levels generally follow a cookbook style in which students are presented with aims and detailed steps for carrying out the experiments. In which case students may or may not learn something about the way scientists do things in the circumstances. It is argued in the literature that this teaching approach is not only an ineffective means of developing students' understanding of science concepts, but also presents a misleading way of how scientists develop scientific knowledge and skills (Lazarowitz and Tamir 1994). For practical work at both the school and high education levels to be real value, the literature suggests it needs to involve an inquiry-based approach to chemistry learning (Nakhleh et al. 2002). If engaged in inquiry-based learning in the laboratory students can come to understand the nature of scientific inquiry by engaging in inquiry themselves. However, students' knowledge about scientific inquiry and the nature of science does not occur automatically once they are placed in a laboratory. Students do not develop an understanding simply through experiment inquiry, instead they need to learn from their experiences in the laboratory under expert guidance (Hume and Coll 2008, 2009). Hence, chemistry teachers need to provide students with experiences in methods of scientific inquiry and reasoning, and in the application of scientific knowledge related to everyday life (National Research Council (NRC) 2000).

Teaching Kinetics in Thailand

Because of its importance for the understanding of various chemical processes, chemical kinetics is of introductory chemistry courses at both secondary school (Grade 9) and higher educational levels in Thailand (Grade 11). At the university level, as might be expected, the kinetics concepts taught are more complex than at the secondary school level.

Most chemistry practical classes in Thai universities are traditional in approach, meaning they are teacher-dominated and that practical classes follow a cookbook style, or consist of teacher demonstrations of practical work. In an effort to change from a teacher-centered approach to learning, there is currently discussion in Thailand as how to change the teaching and learning approaches and strategies in order to facilitate the acquisition of the new types of knowledge. So, the national curriculum, which is now regarded as the educational standard, states that at any level of education, teaching–learning activities must emphasize 'learning to think, to do and to solve problems' (Pravalpruk 1999). IPST has incorporated the inquiry approach in science curricula, and emphasizes an inquiry-based teaching and learning process (Ministry of Education (MoE) 1996).

Development of an Inquiry-Based Approach to Teaching Kinetics

Active learning involves students doing practical experiment in the laboratory, rather than relying purely on classroom teaching, and as noted above the literature suggests inquiry-based learning may help student so understand complex chemistry concepts, including chemical kinetics. Therefore, the recent work was done to improve the teaching of kinetics via inquiry-based learning in the laboratory (Chairam et al. 2009). A key feature of this project was that active teaching and learning must involve students doing practical or laboratory work themselves, rather than just watching the teacher conduct a demonstration in the laboratory or classroom. All students were asked to solve problems given by designing the experimental set up, conducting the experiments in which they try to examine the aspects of chemical kinetics, exploring what happened when the reaction occurs, explaining what happened when changing which affect the rate of reaction for studying chemical kinetics, calculating the experiment data which mathematically integrated to give the rate of reaction, analyzing the graphs and mathematic calculations, and finally discussing in groups about the results in which they might gain confidence in a deep understanding of knowledge.

The experiment developed as an intervention focuses on the kinetics of acid–base reactions (see Appendix B). Acids such as hydrochloric acid (HCl) react quickly with calcium carbonate to produce a salt and water, and releases gaseous carbon dioxide. Other acids such as the acids present in vinegar also react with carbonates.

The reaction is: $CaCO_3(s) + 2HCl(aq) \rightarrow CaCl_2(aq) + H_2O(l) + CO_2(g)$

In the reaction above, how the acid and carbonate react depends on a number of factors, including the concentration of the acid, the particle size of the carbonate, and the temperature of a reaction. The chemical equation can be applied to determine the rate of a reaction by plotting the carbon dioxide generated over time.

The experiment was devised to be more open in nature, so that the students would do things in a way more like scientists. The students were required to conduct an experiment in which they try to examine the aspects of chemical kinetics for the acid–base reaction between eggshells (mostly calcium carbonate) and acids (hydrochloric acid and vinegar). That is, the students had to design the experiment procedure themselves to gain an understanding of the process of scientific inquiry. The Prediction-Observation-Explanation (POE) technique—a learner-centered strategy that arouses students' curiosity (White and Gunstone 1992) was incorporated into the experiment (see Appendix). The questions focused on four POE activities looking at the influence of variables on chemical kinetics: surface area, temperature, concentration and type of acid on the rate of a chemical reaction. In a whole-class setting, the students were asked to predict the results of some events and justify the reasons used to support their prediction. Students then were asked to describe what they observed when a reaction occurs while doing the experiment. Lastly, they were required to explain any conflict between what they have predicted and observed. The POE task is also measure of students' ability to apply knowledge and it also is the powerful instrument in which students can use to interpret real events. The POE activities used helped develop a student-centered approach in this laboratory class in combination with other techniques, such as small-group learning (Johnson and Johnson 2005) and negotiation involving argumentation (Pinnell 1984).

Student's Learning of Kinetics Via Inquiry-Based Learning

The students were required to conduct an experiment in which they examined aspects of reaction kinetics for the reaction between egg shells (mostly calcium carbonate) and acids (hydrochloric acid and vinegar). The students in groups of 4–5 students had to form hypotheses that could be tested by collecting data, conduct scientific experiments that control all but one variable, predict the outcome of the results, collect and record data accurately, and finally explain and interpret their data. Examination of the findings suggested that many of the students were able to provide good experimental design which could test their hypotheses. The experimental procedure was clear and simple, and the students identified three groups of variables (i.e., independent variable, dependent-variable, and controlled variable) for investigation in the experiment. An example illustrates their approach:

Problem—Influence of surface of egg shells to the rate of a reaction

Hypothesis—The different surface of egg shells gives a different rate of a reaction

Independent variable—Influence of surface of egg shells

Dependent variable—The rate of a reaction

Controlled variable—Size of Erlenmeyer flask, type of acid used, concentration of acid, reaction temperature, source of egg shells, laboratory environment

The sample preparation in this experiment involved a sample of egg shells and is seen here an essential stage, since it is a key step to successful completion of the experiment. In this experiment, there is no single method of sample preparation for the solid reactant (egg shells), meaning students have to decide how to prepare the solid sample themselves. The students can learn how to conduct some aspects of this experiment, but they already understood the preparation of solid sample (and acids) before doing the experiment based on previous work. From analysis of students' laboratory documents, it seems they felt the particle size of egg shells should be consistent, and thus they ground the shells to obtain a fine powder with uniform (homogeneous) particle size:

> Sample preparation of egg shells—In this part, the particle sizes of the solid reactant should be made at least into three sizes: big, medium and small (like sands) size. The white layer which covers on the egg shells should be removed before grinding.

The students were required to design their own experimental procedure after they decided the how to investigate the problem:

> Weigh egg shells 0.1 g and then pour them into the flask… Pour HCl (aq) 4.0 mL into the vial and then place it into the Erlenmeyer flask (be careful not do not mix HCl with egg shells)…fill the water into the 25.0 mL burette…Connect the burette and the flask with the U-tube and the rubber stopper…Shake gently the flask, mixing HCl and egg shells together …Observe the volume of CO_2*in the burette and record the results at time…Change the type of acid used from HCl to vinegar and then repeat the experimental procedure.*

The students are introduced to an understanding of the rates of reaction for different examples, and investigating the influence of variables (e.g., surface of solid reactant, concentration of acid, temperature of a reaction, and type of acid) on the rate of a reaction. In general, the students can carry out the investigation of the influence of surface of egg shells for two different particle sizes. Students in each group have to vary the particle size of egg shells themselves, and to explain the influence of surface area of the solid reactant and the rate of reaction, the students commonly reasoned that the change of the rate of a reaction is due to changes in physical dimensions of the solid reactant. After completing the experiment, in whole-class discussion, the experimental data from investigations of kinetics are analyzed by the class to compare the rates of reaction by plotting the relationships between the amounts of carbon dioxide over time using standard computer software (i.e., Solver Parameters in Microsoft Excel):

> The rate of a reaction is dependent on the surface of the solid reactant (egg shells). If the size of egg shells is big, the rate of a reaction is slow. On the other hand, if the size of egg shells is small, the rate of a reaction is fast.

Some students in their groups drew upon analogy of a cube to explain the increase of the surface area of solid reactants. This was perhaps because it was simple for students to understand in their mind as is important feature of the student-generated analogy (Coll et al. 2005) (Fig. 16.1).

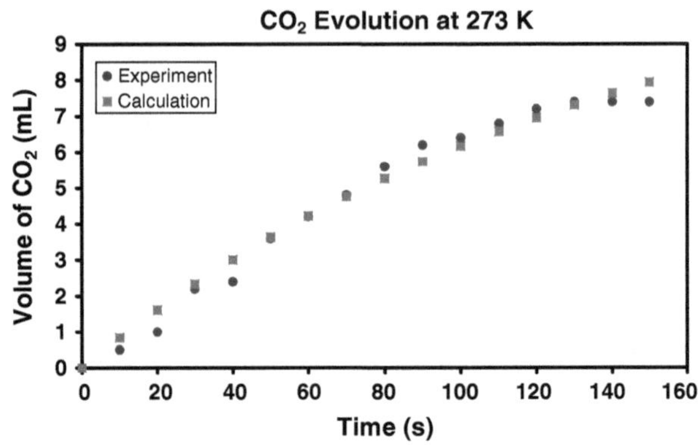

Fig. 16.1 Using solver parameters in Microsoft Excel for plotting data from the experiment at 273 K compared with computer calculation

> If we put four cubes together, the surface area is only 16 cm^2. However, if we separate all four cubes away from each other, the surface area is 24 cm^2.

To investigate the influence of temperature, the students carried out the reaction at least two different temperatures. They start doing a reaction at low temperature and then moved on to higher temperatures, although some start at higher temperatures and moved to lower temperatures. In general, the students who changed the reaction conditions moved from room temperature to a higher temperature, reasoning pragmatically that it is easier to carry out such a change to the temperature using water bath.

To explain the influence of different temperatures on the rate of a reaction, students observed that "the rate of a reaction increases at higher temperatures".

> When increasing the temperature for a reaction, the kinetic energy of reactant molecules increases. So, molecules move faster and more collision. The rate of a reaction increases and then the reaction occurs quickly.

Students also were able to perform the calculations for the rate of a reaction correctly and typically explained their findings in the following way: "At the beginning, the slope of graphs is very sharp, because the rate of a reaction is large. As the reaction progresses, the reaction becomes slower. The rate of a reaction decreases, eventually to zero when the reaction is completed" (see example) (Fig. 16.2).

Conclusions

The students were able to explain changes to the rate of a chemical reaction based on kinetic theory, and drew upon energy and particle theory to explain changes in rates of reaction. They understood how to conduct experiments, and the notion of investigating variables by changing each separately, while maintaining the others constant. The use of computers, and, as for example here, the experimental data

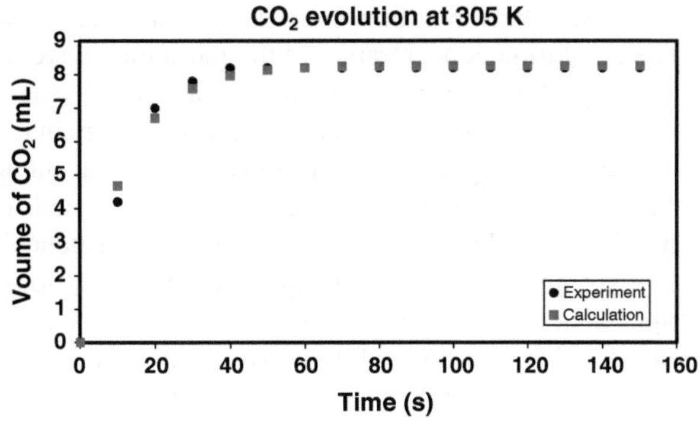

Fig. 16.2 Students' laboratory data—the volume of gaseous carbon dioxide formed at the temperature of 305 K

from the experiments was easily analyzed using Solver Parameters in the tool function of Microsoft Excel, which enabled students to plot the relationship between the production of carbon dioxide and time. The findings reported here suggest science teachers may wish to consider teaching chemical kinetics using simple chemical reactions and materials related to everyday processes, and based on inquiry learning. Science teachers may wish to modify such experiments to suit their own needs and circumstances using, say, other variables, such as catalysts or a size of the reaction flask, for students to investigate how the rate of a reaction differs with other variables.

Teaching Dilution Chemistry Via Interactive Group Learning in the Laboratory

Dilution is an important part of practical introductory chemistry. To do almost any practical work in the laboratory students need to be able to make diluted solutions, or to prepare solutions, including standard solutions (Dunnivant et al. 2002; McElroy 1996; Wang 2000). Understanding dilution requires students to understand a number of related concepts: concentration, solvent, solute, solution, solubility, and the the amount of substance (Mole). Additionally, other related topics like volume, and molecules are implicit in the understanding of dilution—along with the use of chemical equations.

Student Difficulties Learning Dilution Concepts

Dilution concepts are somewhat abstract and difficult to learn, leading to many alternative conceptions such as: confusion about the relationship between the amount of solute and volume of solution (Dahsah and Coll 2008; Devetak et al. 2009; Jansoon et al. 2009), how to prepare a diluted solution (Çalik 2005), the relationship between the solvent and solute (Çalik and Ayas 2005a; Devetak et al. 2009), solution concentration at the particulate level (Devetak et al. 2009), and the meaning of homogenous solutions (Çalik and Ayas 2005b).

The literature suggests that when teaching about dilution, teachers often provide students with algorithmic formulas for solving numerical problems, such as the ubiquitous $C_1V_1 = C_2V_2$. Students often use such equations to solve numerical chemistry problems, but they use these as 'crutches' without understanding their meaning and when to use or not use such equations. So, in order to acquire the 'right' answer, they simply memorize equations and 'plug in' the numbers, rather than develop a solution for the problem using fundamental concepts (Beall and Prescott 1994; Bunce et al. 1991; Lythcott 1990; Robinson 2003). Bunce et al. (1991) suggest that students are able to solve symbolic level problems 'successfully', without applying the chemistry knowledge correctly. However, Dahsah and Coll (2007) point out that when students cannot solve problems, it is often because they misunderstand the related underlying concepts (e.g., solvent, solute, solution, concentration, solubility, and the the amount of substance (Mole), and suggest that teachers need to be sure students understand such concepts before teaching topics like dilution or stoichiometry.

Because dilution topics are related to volume, molecules and use equations, students need to understand concepts from the macroscopic, microscopic and symbolic levels, and to be able to integrate knowledge across these levels (Heyworth 1999; Johnstone 1991; Larkin 1983). To illustrate difficulties in this, in laboratory classes, students may visualize the chemical phenomena at the macroscopic level, and subsequently be required to explain their observations using the submicroscopic and symbolic levels (Gabel et al. 1992). Research suggest that students can be moved from instrumental level to the relational level (i.e., from the macroscopic to the submicroscopic and symbolic levels) if they have the ability to understand and to explain chemistry which is abstract and complex (Treagust et al. 2003).

Teaching Dilution and Related Concepts Using Mental Models, Analogies, and Practical Work

In order to help students understand dilution and related concepts, teachers may make use of analogy and models or colourful demonstrations: diluting cell suspensions, using coloured solutions like orange juice to better show dilution, and the squares-and-points model to visualize the dilution process (Demeo 1996; Heyworth 1999; McElroy 1996; Raviolo 2004). It seems that students are able to better describe their own mental models (i.e., they were able to describe their mental images of the of microscopics entities after using the squares-and-points model when the teacher teaches solubility topics using particles models, with dissolution explained in terms of the distribution of particles (Kabapinar, Leach, and Scott Kabapinar et al. 2004). Likewise, as a consequence of the use of analogy, students are able to draw their own mental models and better use these to explain the process of dissolving, and the difference between dissolving and melting (which is often confused by students). Using models and analogies means students also are able to discuss mass conservation and to solve problems, to explain, and to predict when they constructed their own mental models—consistent with work on other mental model teaching generally (see e.g., Coll and Treagust 2001, 2002, 2003; Eilam 2004; Glynn 1997; Johnson-Laird 1983).

Coll et al. (2005) suggest that student's achievement, in terms of their understanding of mental models, is enhanced when students are given the opportunity to reflect, discuss their mental models in groups and in laboratory situations. Students generally enjoy working in the chemistry laboratory, and practical work can enhance students' interest in science (Nakhleh et al. 2002). Cooperative learning in chemistry laboratory classes within groups, means students can solve more difficult problems and in addition, enjoyed their classes more (Fleming 1995). In particular, special models for teaching using in introductory chemistry laboratory classes—like the Jigsaw method—address a lack of student preparation and poor understanding of chemistry concepts (e.g., for acid/base titration—see Smith et al. 1991). The Jigsaw method is unique in that students prepare a small part of an experiment and share the data and the results from their group with others. Jigsaw-based methods work well for abstract topics like atomic structure (Eilks 2005) and are particularly helpful in shifting students from the macroscopic to microscopic levels of representation (Johnson 1990).

Teaching Dilution Chemistry in Thailand Using the Jigsaw IV Method

A hands-on activity was designed to teach students about dilution and related concepts, drawing upon research about learning, group learning, the value of practical work—all combined to develop an active-learning intervention based on the Jigsaw approach. The hands-on activity consisted of an experiment designed to determine total phenolic compound in green tea beverage samples based on the Folin-Ciocalteu method (see Appendix C). This topic was chosen because it involved the use of green tea—a well-known beverage in Thailand, and thus the intervention sought to link abstract concepts in dilution chemistry to a common everyday material students would be well familiar with.

In this work we wanted to do two things. First was use laboratory activities and collaborative group work to enhance students' enjoyment of leaning dilution chemistry. We wanted to probe students' ability to present their mental models of dilution chemistry at all three levels of representation. Second, we want to explore how well a model like Jigsaw would work in Thailand, where such interactive teaching approaches are rare, despite statements indicating otherwise in national curriculum documents. We thus also wanted to understand if students enjoyed learning in this very new (to them) teaching approach—hence we wanted to evaluate affective variables as well as learning outcomes. The details of the intervention are now described.

According to Holliday (2000), the three important features of Jigsaw IV are the introduction, the quiz, and re-teaching:

1. *Introduction:* The teacher introduces the lesson by means of lectures, literature, questions, problems or showing a movie. The purpose here is to stimulate student interest in the lesson;
2. *Quiz:* The students are evaluated by means of two quizzes:

   The first quiz is designed to check the accuracy and understanding of student in the expert group—this based on the expert sheet and
   The second quiz is designed to check accuracy and understanding of students in the home group—this based on all original material; and

3. *Re-teach:* The teacher re-teaches the material which they think has been misunderstood based on the individual assessment process.

Holliday (2002) goes on to say that class activities can be sorted into nine processes.

1. *Introduction.* The teacher introduces the principle and experiment to the students in a plenary session, and assigns students to a home group, containing six students. The members of each home group are divided into expert groups;
2. *Expert sheets assigned to expert groups*;
3. *Answer expert questions prior to returning to home group.* The students are asked questions based on their expert sheet to check their understanding prior to returning to their home group;
4. *Quiz on material in the expert groups checking for accuracy.* The teacher administers quizzes to assess the validity of their responses;
5. *Return to home groups to share their information with their group.* The students return to their home group to teach their peers, and to share information with each other in their home group;
6. *Quiz on material shared checking for accuracy.* The students are asked questions based on all original material;
7. *Review process.* The teacher reviews and clarifies any concepts which it appeared the students did not understand;
8. *Individual assessment and grade.* Each student is reassessed using a post-test; and
9. *Re-teach.* The teacher re-teaches any topics found to be difficult based on the post-test assessment.

So in summary in Jigsaw students are assigned to study specific topics in an expert group, they become the expert on their topic, and subsequently they teach all their home group members. This means they have the opportunity to teach and learn in their groups, they are able to share their ideas, they develop their self-confidence, cooperation and motivation (Barbosa et al. 2004).

In a new experiment, a sample of first-year (or freshman) undergraduate chemistry students at a large Thai university was required to determine total phenolic compounds in green tea beverage samples based on the Folin-Ciocalteu method (the industry standard method). The activity was designed based on the processes identified by Holliday (2002) for use with the Jigsaw IV method (see Appendix).

Evaluation of the laboratory activity (based on observation of laboratory classes, examination of student's work, surveys, and questionnaires) indicates that the students could understand chemistry concepts better if they understood dilution topics at the three 'thinking levels' or levels of representation. In addition, they were able to integrate the three levels of representation and understand the similar concept, so they understand as relational understanding. This suggests that the students were able to achieve a greater depth of understanding (Treagust et al. 2003) (Table 16.1).

Table 16.1 Students' attitude toward chemistry, prior learning approaches to dilution and related concepts, and knowledge of practical chemistry (N = 244)

Item		SA + A (%)	SD + D (%)
1	I think chemistry is a very interesting subject	96	4
2	I knew about dilutions before doing this experiment	91	9
3	I felt I understood why we need to know how to dilute solutions before doing this experiment	80	20
4	I felt I understood how to make dilutions of stock solutions before doing this experiment	80	20
5	I felt I understood how to calculate concentrations before doing this experiment	71	29
6	I had already used the formula $C_1V_1 = C_2V_2$ into calculate dilutions before doing this experiment	80	20
7	I knew what the formula $C_1V_1 = C_2V_2$ means before doing this experiment	64	36
8	Before doing this experiment, I would calculate the concentration of solutions by another method (open response with description of other method also solicited)	36	64
9	I was familiar with the technique of UV–Visible spectroscopy before doing this experiment	29	71
10	I was familiar with calibration graphs before doing this experiment	53	47
11	Before doing this experiment, I knew why we need to draw calibration graphs	67	33
12	I was familiar with phenols before doing this experiment	34	66
13	I was familiar with green tea before doing this experiment	98	2
14	I used to drink green tea beverages before doing this experiment	96	4
15	I knew about the advantages and disadvantage of green tea beverages before doing this experiment	84	16

Key: *SA* strongly agree; *A* agree; *D* disagree; *SD* strongly disagree

Student perceptions of their learning of dilution and related concepts with the Jigsaw IV approach are reported in Table 16.2

Table 16.2 Students' perceptions of learning dilution chemistry with the Jigsaw IV method ($N = 244$)

Item	SA + A (%)	SD + D (%)
1 In this experiment, I felt I learned how to make dilutions of stock solutions	96	4
2 In this experiment, I felt I learned how to calculate the concentration of solutions	90	10
3 In this experiment, I felt I understood clearly about the concentration of solutions	82	18
4 In this experiment, I felt I learned how to use the UV–Visible spectrometer	94	6
5 In this experiment, I felt I learned how to draw a calibration curve	87	13
6 In this experiment, I felt I learned how to calculate the concentration of total phenols in green tea beverages	89	11
7 In this experiment, I felt I learned more about total phenols in green tea beverages	81	19
8 In this experiment, I liked using the Jigsaw IV Method(open response with description things liked and not liked also solicited)	49	51
9 In this laboratory, I felt happy and relaxed (open response with description things liked and not liked also solicited)	38	62

Key: *SA* strongly agree; *A* agree; *D* disagree; *SD* strongly disagree

Students' laboratory reports were examined thematically for each of the six questions that related to the experiment looking for evidence of representation at the three thinking levels of chemistry. We consider each item in turn

Item 1: What is the *concentration* of total phenolic compound in the blue solution measured by the spectrometer?

Item 2: What is the *amount* of total phenolic compounds in 25 mL of the blue solution?

Item 3: What is the *amount* of total phenolic compounds in 1 mL of 10 % green tea beverage?

Item 4: What is the *amount* of total phenolic compounds in 100 mL of 10 % green tea beverage?

Item 5: What is the *amount* of total phenolic compounds in 10 ml of green tea beverage?

Item 6: What is the *amount* of total compound in 500 mL of green tea beverage sample?

The data were compared before and after learning 'Thinking levels of chemistry' as shown in Figs. 16.3 and 16.4.

The Jigsaw IV approach was applied to these practical classes, and the students worked together as a group, and employed more interactive learning strategies. In particular, each student became an 'expert' for a specific topic, and subsequently taught this to their home group. We found that students were not all positive about their Jigsaw IV experiences. However, in contrast, they were strongly confident that the experiment helped them understand dilution chemistry, concentration, how to

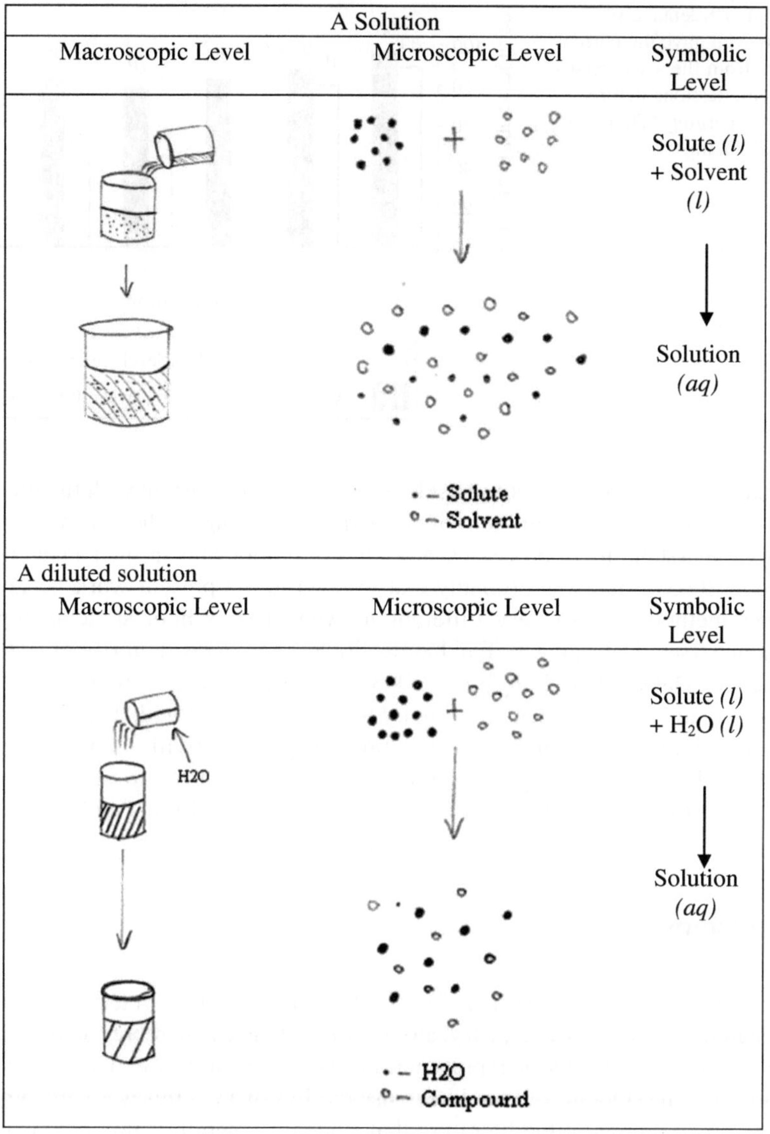

Fig. 16.3 An example of student work

use the UV–Visible spectrometer, the use of calibration curves, how to calculate total phenolic compounds in green tea beverages and gain the knowledge of green tea. Overall fewer than half of the students said they liked the method, and were happy and relaxed in their class. However, some students said they did not really understand the Jigsaw IV method. It seems that a reasonable proportion of the students did not like to use the Jigsaw IV method because they did not really

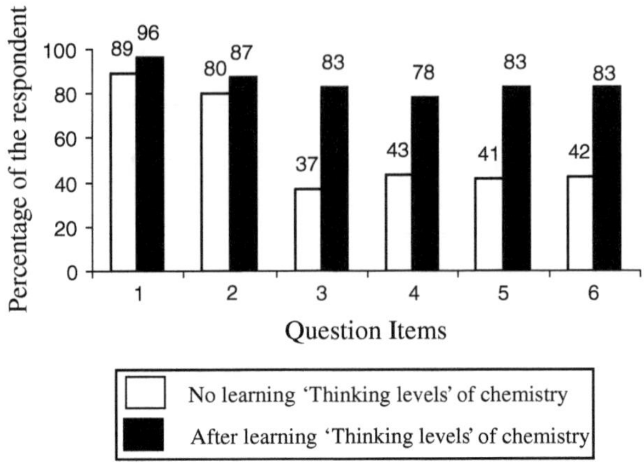

Fig. 16.4 Percentage of students who gave the correct answers from the six reverse-sequence questions before and after learning 'Thinking level' of chemistry

understand the purpose of method itself. This is in fact consistent with the literature, which suggest a common problem of cooperative learning methods lies in students actually understanding of processes, and roles of participants in such a new method (Balfakih 2003). This was probably exacerbated in the present work because the learning method was so very different to what Thai school students typically experience (see Dahsah and Faikhamta 2008), and indeed university students experience (Dahsah and Coll 2007, 2008). In addition, the participants felt they spent too much of time when learning by the Jigsaw IV method. The teachers also felt that spending too much of time preparing the activities and procedures, something the literature suggest is a common perceived barrier to new, particularly constructivist-based learning approaches like Jigsaw (see Colosi and Zales 1998).

Conclusions

The research reported here points to useful learning outcomes in terms of student's understanding of dilution chemistry, along with their ability to represent their mental models at all three levels of representation. However, it seems that a reasonable proportion of the students did not like to use the Jigsaw IV—this was somewhat of a surprise-given reports in the literature that students generally enjoy practical work and more active learning strategies (Johnson and Johnson 2005), and indeed cooperative learning strategies such as group work (Lazarowitz and Hertz-Lazarowitz 1998). There are two reasons why these students might not have enjoyed Jigsaw as much as anticipated. First, is the Thai education system where students are much more accustomed to passive learning in which the teacher gives clear directions and controls the learning environment. Coll et al. (2002) note that this also is true even in higher education where students are expected to become more independent learners, and especially academically able students, prefer the teacher to exercise control over the classroom and learning activities, because this results in greater clarity about what is needed to be done to succeed in assessment tasks. Second, the literature suggests one common problem of cooperative learning methods lies in students actually understanding the processes, and roles of participants (Balfakih 2003). If the purpose of the new approach is not made explicit, it seems students and indeed teachers may focus on the new activity and fail to grasp its purpose and thereby value in the learning process. Dahsah and Coll (2007) report that learner-centered teaching approaches such as Jigsaw are very new for Thai teachers, who typically only pay lip service to learner-centered education. If this is the case, then it also is possible that the teachers did not really understand the purpose of the intervention; only seeing it as a convenient way of assuring their superiors that they were indeed engaged in learner-centered education.

It summary, it seems there were some useful gains in student's learning, but that teachers and students both need more experience in cooperative group learning before they become familiar enough to appreciate improvements in learning practical chemistry (Charania et al. 2001). In particular they need to understand the purpose of using cooperative learning approaches such as Jigsaw meaning they can indeed deliver a learner-centered education in the way required by Thai education authorities. Hume and Coll (2008) suggest that any new teaching approach requires time for all parties to become accustomed to it before its full potential can be realized.

Active Learning of Chemistry in Thailand: Some Conclusions from Local Research

All three of these studies sought to help student to learn chemistry better and to enjoy their leaning but fostering an active-learning environment. The driving force behind the interventions was a desire to develop learner-centered instruction that is consistent with the aims of the Thai science curriculum. As such the interventions consisted on hands on activities, such as laboratory work, collaborative group learning, argumentation and analogy. Specific pedagogies included POE, inquiry-based learning (IBL), interviews about-instances and interviews about events (IAE/IAI), and Jigsaw IV. These are plainly more active pedagogies than is common in most Thai classrooms or laboratories (at any level of schooling), and the research findings point to some gains in terms of learning. There is reasonable evidence that learning outcomes were enhanced. Given that such pedagogies are not common and thus represent a new experience for Thai learners, it is not surprising the interventions need some further work; this is subject to on-going research in our research group. It is also not surprising that in many case teachers and students were initially uneasy about these approaches—even experienced teachers. Deviation from normal practice can be alarming for students, especially high-achieving students who have learnt how to succeed in a traditional, teacher-dominated, highly structured classroom environment. Likewise, teachers must cede some control of the classroom, and may worry that they will be able to complete the curriculum. It is positive that as the teachers became accustomed to leaner-centered, more active learning, they became more confident and at ease with these new approaches. This along, with the benefits in terms of learning outcomes, is encouraging. Thailand has committed to more active, learner-centered teaching approach and these studies indicate that this is possible but that it will take some time for all stakeholders to become comfortable with this.

Appendix A

Stoichiometry Learning Unit: 2

Subject: Chemistry	**Level:** Grade 10
Topic: Average atomic mass	**Time:** 1 period (50 min)

Learning Outcome

1. Students should be able to define the meaning of an average atomic mass;
2. Students should be able to do the experiment and calculate average mass of objects
3. Students should be able to calculate atomic mass of an element
4. Students should be able to describe how scientists determine average atomic mass using mass spectrometer.

Science Concept

The **average atomic mass** of an element is the average atomic mass for the naturally occurring element, expressed in atomic mass units. The scientist uses mass spectrometer to determine the isotope of element and average atomic mass.

Learning Activities

Express and Share Ideas

1. Explore students idea about number of basic particles in atoms and isotope (Worksheet I)
2. Demonstration using red and green balls to explore students' prior knowledge about atomic number, mass number, and isotope. Discuss the responses in class. (Isotope demonstration)
3. Predict how scientist determines the atomic mass of the element that has isotope. Discuss in group and in class

Challenge Ideas

4. Do analogy experiment about average mass of beans (Worksheet II: Average Mass Experiment)

Accommodate Ideas

5. Discuss the analogy experiment and link to the concept of average atomic mass

Apply Ideas

6. Calculate an average atomic mass of an elements (Worksheet III)
7. Search for information about mass spectrometer and how scientists use it to determine atomic mass. Present in next class.

Instructional Materials

Worksheets and Demonstration equipment

Assessments

1. Students' response; discussion, presentation both in group and in class
2. Do experiment
3. Group activity
4. Worksheet
5. Searching and Report
6. Students' Journal.

Worksheet I

Atomic number, Mass Number, and Isotope

1. Complete the following table

Symbol	Number of Proton(s)	Number of Neutron(s)	Number of electron(s)	Atomic Number	Mass Number
$^{1}_{1}H$					
$^{2}_{1}H$					
$^{3}_{1}H$					
$^{12}_{6}C$					
$^{13}_{6}C$					
$^{14}_{6}C$					
$^{14}_{7}N$					
$^{15}_{7}N$					

2. Are there any Isotope shown in the table from item 1? Explain
3. What is atomic number?
4. What is mass number?
5. What is isotope?
6. If an element that has isotope, how do we define the atomic mass of that element?

Worksheet II

Average Mass Experiment

Instruction: Group of three students find out the average mass of beans

Pre Questions:

1. What are average weight of boys and girls in our class?
2. What is average weight of the student in a class from the information in item 1?

3. How to find out the average mass of seed of bean in a beaker?
4. If we have three types of bean, and know average mass of each bean, how do we determine the average mass of bean?

Materials:
Three beakers, black beans, soy beans, green beans, digital balance

Procedures:

- Weight mass of each bean in the given amount
- Count the number of each bean in the given amount (about 40–100 seeds)
- Calculate average mass of each seed
- Calculate percentage of each bean compare to all beans

  $$\text{e.g percent of green bean} = \frac{\text{number of green bean}}{\text{number of all beans}} \times 100$$
- Calculate average mass of one seed of bean

$$\text{Average mass of beans} = \frac{(\text{mass of green bean} \times \%\ \text{ofgreen bean}) + (\text{mass of soy bean} \times \%\ \text{of soy bean}) + (\text{mass of blackbean} \times \%\ \text{of black bean})}{100}$$

Source: http://www.ndsu.edu/ndsu/goswald/chem117/labs/IsotopeLab.pdf

Worksheet III
Average Atomic Mass

1. The chemistry score (100 points in total) divided into three part; 50 points for test, 25 points for experiment, and 25 points for homework. Aree got 85 % from test, 77 % from experiment, and 91 % from homework, what is Aree's chemistry score?
2. What is the average atomic mass of Silicon

Isotope	Atomic Mass	Percent in nature
Silicon-28	27.98	92.21
Silicon-29	28.98	4.70
Silicon-30	29.97	3.09

3. Carbon has two isotopes which are C-12 and C-13. The atomic mass of C-12 and C-13 are 12.000 and 13.003, respectively. If the average atomic mass of carbon is 12.011 what is the ratio of each isotope?
4. The results from mass spectrometer indicated that Ar composted of three isotope which are $^{36}_{18}Ar$, $^{38}_{18}Ar$, and $^{40}_{18}Ar$. The amount of each isotope is 0.1, 0.3, and 99.6 %, respectively. What is the average mass of Ar?

Isotope Demonstration
Objective: Explain the meaning and determine atomic number and atomic mass of isotope of element
Material: Red balls, Green balls, round-bottom flask, periodic table

Procedure:

1. Tell students that using red ball represents proton, and green balls represent neutron, and round- bottom flask is a nucleus of atom
2. Ask students "what word represent number of proton?" and "how the number of proton important?"
3. Put one red ball in a flask, ask students "what element that the model represents?" and "what is atomic mass, and atomic number?" *(Hydrogen; 1; 1)*
4. Add two green balls, "what happen to this model, still be the same element? Why?," and "what is the symbol of this?" ($^{3}_{1}H$)
5. Add one red ball, "what happen to this model, still be the same element? Why?," and "what is the symbol of this?" ($^{4}_{2}He$)
6. Add more balls and ask the students to make sure that they understand about atomic mass, atomic number, and isotope.

Student's Journal

1. What did you learned from this class?
2. Any question do you have?
3. Could you apply what you learned to your daily life, how?
4. What activities do you like the most?
5. What activities you do not like?
6. Any comment and suggestion about the teaching and learning

Appendix B

Chemical Kinetics

This experiment focuses on the kinetics of acid–base reactions. The concept of chemical kinetics of this reaction is often taught in secondary or tertiary education levels. Whilst concrete which buildings are made of is chemically different to calcium carbonate, the overall idea is similar in that acids destroy carbonates—and this experiment uses materials that are a bit easier for us to handle in the laboratory class. Acids such as hydrochloric acid (HCl) will react quickly with calcium carbonate to produce a salt, water and release gaseous carbon dioxide. Other acids such as the acids present in vinegar also react with carbonates.

The reaction is: $CaCO_3(s) + 2HCl(aq) \rightarrow CaCl_2(aq) + H_2O(l) + CO_2(g)$

In the reaction above, how the acid and carbonate react may depend on a number of factors which we want you to investigate. Things you can consider are: the concentration of the acid, the particle size of the carbonate, the temperature of a reaction, and any other factors you can think of. This chemical equation can be applied to determine the rate of a reaction by plotting the relationships between the production of carbon dioxide over time. The experiment is first-order in its kinetics with respect to calcium carbonate and acid. The experimental data from kinetics investigations can be analyzed using Microsoft Excel Solver.

Inquiry-Based Learning

Teachers indicate the students a POE in an inquiry-based experiment in teaching and learning chemical kinetics: acid–base reactions. The use of a POE focuses on the student's understanding of a laboratory. Students need to practice using the ideas themselves to gain the ways of thinking by requiring written responses for this experiment. Students are given to design the experiment for studying how variables affect the rate of a reaction.

Prediction-Observation-Explanation

Prediction-Observation-Explanation, POE, probes student understanding by requiring students to carry out three tasks. It is most important to ensure that students are being asked to make a POE. In the whole classes, students have to:

- predict the outcome of some events, and justify reasons students have to support their prediction,
- describe what students see when the reaction occurs while doing the experiment, students have to write down their observation, and
- reconcile any conflict between what students predicted and what students observed.

Example: Predict how the surface of solid reactant, calcium carbonate, might affect the rate of a reaction, when we change the particle size from either

☐ small particle sizes to larger particle sizes

or

☐ large particle sizes to smaller particle sizes.

Prediction: When reacting with the same concentration of acid at the same temperature:

☐ the rate of a reaction increases

☐ the rate of a reaction decreases

☐ the rate of a reaction does not change.

Explanation for Prediction:

Observation:

Reconciliation of Prediction and Observation:

The experimental design used in this class of inquiry-based learning seeks to enhance students processes of scientific inquiry and to enhance their understanding of chemical kinetics. Here we use POE activities in this laboratory class in combination with several other tools. First is argumentation and argumentative practice. This means each students needs to defend or ague for the rightness of his or her predictions, observations, and explanations. This type of activity is a central activity of scientists and is used within research groups, in this experiment your assigned group. Here we emphasize the knowledge of chemical kinetics by sharing individual ideas between teachers and students in the groups.

Argumentation

The rationale of argumentation in this study is the contribution of the scientific arguments to the construction of scientific knowledge. The arguments can be seen to take place as an *individual* activity, through thinking and writing, or as a *social* activity to take place within a group, a negotiated social act within a specific community. The question that needs to be asked is not only *what* phenomenon is, but also *how* it related to events, and *why* it is important. The classroom practice does provide the opportunity to develop student's abilities to construct arguments. It is important to ensure that all students are asked to:

- indicate both the prediction of the outcomes and provide reasons to support the prediction.
- explore what happened, when the reaction occurs. All students have to write down their individual observations based on some personal reasoning.
- explain what happened, when students change variables which affect the rate of a reaction for studying chemical kinetics.
- discuss in your group, for example, students represent individual idea for few minutes through promoting appropriate classroom activities. Students might gain confidence in a deep understanding of knowledge.

Importance of Group Work

The teaching and learning approach in this experiment places emphasis on the discussion or argumentation described above in group work for promoting the negotiation and argument in order to develop the student's conceptual understanding. Teachers here in this experiment will try to encourage students to predict, observe, and explain what they are doing in the experiment in a group setting as well as in whole-class discussion. In the whole laboratory classes, students are also given the opportunity dealing with a particular problem in a group work.

Appendix C

Study Basic Chemistry with Green Tea Beverages

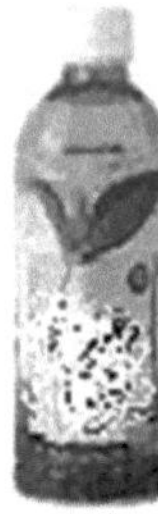

Why drinking green tea could prevent cancer

Epidemiological studies suggest that the consumption of green tea may help prevent cancers in humans; also, breast and prostate cancers in animal models are reduced by green, but not black, tea[1]. Here we offer a possible explanation. We have inferred (using molecular modelling) and subsequently demonstrated that one of the major ingredients of green tea inhibits urokinase, an enzyme crucial for cancer growth.

Tea is drunk in three forms: black (78%), green (20%) and oolung (2%). Green tea contains many polyphenols known as cathechins, including epigallocathechin-3 gallate (ECCC), epigallocathechin (ECC) and epicathechin-3 gallate (ECC). The brewing of black tea oxidizes the cathechins, destroying any beneficial effects[1]. Several mechanisms of anticancer activity of cathechins have been postulated, but none seems universal for all cancers[1-4].

Human cancers need proteolytic enzymes to invade cells and form metastases. One of these enzymes is urokinase (uPA). Inhibition of uPA can decrease tumour size or even cause complete remission of cancers in mice[4,5]. The known uPA inhibitors are unlikely to be used in anticancer therapy because of their weak inhibitory activity or high toxicity.

We have searched for new uPA inhibitors by computer modelling using the active site of uPA as a template. Coordinates of human uPA were kindly provided by C. Philips*; National Cancer Institute,

Jankun et al. (1997)

Green tea contains phenolic compounds. The phenolic compounds in green tea are the four flavanol: epicatechin, epicatechin gallate, epigallocatechin, and epigallocatechin gallate. The total phenolic compounds have been determined by the *Folin-Ciocalteu method*. This is a colorimetric redox reaction that measures all phenolic compounds. The Folin-Ciocalteu reagent is a solution of polymeric complex ions formed from phosphomolybdic acid ($H_3PMo_{12}O_{40}$) and phosphotungstic acid ($H_3PW_{12}O_{40}$).

In an alkaline solution, which is adjusted by sodium carbonate solution to pH 10, phenol was dissociated to phenolate anion. Folin-Ciocalteu is reduced to blue complex during phenolic compound oxidation. The absorption is measured at 760 nm.

$$\text{phenol} + Na_2CO_3 + \text{FC reagent} \rightarrow \text{blue complex}$$

The procedure is used to measure the relative phenolic compound contents in green tea, using gallic acid as a standard. The results are typically expressed as *gallic acid equivalents (GAE).*

Purpose

For this experiment, the objectives are:

1. To study the dilution method and the concentration of solutions;
2. To study the calibration curve; and
3. To determine the total phenolic compound in green tea beverages by UV–Vis spectrometer.

Materials and Reagents

For this experiment, the materials and reagents are:

1. 25-, 50-, and 100-ml volumetric flask;
2. 5-ml cylinder;
3. ml pipette;
4. 100-ml beaker;
5. Spectronic 20;
6. Water bath;
7. Balance;
8. Gallic acid;
9. Sodium carbonate; and
10. Folin-ciocalteu reagent.

Experiment procedures:

Part A: Prepare standard solution and create a calibration curve

1. Make up 0, 50, 100, 150, and 200 ppm solutions of gallic acid from the 1000 ppm gallic acid stock solution.
2. Add 1.0 mL aliquot of each gallic acid standard solution into beakers No.1, No.2, No.3, No.4, and No.5; add the following in order to each beaker:
   - 5 mL of 10 %v/v FC reagent and wait 3 min
   - 2 mL of 15 %w/v Na_2CO_3
3. Incubate the mixed solution for 15 min at 50 °C and transfer to 25-mL volumetric flask. Adjust the volume to exactly 25 mL with distilled water.
4. Record the UV absorbance at 760 nm by Spectronic 20.
5. Create a calibration curve with 0, 50, 100, 150, and 200 ppm gallic acid.

Part B-1: Determine total phenolic compound in green tea beverage sample

1. Filter the green tea beverage through paper and **dilute to 10 % with water**.
2. Add 1.0 mL aliquot of sample solution into beakers No.1, No.2, and No.3 and add the following in order to each beaker:
   - 5 mL of 10 % v/v FC reagent and wait 3 min
   - 2 mL of 15 % w/v Na_2CO_3
3. Incubate the mixed solution for 15 min at 50 °C and transfer to 25-, 50-, and 100-mL volumetric flasks, and adjust volume to exactly 25, 50, and 100 ml with distilled water.
4. Record the UV absorbance at 760 nm by Spectronic 20.

Part B-2: Determine total phenolic compound in green tea beverage sample

1. Filter the green tea beverage through paper and **dilute to 10 % with water**.
2. Add 1.0, 2.0, and 3.0 ml aliquot of sample solution into beakers No.4, No.5, and No. 6 respectively, and add the following in order to each beaker:

- 5 ml of 10 % v/v FC reagent and wait 3 min
- 2 ml of 15 % w/v Na_2CO_3

3. Incubate the mixed solution for 15 min at 50 °C and transfer to 25 mL volumetric flask and adjust volume to exactly 25 mL with distilled water.
4. Record the UV absorbance at 760 nm by Spectronic 20.

Calculation

Calculate the total phenolic compound (mg of GAE) in one bottle of green tea beverages

500 mL

Laboratory report
Study Basic Chemistry with green tea beverages

Part A: Prepare standard solution and create a calibration curve

1. Make up a 0, 50, 100, 150, and 200 ppm gallic acid solution from 1000 ppm gallic acid, and record the information in the table below.

	0 ppm	50 ppm	100 ppm	150 ppm	200 ppm
Volume of 1000 ppm gallic acid (mL)					
Volume of solutions (mL)					

2. Create and draw a calibration curve, using the information in the table below (Fig. A.1).

Notes

....
....
....

Part B-1: Determine total phenolic compound in green tea beverage

	10 % sample (ml)	10 % FC (ml)	15 % Na_2CO_3 (ml)	V_{tot} (ml)	A	C ppm of GAE
Beaker No.1	1.00	5	2	25		
Beaker No.2	1.00	5	2	50		
Beaker No.3	1.00	5	2	100		

* V_{tot} Total of solution volumes

Fig. A.1 Gallic acid calibration curve

	Concentration (ppm)	A
1		
2		
3		
4		
5		

Notes**..**

Part B-2: Determine total phenolic compound in green tea beverage

	10 % sample (ml)	10 % FC (ml)	15 % Na_2CO_3 (ml)	V_{tot} (ml)	A	C ppm of GAE
Beaker No.4	1.00	5	2	25		
Beaker No.5	2.00	5	2	25		
Beaker No.6	3.00	5	2	25		

* V_{tot} = Total of solution volumes

Notes**..**

Calculation

Calculate the total phenolic compound (mg of GAE) in one bottle of green tea beverages

Macroscopic level ☞ Can be observed from experiment.

(Laboratory level)

Submicroscopic level ☞ Can be explained by the molecules at the macroscopic level.

(Molecular level)

Symbolic level ☞ Can be represented by the equation.

Part B-1 Determine total phenolic compound in green tea beverage sample

1.1 Beaker No.1

(1) What is the equivalent concentration of total phenolic compound in the blue solution measured by Spectronic 20?

(2) What is the equivalent amount of total phenolic compound in 25 ml of the blue solution?

(3) What is the equivalent amount of total phenolic compound in 1 ml of 10 % green tea beverage?

(4) What is the equivalent amount of total phenolic compound in 100 ml of 10 % green tea beverage?

5) What is the equivalent amount of total phenolic compound in 10 ml of green tea beverage?

(6) What is the equivalent concentration of total compound in 500 ml of green tea beverage sample?

1.2 Beaker No.2

(1) What is the equivalent concentration of total phenolic compound in the blue solution measured by Spectronic 20?

(2) What is the equivalent amount of total phenolic compound in 25 ml of the blue solution?

(3) What is the equivalent amount of total phenolic compound in 1 ml of 10 % green tea beverage?

(4) What is the equivalent amount of total phenolic compound in 100 ml of 10 % green tea beverage?

(5) What is the equivalent amount of total phenolic compound in 10 ml of green tea beverage?

(6) What is the equivalent concentration of total compound in 500 ml of green tea beverage sample?

1.3 Beaker No.3

(1) What is the equivalent concentration of total phenolic compound in the blue solution measured by Spectronic 20?

(2) What is the equivalent amount of total phenolic compound in 25 ml of the blue solution?

(3) What is the equivalent amount of total phenolic compound in 1 ml of 10 % green tea beverage?

(4) What is the equivalent amount of total phenolic compound in 100 ml of 10 % green tea beverage?

(5) What is the equivalent amount of total phenolic compound in 10 ml of green tea beverage?

(6) What is the equivalent concentration of total compound in 500 ml of green tea beverage sample?

Part B-2 Determine total phenolic compound in green tea beverage sample

2.1 Beaker No.4

(1) What is the equivalent concentration of total phenolic compound in the blue solution measured by Spectronic 20?

(2) What is the equivalent amount of total phenolic compound in 25 ml of the blue solution?

(3) What is the equivalent amount of total phenolic compound in 1 ml of 10 % green tea beverage?

(4) What is the equivalent amount of total phenolic compound in 100 ml of 10 % green tea beverage?

5) What is the equivalent amount of total phenolic compound in 10 ml of green tea beverage?

(6) What is the equivalent concentration of total compound in 500 ml of green tea beverage sample?

2.2 Beaker No.5

(1) What is the equivalent concentration of total phenolic compound in the blue solution measured by Spectronic 20?

(2) What is the equivalent amount of total phenolic compound in 25 ml of the blue solution?

(3) What is the equivalent amount of total phenolic compound in 1 ml of 10 % green tea beverage?

(4) What is the equivalent amount of total phenolic compound in 100 ml of 10 % green tea beverage?

(5) What is the equivalent amount of total phenolic compound in 10 ml of green tea beverage?

(6) What is the equivalent concentration of total compound in 500 ml of green tea beverage sample?

2.3 Beaker No.6

(1) What is the equivalent concentration of total phenolic compound in the blue solution measured by Spectronic 20?

(2) What is the equivalent amount of total phenolic compound in 25 ml of the blue solution?

(3) What is the equivalent amount of total phenolic compound in 1 ml of 10 % green tea beverage?

(4) What is the equivalent amount of total phenolic compound in 100 ml of 10 % green tea beverage?

(5) What is the equivalent amount of total phenolic compound in 10 ml of green tea beverage?

(6) What is the equivalent concentration of total compound in 500 ml of green tea beverage sample?

Conclusions

References

Balfakih, N. M. A. (2003). The effectiveness of student team-achievement division (STAD) for teaching high school chemistry in the United Arab Emirates. *International Journal of Science Education, 25*(5), 605–624.

Barbosa, R., Jofili, Z., & Watts, M. (2004). Cooperating in constructing knowledge: Case studies from chemistry and citizenship. *International Journal of Science Education, 26*(8), 935–949.

Beall, H., & Prescott, S. (1994). Concepts and calculations in chemistry teaching and learning. *Journal of Chemical Education, 71*(2), 111–112.

Bergquist, W., & Heikkinen, H. (1990). Student ideas regarding chemical equilibrium. *Journal of Chemical Education, 67*(12), 1000–1003.

BouJaoude, S., & Barakat, H. (2000). Secondary school students' difficulties with stoichiometry. *School Science Review, 81*(296), 91–98.

Bunce, D. M., Gabel, D. L., & Samuel, K. B. (1991). Enhancing chemistry problem-solving achievement using problem categorization. *Journal of Research in Science Teaching, 28*(6), 505–521.

Cain, L. (1986). S' Mores-demonstration of stoichiometry relationships. *Journal of Chemical Education, 63*(12), 1048–1049.

Camacho, M., & Good, R. (1989). Problem solving and chemical equilibrium successful versus unsuccessful performance. *Journal of Research in Science Teaching, 26*(3), 251–272.

Carr, M. (1984). Model confusion in chemistry. *Research in Science Education, 14*(2), 97–103.

Çalik, M. (2005). A cross-age study of different perspectives in solution chemistry from junior to senior high school. *International Journal of Science and Mathematics Education, 3*(4), 671–696.

Çalik, M., & Ayas, A. (2005a). A comparison of level of understanding of eighth-grade students and science student teachers related to selected chemistry concepts. *Journal of Research in Science Teaching, 42*(6), 638–667.

Çalik, M., & Ayas, A. (2005b). A cross-age study on the understanding of chemical solution and their components. *International Education Journal, 6*(1), 30–41.

Chang, R. (2003). *General chemistry*. Maidenhead, UK: McGraw-Hill Higher Education.

Charania, N., Kausar, F., & Cassum, S. (2001). Playing jigsaw: A cooperative learning experience. *Journal of Nursing Education, 40*(9), 420–421.

Chairam, S., Somsook, E., & Coll, R. K. (2009). Enhancing Thai students' learning of chemical kinetics. *Research in Science and Technological Education, 27*(3), 95–115.

Chiu, M.-H. (2001). Algorithmic problem solving and conceptual understanding of chemistry by students at a local high school in Taiwan. *Proceedings of the National Science Council, Republic of China, Part D, 11*, 20–38.

Choi, M. M. F., & Wong, P. S. (2004). Using a datalogger to determine first first-order kinetics and calcium carbonate in eggshells. *Journal of Chemical Education, 81*(10), 859–861.

Coll, R. K., France, B., & Taylor, I. (2005). The role of models and analogies in science education: Implications from research. *International Journal of Science Education, 27*(2), 183–193.

Coll, R. K., Taylor, I., & Fisher, D. L. (2002). An application of the questionnaire on teacher interaction and college and university classroom environment inventory in a multicultural tertiary context. *Research in Science and Technological Education, 20*(2), 165–183.

Coll, R. K., & Treagust, D. F. (2001). Learners' mental models of chemical bonding. *Research in Science Education, 31*(3), 357–382.

Coll, R. K., & Treagust, D. F. (2002). Exploring tertiary students' understanding of covalent bonding. *Research in Science and Technological Education, 20*(2), 241–267.

Coll, R. K., & Treagust, D. F. (2003). Learners' mental models of metallic bonding: A cross-age study. *Science Education, 87*(5), 685–707.

Colosi, J. C., & Zales, C. R. (1998). Jigsaw cooperative learning improves biology lab courses. *BioScience, 48*(2), 118–124.

Coştо, B. (2007). Comparison of students' performance on algorithmic, conceptual, and graphical chemistry gas problems. *Journal of Science Education and Technology, 16*(5), 379–386.
Dahsah, C. (2007). *Teaching and learning using conceptual change to promote Grade 10 student understanding and numerical problem solving skills in stoichiometry*. Unpublished PhD thesis. Thailand: Kasetsart University.
Dahsah, C., & Coll, R. K. (2007). Thai Grade 10 and 11 students' conceptual understanding and ability to solve stoichiometry problems. *Research in Science and Technological Education, 25*(2), 227–241.
Dahsah, C., & Coll, R. K. (2008). Thai Grade 10 and 11 students' understanding of stoichiometry and related concepts. *International Journal of Science and Mathematics Education, 6*(3), 573–600.
Dahsah, C., & Faikhamta, C. (2008). Science education in Thailand: Curriculum reform in transition. In R. K. Coll & N. Taylor (Eds.), *Science education in context: An international examination of the influence of context on science curricula development and implementation* (pp. 291–300). Rotterdam: Sense.
Dahsah, C., Coll, R. K., Cowie, B., Sung-ong, S., Yutakom, N., & Sanguanruang, S. (2009). Enhancing Grade 10 Thai students' stoichiometry understanding and ability to solve the problems via a conceptual change perspective. *Journal of Science and Mathematics Education in Southeast Asia, 31*(1), 1–43.
Dominic, S. (1996). What's a mole for? *Journal of Chemical Education, 73*(6), 309–310.
Dori, Y. J., & Hameiri, M. (1996). The mole environment: Development and implementation of studyware. *Journal of Chemical Information and Computer Sciences, 36*(4), 625–628.
Dori, Y. J., & Hameiri, M. (1998). The mole environment studyware: Applying multidimensional analysis to quantitative chemistry problems. *International Journal of Science Education, 20*(3), 317–333.
Demeo, S. (1996). Mathematically modeling dilution. *The Chemical Educator, 1*(1), 1–5.
Devetak, I., Vogrinc, J., & Glažar, S. A. (2009). Assessing 16-year-old students' understanding of aqueous solution at submicroscopic level. *Research in Science Education, 39*(2), 157–179.
Dunnivant, F. M., Simon, D. M., & Willson, S. (2002). The making of a solution: A simple but poorly understood concept in general chemistry. *The Chemical Educator, 7*(4), 207–210.
Eilam, B. (2004). Drops of water and of soap solution: Students' constraining mental models of the nature of matter. *Journal of Research in Science Teaching, 41*(10), 970–993.
Eilks, I. (2005). Experiences and reflections about teaching atomic structure in a jigsaw classroom in lower secondary school chemistry lessons. *Journal of Chemical Education, 82*(2), 313–319.
Fleming, F. F. (1995). No small change: Simultaneously introducing cooperative learning and microscale experiments in an organic lab course. *Journal of Chemical Education, 72*(8), 718–729.
Furio, C., Azcona, R., & Guisasola, J. (2002). The learning and teaching of the concepts 'amount of substance' and 'mole': A review of the literature. *Chemistry Education: Research and Practice in Europe, 3*(3), 277–292.
Gabel, D. L., Briner, D., & Haines, D. (1992). Modelling with magnets. *The Science Teacher, 59*(3), 58–63.
Gabel, D. L., & Bunce, D. M. (1994). Research on problem solving: Chemistry. In D. L. Gabel (Ed.), *Handbook of research on science teaching and learning: A project of the National Science Teacher Association* (pp. 301–326). New York: Macmillan.
Gabel, D. L., & Sherwood, R. D. (1984). Analyzing difficulties with mole-concept tasks by using familiar analog tasks. *Journal of Research in Science Teaching, 21*(8), 843–851.
Gabel, D. L., Samuel, K. V., & Hunn, D. (1987). Understanding the particulate nature of matter. *Journal of Chemical Education, 64*(8), 695–697.
Glynn, S. (1997). Drawing mental models. *Science Teacher, 64*(1), 30–32.
Holliday, D. C. (2000). *The development of Jigsaw IV in a secondary social studies classroom*. Lanham, MD: University Press of America.

Holliday, D. C. (2002). *Jigsaw IV: Using student/teacher concerns to improve Jigsaw III*. Lanham, MD: University Press of America.
Hegarty-Hazel, E. (1990). *The student laboratory and the science curriculum*. London: Routledge.
Heyworth, R. M. (1999). Procedural and conceptual knowledge of expert and novice students for the solving of a basic problem in chemistry. *International Journal of Science Education, 21*(2), 195–211.
Huddle, P. A., & Pillay, A. E. (1996). An in-depth study of misconceptions in stoichiometry and chemical equilibrium at a South African university. *Journal of Research in Science Teaching, 33*(1), 65–77.
Hume, A., & Coll, R. K. (2008). Student experiences of carrying out a practical science investigation under direction. *International Journal of Science Education, 30*(9), 1201–1228.
Hume, A., & Coll, R. K. (2009). Assessment of learning, for learning, and as learning: New Zealand case studies. *Assessment in Education: Principles, Policy & Practice, 16*(3), 283–304.
Institution for Promoting Science and Technology [IPST]. (2003a). *Chemistry Textbook 2* (2nd ed). Bangkok: Curusaphaladphoa.
Institution for Promoting Science and Technology [IPST]. (2003b). *The Teacher Manual of Chemistry Textbook 2* (1st ed.). Bangkok: Curusaphaladphoa.
Jankun, J., Selman, S. H., Swiercz, R., & Skrzypczak-Jankun, E. (1997). Why drinking green tea could prevent cancer. *Nature, 387*(6633), 561.
Jansoon, N., Coll, R. K., & Somsook, E. (2009). Understanding mental models of dilution in Thai students. *International Journal of Environmental and Science Education, 4*(2), 147–168.
Johnson-Laird, P. N. (1983). *Mental models: Towards a cognitive science of language, inference and consciousness*. Cambridge, UK: Cambridge University Press.
Johnson, A. W. (1990). The year-long first course in organic chemistry. *Journal of Chemical Education, 67*(4), 299–303.
Johnson, D. W., & Johnson, R. T. (2005). Learning groups. In S. A. Wheelan (Ed.), *The handbook of group research and practice* (pp. 441–461). Thousand Oaks, CA: Sage.
Johnstone, A. H. (1991). Why is science difficult to learn? Things are seldom what they seem. *Journal of Computer Assisted Learning, 7*(1), 75–83.
Justi, R. (2003). Teaching and learning chemical kinetics. In J. K. Gilbert, O. D. Jong, R. Justi, D. F. Treagust, & J. H. Van Driel (Eds.), *Chemical education: Towards research-based practice* (pp. 69–94). London, The Netherlands: Kluwer.
Kabapinar, F., Leach, J., & Scott, P. (2004). The design and evaluation of a teaching-learning sequence addressing the solubility concept with Turkish secondary school students. *International Journal of Science Education, 26*(5), 635–652.
Kolb, D. (1978). The mole. *Journal of Chemical Education, 55*(8), 728–732.
Kousathana, M., & Tsaparlis, G. (2002). Students' errors in solving numerical chemical-equilibrium problems. *Chemistry Education: Research and Practice in Europe, 3*(1), 5–17.
Krishnan, S. R., & Howe, A. C. (1994). Developing an instrument to assess conceptual understanding. *Journal of Chemical Education, 71*(7), 653–655.
Larkin, J. H. (1983). The role of problem representation in physics. In A. L. Stevens & D. Gentner (Eds.), *Mental models* (pp. 75–99). Hillsdale, NY: Erlbaum.
Lazarowitz, R., & Hertz-Lazarowitz, R. (1998). Cooperative learning in the science curriculum. In B. J. Fraser & K. G. Tobin (Eds.), *International handbook of science education* (pp. 449–469). Dordrecht: Kluwer.
Lazrowitz, R., & Tamir, R. (1994). Research on using laboratory instruction in science. In D. L. Gabel (Ed.), *Handbook of research on science teaching and learning: A project of the National Science Teachers Association*. New York: Macmillan.
Lin, Q., Kirsch, P., & Turner, R. (1996). Numeric and conceptual understanding of general chemistry at a minority institution. *Journal of Chemical Education, 73*(12), 1003–1005.

Lythcott, J. (1990). Problem solving and requisite knowledge of chemistry. *Journal of Chemical Education, 67*(3), 248–252.
Ministry of Education (MoE). (1996). *Education reform at the Ministry of Education Thailand*. Bangkok: Ministry of Education.
Ministry of Education (MoE). (2009). *Core basic education curriculum B.E. 2551*. Retrieved October 9, 2009, from http://www.curriculum2551.com/Download/2551.pdf (in Thai)
National Research Council (NRC). (2000). *Inquiry and the national science education standards: A guide for teaching and learning*. Washington, DC: National Academy Press.
McElroy, L. J. (1996). Teaching dilutions. *Journal of Chemical Education, 73*(8), 765–766.
Mettes, C. T. C. W., Pilot, A., Roossink, H. J., & Kramers-Pals, H. (1980). Teaching and learning problem solving in science. *Journal of Chemical Education, 57*(7), 882–885.
Nakhleh, M. B. (1993). Are our students conceptual thinkers or algorithmic problem solvers? Identifying conceptual students in general chemistry. *Journal of Chemical Education, 70*(1), 52–55.
Nakhleh, M. B., & Mitchell, R. C. (1993). Concept learning versus problem solving. *Journal of Chemical Education, 70*(2), 190–192.
Nakhleh, M. B., Polles, J., & Malina, E. (2002). Learning chemistry in a laboratory environment. In J. Gilbert, O. D. Jong, R. Justi, D. Treagust, & J. H. v. Driel (Eds.), *Chemical education: Towards research-based practice* (pp. 69–94). Dordrecht, The Netherlands: Kluwer.
Niaz, M. (1995). Progressive transitions from algorithmic to conceptual understanding in student ability to solve chemistry problems: A Lakatosian interpretation. *Science Education, 79*(1), 19–36.
Office of the National Education Commission [ONEC]. (2003). *National Education Act B.E. 2542 (1999) and Amendments (Second National Education Act B.E. 2545 (2002)*. Bangkok: Pimdeekarnpim.
Office of the Education Council [OEC]. (2006). *Education in Thailand 2005/2006*. Bangkok: Amarin.
Parkash, B., & Kumar, A. (1999). Chemical kinetics illustrated by an improvised experiment. *School Science Review, 80*(292), 114–117.
Pinnell, G. S. (1984). Communication in small group settings. *Theory into Practice, 23*(3), 246–254.
Pravalpruk, S. (1999). Learning and assessment in the science classroom in Thailand. *Assessment in Education, 6*(1), 75–82.
Raviolo, A. (2004). An analogic model for understanding the preparation of volumetric solutions. *The Chemical Educator, 9*(4), 211–215.
Robinson, W. R. (2003). Chemistry problem-solving: Symbol, macro, micro, and process aspects. *Journal of Chemical Education, 80*(9), 978–982.
Sawrey, B. A. (1990). Concept learning versus problem solving: Revisited. *Journal of Chemical Education, 67*(3), 253–254.
Schmidt, H. J. (1991). A label as a hidden persuader: Chemists' neutralization concept. *International Journal of Science Education, 13*(4), 459–471.
Schmidt, H. J. (1994). Stoichiometry problem solving in high school chemistry. *International Journal of Science Education, 16*(2), 191–200.
Schmidt, H. J., & Jigneus, C. (2003). Students' strategies in solving algorithmic stoichiometry problems. *Chemistry Education: Research and Practice, 4*(3), 305–317.
Smith, M. E., Hinckley, C. C., & Volk, G. L. (1991). Cooperative learning in the undergraduate laboratory. *Journal of Chemical Education, 68*(5), 413–415.
Stephans, J. (1994). *Targeting students' science misconceptions: Physical science activities using conceptual change model*. Riverview, FL: The Idea Factory.
Tinger, J. B., & Good, R. (1990). Effects of cooperative grouping on stoichiometric problem solving in high school chemistry. *Journal of Research in Science Teaching, 27*(7), 671–683.
Treagust, D. F., Chittleborough, G. D., & Mamiala, T. L. (2003). The role of submicroscopic and symbolic representations in chemical explanations. *International Journal of Science Education, 25*(11), 1353–1368.

Wang, M. R. (2000). An introductory laboratory exercise on solution preparation: A rewarding experience. *Journal of Chemical Education, 77*(2), 249–250.
Wellington, J. (1998). Practical work in school science. In J. Wellington (Ed.), *Practical work in school science: Which way now?* (pp. 35–51). London: Routledge.
White, R., & Gunstone, R. F. (1992). *Probing understanding*. London: Falmer Press.
Woolnough, B. E. (1991). *Practical science: The role and reality of practical work in school science*. Buckingham, UK: Open University Press.

Chapter 17
Active Learning in Computerized Chemical Education Environments

Yehudit Judy Dori, Miriam Barak and Miriam Carmi

Introduction

In light of the importance of integrating visualization tools and connecting the learning material to students' daily life experiences, the Chemistry Committee of the Israeli Ministry of Education initiated a reform in high school chemistry curriculum, which took place during the last decade (Barnea et al. 2010). As a result, new learning units were developed, introducing inquiry-based leaning, interdisciplinary connections, and the use of advanced technologies. Some of the new learning units introduced two active learning environments: Case-based Computerized Laboratories (CCL) and Computerized Molecular Modeling (CMM) (Barak and Hussein-Farraj 2013; Dori and Kaberman 2012; Dori and Sasson 2008; Kaberman and Dori 2009a).

In the CCL learning environment, teams of two to three students experienced both visual and textual representations of the learning materials. The visual representations included hands-on laboratory experiments and real-time graph construction and interpretation. The textual representations included case studies, also known as case narratives, which are authentic stories that are connected to students' daily lives.

In the CMM learning environment, pairs of students were introduced to CMM systems that allowed them to view, manipulate, and measure virtual molecules, as

Y. J. Dori (✉) · M. Barak · M. Carmi
Department of Education in Science and Technology, Technion—Israel Institute of Technology, Haifa, Israel
e-mail: yjdori@technion.ac.il

Y. J. Dori
Electrical Engineering and Computer Science Department, Massachusetts Institute of Technology, Cambridge, MA, USA

M. Carmi
The Hebrew Gymnasium "Herzeliya", Tel Aviv, Israel

I. Devetak and S. A. Glažar (eds.), *Learning with Understanding in the Chemistry Classroom*, DOI: 10.1007/978-94-007-4366-3_17,

well as to modify or construct new ones. Using Web-based modeling tools, students were able to construct virtual molecules and predict their spatial structure.

This chapter describes three studies that examined the implementation of the two active learning environments among chemistry students and teachers in the context of chemistry education in Israel. The first study examined the assimilation of a laboratory unit which is the fifth one out of five learning units. In this unit we combined CCL and CMM for teaching honors high school chemistry students. The second study, which was also conducted among the 12th grade chemistry students, examined the assimilation of a new biochemistry learning unit and investigated students' understanding of protein structure and function while learning in a CMM environment. The third study examined chemistry teachers' concerns and Pedagogical Content Knowledge (PCK) while implementing the CCL active learning unit.

Background

Active Learning: Theory and Implementation

Three decades ago, researchers noted that active learning involved risks for faculty, such as (1) limited class time, (2) a possible increase in preparation time, (3) the potential difficulty of using active learning in large classes, (4) lack of needed materials, equipment, or resources, (5) fear that students would not participate or learn sufficient content, and (6) loss of control (Bonwell and Eison 1991). Science educators and scientists currently agree that students must do more than just listen; they must ask questions, design experiments, analyze data, and be engaged in solving problems and higher order thinking assignments.

Motivated by a desire to change the prevalent passive teaching mode and to involve students in technology-enhanced active learning, several studies describe the integration of innovative learning environments as part of a reform in the science curriculum (Barak and Dori 2005; Dori and Belcher 2005; Hopson et al. 2001; Meyers and Jones 1993). Contemporary innovative learning environments have based their theoretical framework on constructivism, a "theory of knowledge with roots in philosophy, psychology and cybernetics" (von Glasersfeld 1995, p. 162) and on students' active engagement in the science laboratory setting (Hofstein and Lunnetta 1982, 2004; Lazarowitz and Tamir 1994).

Constructivism puts the construction of knowledge in one's mind as the centerpiece of the educational process. In constructivist learning environments, learners are encouraged to create their own mental framework and formulate their own conceptual models. Constructivism calls for the elimination of a standardized curriculum, the endorsement of hands-on problem solving, and the promotion of active learning (Bruner 1990). Nevertheless, active learning is not a new idea. At the beginning of the twentieth century, active learning was widely promoted among progressive educators. Active learning is consistent with the idea that students must actively process information in order to learn in a meaningful way. In active learning, students are involved in more than listening passively, as emphasis is placed less on transmitting information and more on developing their cognitive and motor skills (Keyser 2000).

Well-delivered lectures are valuable and are common in the academia. However, the thinking required while attending a lecture is low-level comprehension that goes from the ear to the writing hand (Towns and Grant 1997). In their summary of research on the use of lectures, Johnson and colleagues (1998) maintained that students' attention to what the instructor said decreased as the lecture progressed. The researchers found that lectures presume the listener is oriented toward auditory learning, and that they tend to promote only lower level learning of factual information (Johnson et al. 1998). Contrary to that, active learning environments encourage students to be engaged in solving problems, discussing ideas, providing feedback, and teaching each other, which requires higher order thinking (Johnson et al. 1998; Towns and Grant 1997).

Active learning puts the responsibility of organizing the learning in the hands of the learners (Keyser 2000; Niemi 2002) and allows for a diverse range of learning styles (Johnson et al. 1998). Requiring students to actively solve problems, talk about what they learned, and reflect upon their thoughts is most important for effective teaching and learning (Niemi 2002). Since learning is considered as something a learner does (not something that is done to the learner), active learning can support the construction of meaning among students (Johnson et al. 1998; Niemi 2002). Integrating active learning strategies as part of formal learning sessions can advance students' learning as well as address the concerns of instructional change (Niemi 2002).

Computerized Modeling and Simulations

Modeling and simulations are used in chemistry research for describing, explaining, and exploring phenomena, processes, and abstract ideas (Dori and Barak 2001). Models stimulate their creators and viewers to pose questions that take them beyond the original phenomenon, and therefore, might assist them in formulating new hypotheses. These capabilities have opened the way for advanced research in chemistry. Computational chemistry, modeling and simulations, which are well-integrated into chemistry research, are finding their way into chemistry teaching.

Static graphics of chemical structures found in textbooks may help learners to form two-dimensional (2D) mental images, but computers can provide three-dimensional (3D) visualizations. Computerized Molecular Modeling (CMM) tools enable dynamic, interactive, 3D simulations of molecular formulae and their spatial structure. CMM allows students to view, rotate, and measure virtual molecules, as well as to modify them and construct new ones. These visualization tools help translate abstract ideas into concrete ones, helping students understand chemical concepts (Barak and Dori 2005; Barnea and Dori 1999; Dori and Barak 2001). Williamson and Abraham (1995) studied the effect of computer animations on college students' mental models of chemical phenomena. The researchers indicated that the animations helped students understand the subject matter better while improving their ability to construct dynamic mental models of chemical processes.

Majors in the Department of Chemistry and Biochemistry at California State University Fullerton were introduced to chemical computation early in the undergraduate curriculum in an electronic classroom equipped with networked Silicon Graphics workstations (Kantardjieff et al. 1999). Lipkowitz et al. (1999), used computational chemistry as a bridge between the disciplines of geology and chemistry in an undergraduate geology course on the topic of mineralogy. In both studies students were engaged in exploration activities whereby they learned how to use modern software packages as tools to understand chemistry and nature.

Dori et al. (2003) described a Web-based general chemistry course that encouraged its students to participate in a CMM project. Their findings indicated that technology-enhanced teaching positively affects students' achievement, provided the students are actively engaged in constructing the computerized models. These results are in line with the findings of Donovan and Nakhleh (2001), who concluded that the Website used in their general chemistry course was instrumental in visualizing and understanding chemistry. Cox et al. (2003) developed interactive physics-based curricular materials that helped their students learn concepts of thermodynamics with a particular focus on kinetic theory models. The simulations helped students visualize ideal gas particle dynamics and develop a conceptual framework for problem solving. Indeed, among the many advantages of using innovative technologies in chemical education, CMM and simulations are significantly important for students' chemical understanding and spatial ability (Barak and Dori 2005; Barnea and Dori 1999; Dori and Barak 2001).

Computer-Based Laboratory and Real-Time Graphing

Computers usage as laboratory tools may offer a fundamentally new way for aiding students' construction of science concepts. Designated software and probes can help students collect, record, and graph data. They can provide opportunities for asking and refining questions, making predictions, designing plans and/or experiments, collecting and analyzing data, communicating and debating ideas, drawing conclusions, and asking new questions (Linn et al. 1987). Examples of probes used to collect laboratory data include temperature, motion, force, pH, sound, light, conductivity, and pressure. Real-time graphing, formerly often referred to as microcomputer-based laboratories (MBL), offers a dynamic representation of the relationships between at least two variables, such as time, pH, and temperature. Using real-time graphing, data can be collected by various probes and then stored in a computer or a calculator. Real-time graphing allows for frequent repetition and provides opportunities to experience graphically chemical and physical phenomena. The ability to access data over time intervals of varying durations and the power to rapidly process and display the collected data leaves more time for students to solve test problems, propose hypotheses, manipulate variables, generate knowledge, explore relationships, and employ higher-order thinking skills (Dori and Sasson 2008; Russell et al. 2004).

Using an interactive software program, Virtual Chemistry Laboratory (VCL), Martínez-Jiménez and colleagues (2003) conducted an experiment to assess the software's influence on students' understanding of basic organic chemistry laboratory techniques. They concluded that the use of VCL helped students gain better understanding of the techniques and basic concepts used in laboratory work. The VCL contributed in particular to the progress of students with the greatest learning difficulties. Along this line, Stratford et al. (1998), found that computer-based laboratories enabled students to connect between multiple representations of scientific phenomena and processes.

Dori and her research group (Dori and Sasson 2008; Dori et al. 2004; Kaberman and Dori 2009a) presented a project aimed at integrating computerized, hands-on experiments into chemistry teaching in order to foster students' higher order thinking skills, to teach in an up-to-date environment, and to motivate and stimulate the students. They found a significant improvement in students' question posing, scientific inquiry, modeling, and graphing skills, as well as in students' satisfaction from the computerized learning environment and the case-based inquiry approach.

Overall, studies have indicated that computer-based laboratories and real-time graphing can serve as a platform for incorporating inquiry strategies and fostering conceptual understanding and transfer between chemical representations (Kaberman and Dori 2009b; Wu et al. 2001).

Science Teachers and Their Role in Enhancing Active Learning Environments

The success of a science education reform depends on the science teachers' knowledge, skills and practice (Fullan and Hargreaves 1992; Fullan 2002). A major component of this knowledge is PCK. PCK development is embedded in the classroom practice (Van Driel et al. 1998). Originally introduced by Shulman (1986, 1987), PCK was defined as a teacher's integration of content and pedagogy knowledge in a specific domain and its implementation in the teacher's instruction. Van Driel and De Jong (2001) called for a multi-method approach for investigating PCK. Loughran et al. (2004) developed tools for uncovering science teachers' PCK.

Theories and models of teacher learning already exist, but many of these prescribe how teachers should learn, neglecting how the process actually occurs. Several studies examined teachers' learning in their workplace. Lohman, and Woolf (2001) examined learning activities initiated by experienced public school teachers in order to develop their professional knowledge. They found three different types of self-initiated learning activities: knowledge exchanging through sharing and collaborating with colleagues, experimenting by reflection in action, and environmental scanning. Henze et al. (2009), used the storyline method as a narrative research instrument for investigating ways by which experienced science teachers learn in the workplace. They identified two types of teachers. Type I represents a revolutionary course of development in a teacher's engagement in mainly individual activities in the working context. Type II symbolizes an evolutionary development in a teacher's participation in both individual and collaborative activities. They also mentioned the possible connection to teachers' model of concerns developed by Fuller (1969).

Our previous study investigated beliefs and concerns of chemistry teachers after 1 and 2 years of experiencing—the Case-based Computerized Laboratory (CCL) as a new unit of the chemistry curriculum in the context of the chemical education reform in Israel (Dori et al. 2005). The successful implementation of educational change is related to the level and type of individuals' concerns regarding a new program or innovation that is relevant to their daily job. Recognizing the need to study the change process that teachers undergo while implementing CCL, we used the Concerns Based Adoption Model (CBAM), which was administered during summer CCL workshops and at each end-of-year meeting. The model assumes that concerns of people who consider and/or experience a change can be investigated by characterizing different stages of the adoption process (Horsley and Loucks-Horsley 1998).

The Stages of Concerns Questionnaire (SoCQ) is designed to examine teachers' concerns of three types according to CBAM: self concerns, task concerns, and impact concerns. Self concerns refer to (1) awareness concern, i.e., having little knowledge of the program and no interest in taking part in the implementation, (2) informational concern, i.e., showing willingness to learn more about the nature of the program, and (3) personal concerns, i.e., raising concerns regarding one's role and change of status due to the adaption of the new program. Task concerns relate to (1) management concerns—having concerns regarding organizing, scheduling, and time demands during the implementation process. Impact concerns refer to several stages: (1) the consequence stage, which includes concerns regarding the relevance of the program for the students, evaluating students' outcomes, and issues of competencies, (2) the collaboration stage of focusing on cooperation and coordination with other teachers regarding the use of the program, and (3) the refocusing stage, in which one explores additional possible benefits from the program and raising ideas about alternatives to the proposed or existing form of the program.

The Three Studies: An Overall View

The goal of the three studies was to investigate the effect of the CCL and CMM active learning environments on students' higher order thinking skills and perceptions. The first and second studies investigated 12th grade chemistry students from different high schools in Israel who studied in these active learning environments. In the third study, we examined how different factors, such as teachers' beliefs and concerns, in addition to in-service training and support, contribute to the complex process of implementing the CCL curriculum.

Study I: CCL as an Active Learning Environment

The objective of the CCL as an active learning environment study was to investigate the effect of the CCL active learning environment on students' perceptions of the science laboratory.

Research Setting

Laboratory in science education in general and in chemistry education in particular is an important learning environment that facilitates improving social relations in class, creating positive attitude toward science, and encouraging thinking skills. In the last decade, the CCL learning unit, developed at the Technion, combines activities that encourage inquiry studies and development of skills at different thinking levels. The unit integrates reading scientific texts describing case studies, application of sensors and real-time graphing while working in teams of two to three students and guided and open-ended inquiry experiments (Dori et al. 2005).

The CCL active learning environment was tested among 12th grade chemistry honor students who studied this 90-h unit as part of the matriculation examination requirements. The research examined how students in both the Jewish and Arab sectors—the experimental group—view the CCL learning environment (see more details in Abed and Dori 2013). The control groups consisted of 12th grade chemistry honor students who studied a confirmatory, close-ended laboratory as a half unit (45 h), and students who studied the traditional chemistry course without any laboratory unit. The underlying assumption of the research was that cultural differences between the two sectors, an active learning environment versus a confirmatory laboratory, and gender differences, may affect students' perceptions toward the laboratory learning environment (Marjieh 2007).

Research Questions

The research questions were the following.

1. What are the students' perceptions of the learning environment in the Case-based Computerized Laboratory environment in 12th grade?
2. Are there any differences between perceptions of students who study in an active laboratory—CCL—environment versus traditional laboratory, and if so, what are these differences?

Research Participants

The participants included high school students who studied chemistry at a five unit—honors level. The experimental group included 383 chemistry students studying the CCL unit: 224 students from the Arab sector and 159 students from the Jewish sector. Additionally, 176 students, who studied the confirmatory, close-ended laboratory, served as one of the control groups. The second control group included 183 students who studied the traditional course without any laboratory unit.

Assessment Method

The research instruments included two SLEI—Science Laboratory Environment Inventory—questionnaires and an open feedback question. The SLEI questionnaire was developed and validated by Fraser et al. (1993), and was translated to Hebrew and validated by Hofstein et al. (1996). In our study, the questionnaire was translated also into Arabic by a chemistry teacher who was also a master student and the translation was validated by a doctoral student in science education who specialized in bilingual teaching and learning as well as in translation from Hebrew into Arabic and vice versa (Abed and Dori 2007). Both the translator and the validator of the translation implemented the CCL unit as chemistry teachers and Arabic was their mother's tongue.

The original SELI questionnaire relates to eight categories: student cohesiveness, teacher supportiveness, involvement, open-endedness, integration, rule clarity, physical environment, and laboratory organization. We added another category—attitudes toward the science laboratory. The students received two versions of the SLEI questionnaire: actual and preferred. An open feedback question was added in order to investigate the students' reasoning for their perceptions. Further, observations were made in some of the experimental classes in order to document learning in the laboratory and to validate the findings of these questionnaires.

Findings

Our findings in previous studies (Abed and Dori 2013; Dori and Kaberman 2012; Dori and Sasson 2008; Kaberman and Dori 2009a) indicated that the scores of the experimental group students improved significantly in question posing, inquiry, graphing, and modeling skills from the pretest to the post-test. The net gain scores of the experimental group students were significantly higher than those of their comparison peers in all the examined skills.

We used quantitative analysis of students' perceptions of the CCL learning environment in the Jewish and Arab sectors which is summarized in Table 17.1 (Marjieh 2007).

We found that the students in the Jewish sector perceived the learning environment in the CCL more positively than the students in the Arab sector in the categories of student cohesiveness and physical environment in both actual and preferred situations (Marjieh 2007). Students in the Arab sector perceived the laboratory learning environment more positively than the experimental group of students in the Jewish sector in the categories of integration, clarity of rules, involvement, open-endedness of the environment, and attitudes about the science laboratory in the actual situation as well as in the preferred one.

The comparison between the perceptions of the Arab students in the CCL learning environment and the traditional laboratory showed that the students in the computerized laboratory perceived the learning environment more positively in the categories of involvement, integration, and organization in both actual and preferred

Table 17.1 Jewish and Arab chemistry students' perceptions toward the CCL learning environment

Category	CCL—Jewish students (N = 159)		CCL—Arab students (N = 224)			
	Actual	Preferred	Actual	Preferred	Actual	Preferred
	Mean	Mean	Mean	Mean	t	t
Teacher supportiveness	4.43	4.44	4.34	4.35	1.38	1.58
Involvement	3.66	3.81	3.95	4.12	5.67***	5.43***
Student cohesiveness	3.90	3.98	3.67	3.71	4.60***	4.91***
Open-endedness	2.57	2.91	2.88	3.17	6.62***	4.69***
Integration	3.70	3.79	4.09	4.03	5.66***	3.43***
Laboratory organization	4.01	4.21	4.34	4.11	1.19	1.51
Rule clarity	3.74	3.90	3.98	4.12	4.1***	3.24**
Physical environment	3.88	4.08	3.59	3.92	4.84***	2.41*
Attitudes toward science lab	3.82	4.06	4.31	4.38	6.54***	4.44***

* $p < 0.05$; ** $p < 0.01$; *** $p < 0.0$

situations, as well as rule clarity and physical environment in the preferred situation. Students in the traditional laboratory setting perceived the learning environment more positively in the category of open-endedness in actual situation.

The perceptions of the male and female students in the CCL environment in both sectors are more similar to each other than the perceptions of the male and female students at the traditional laboratory environment. The female students in the traditional laboratory perceived the learning environment more positively than the male students in the categories of teacher supportiveness, involvement, student cohesiveness, organization, and rule clarity. In the computerized laboratory environment, female students have a more positive perception than male students only in two categories: student cohesiveness among students and their views regarding laboratory.

The response of the experimental CCL group of students in both sectors to the open feedback question showed satisfaction. One of the female students described the popcorn experiment as part of the *Energy* topic: "*It is a great experiment! We study chemistry* [with temperature sensors] *while enjoying ourselves and the results appear on the computer screen… We love to work in teams and help each other.*" The students indicated that they had enjoyed learning, they were interested in the CCL unit, benefited, and felt success due to their self-perception as independent learners.

Study II: CMM Active Learning Environment for Enhancing Students' Understanding of Protein Structure and Function

The objective of the second study on the CMM active learning environment was to examine whether, and to what extent, learning via CMM affect students' understanding of protein structure and function. The study also aimed at examining the student's ability to transfer across the four levels of chemistry understanding: microscopic, macroscopic, symbol, and process (Dori and Hameiri 2003; Barak and Dori 2005).

Research Setting

A new learning unit—*Biochemistry: The chemistry of proteins and nucleic acids* was introduced as part of a new curriculum in honors chemistry. The goal of the learning unit was to enhance students' understanding of biomolecular structure and function in the context of the human body. It takes an interdisciplinary approach and uses CMM for the illustration of structure and function of biological

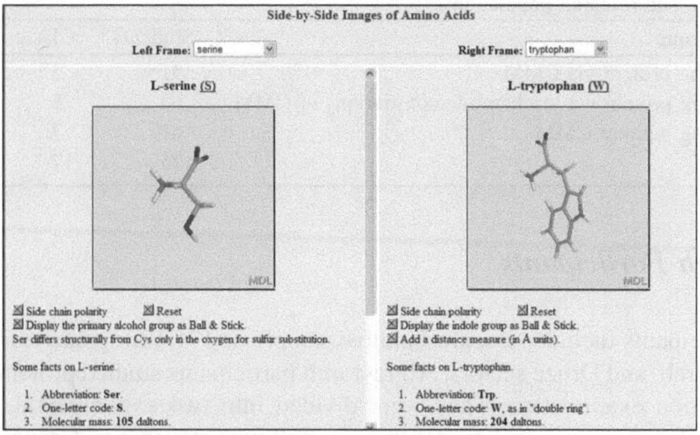

Assignment: Compare between Tryptophan and Serine amino acids

1. Change the models' display to 'ball & stick' and 'space-filling' representations. What does each display represent and what are the chemical properties that each display emphasizes?
2. Draw molecular formula for each amino acid.
3. What is the side-chain functional group of each amino acid?
4. Which one of the two amino acids has a polar side-chain? Why is it considered polar?

Fig. 17.1 An example of the CMM learning environment and amino acids' assignment

molecules. The learning unit includes four chapters, each containing CMM assignments: (1) Introduction to life science, (2) From amino acids to proteins, (3) From nucleotides to nucleic acids, and (4) From nucleic acids to proteins.

The CMM learning environment included Web-based applications, available freely on the Internet. The side-by-side images of amino acids, developed by Carnegie Mellon University (http://www.bio.cmu.edu/courses/biochemmols) is an exemplary website that was used as a platform for the students' activities (see Fig. 17.1 for the screen shots and the appropriate assignment our students received).

The CMM-based learning environment and the specially designed assignments (written in Hebrew or Arabic) encouraged students to manipulate 3D molecular models of amino acids, proteins, DNA, and RNA. The assignments included questions that focused on the connections between a molecule's 3D structure and its function in the human body. The students studied the learning unit for about 9 weeks and were engaged in CMM activities 2 h per week on average, either by teacher demonstrations in a regular classroom setting or via hands-on practice in a computer cluster.

Table 17.2 The research population

Research group	Students	Classes	Schools
A—Learning protein via CMM	51	5	5
B—Learning protein via teachers' demonstrations of CMM	63	5	3
C—Learning without CMM	61	7	4
Total	175	17	12

Research Participants

The participants included a representative sample of 175 12th grade students from Jewish, Arab, and Druze sectors. All research participants studied proteins for their matriculation examination. They were divided into two experimental, groups A and B, and one control group, group C, as follows: students in Group A studied protein structure and function via hands-on CMM activities, Group B students studied protein structure and function via teacher's demonstration of CMM, and students in Group C experienced traditional learning of protein structure and function without the use of CMM.

The research population sorted by comparison groups is presented in Table 17.2.

Assessment Method

The mixed methods research model (Johnston and Onwuegbuzie 2004) was employed by using both quantitative and qualitative methodologies in the analysis and interpretation of data. The research was based on pre- and post-questionnaires that included five main questions. The first question was aimed at testing students' ability to transfer across the three modes of molecular representation: 3D, 2D, and textual-symbolic. The second question tested students' ability to transfer across the macro, micro, symbol, and process levels of chemistry understanding. The fourth, fifth, and sixth questions were aimed at indicating students' knowledge, understanding, and implementation of protein structure and function, respectively.

The questionnaires were administrated to the three groups of students before and after they studied the proteins topic. The quantitative data were analyzed to compare between the research groups using GLM statistical procedures, such as paired sample t-tests and ANCOVA tests.

Findings

Mean scores and standard deviations of the three research groups in the pre- and post-questionnaires are presented in Table 17.3.

Table 17.3 Mean scores and standard deviations of pre-and post-questionnaires

Research group	N	Pre-test		Post-test		t	p
		Mean*	SD	Mean*	SD		
A. Learning via hands-on CMM	51	4.33	2.03	6.45	2.05	5.24	<0.01
B. Learning via teacher's demonstrations of CMM	63	3.81	1.33	5.24	2.14	5.14	<0.01
C. Learning without CMM	61	4.24	1.78	4.43	2.10	0.72	N.S
All the Population	175	4.20	1.75	5.23	2.27	5.24	<0.01

* Mean score is for a range of 0–10 points

As Table 17.3 shows, the mean scores of the pre-questionnaires in all three groups were around 4.00 (out of 10.00), indicating low biochemistry knowledge among students. In the post-questionnaires, group A students, who studied with the use of CMM, achieved the highest mean scores on the post-questionnaire. An ANCOVA test that compared between the research groups indicated a statistically significant difference ($F_{(2,171)} = 16.53, p < 0.01$). Post hoc Sidak test showed that the post-questionnaire mean score of the traditional learning group C was significantly lower than that of the other two research groups. These results suggest that learning protein structure and function in the CMM environment improved students' conceptual understanding, their ability to transfer across the four levels of chemistry understanding, and their ability to transfer across different types of molecular representation, i.e., from 3D to 2D and to textual representation and vice versa.

Examining students' ability to transfer across different types of molecular representations, findings indicated that the post-questionnaire scores of the experimental group A—learning via hands-on CMM (Mean = 7.81, SD = 2.46) were higher than their pre-questionnaire scores (Mean = 5.24, SD = 3.29). These differences were found to be statistically significant ($t_{(50)} = 4.85$, $p < 0.01$). Similarly, the post-questionnaire scores of the experimental group B—learning via teacher's demonstrations of CMM (Mean = 6.11, SD = 3.09) were higher than their pre-questionnaire scores (Mean = 3.61, SD = 2.45). These differences were also found to be statistically significant ($t_{(62)} = 5.34$, $p < 0.01$). Contrary to the two experimental groups, no statistically significant difference was found between the pre- and post-questionnaire scores of the control group C—learning without CMM. These results suggest that students who did not use CMM did not improve their ability to transfer across different types of molecular representations.

Analysis of students' drawings of amino acids models, which was a part of an assignment in the questionnaires, showed that students who experienced hands-on CMM (Group A) drew more accurate models. In their drawings, no atoms were missing, their drawings had a 3D perspective, as they drew shadows for illustrating depth, they depicted correct angles, and they differentiated atoms by colors and/or proportional size. On the other hand, students who studied with teacher's demonstrations of CMM (group B) and students from the control group (group C) provided mostly incorrect drawings with no 3D perspective. We found that most of

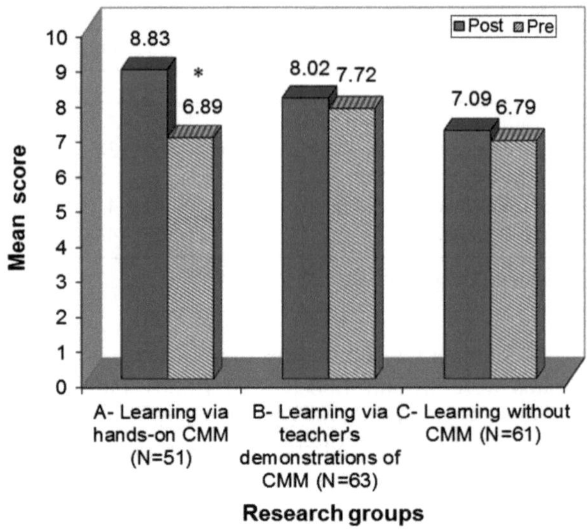

Fig. 17.2 Students' ability to transfer across the chemical understanding levels in the pre- and post-questionnaire, * $p < 0.05$

the students in research groups B (80 %) and C (95 %) chose not to answer this question or answered it incorrectly.

Our results suggest that it is not enough to present molecular models by showing their drawings in textbooks or by teacher's demonstrations. It is important that students experience the use of CMM, construct the computerized models, turn them around, calculate angles and distances between atoms, and change their representation forms.

The mean scores of the pre- and post-questionnaires on the question that examined student's ability to transfer across the chemical understanding levels are presented in Fig. 17.2.

Paired *t* test showed statistically significant difference between pre- and post-scores of students who studied via CMM ($t_{(50)} = 2.80$, $p < 0.01$). There was no statistically significant difference between pre- and post-scores of students who studied biochemistry by teacher's demonstration or students from the control group. An ANCOVA test that compared between the research groups indicated a statistically significant difference ($F_{(2,170)} = 3.39$, $p < 0.05$). Post Hoc Sidak test showed that the post-questionnaire mean score of the traditional learning group was significantly lower than that of the other two research groups. These results suggest that in the process of constructing and manipulating computerized molecular models, students were able to move across the four chemical understanding levels: macro, micro, symbol, and process. Indeed, in order to complete their CMM tasks, these students had to repeatedly review the learning material, thoroughly understand the structure of the molecules and correctly apply chemical principles that were taught in the classroom.

In both the pre- and post- questionnaires, students were asked to answer the following question:

The following condensation reaction occurs between two amino acids. Complete the process, circle the bond that is created that and name it.

Following Bloom's taxonomy (1956), this question was classified as a **knowledge level** question that requires knowledge in the **process** and **symbol** levels.

Findings indicated that the mean score of the post-questionnaires of groups A and B were significantly higher than their prequestionnaire mean score ($t_{(50)} = 2.12$, $p < 0.05$, and $t_{(62)} = 5.41$, $p < 0.01$, respectively). However, no statistically significant difference was found between the pre- and post-questionnaire results of the control group C. These results suggest that experiencing hands-on CMM activities or teachers' demonstrations of CMM both have a positive effect on students' knowledge. ANCOVA test comparing among the three groups indicated statistically significant differences ($F_{(2,\ 170)} = 3.392$, $p < 0.05$). Post Hoc Sidak test showed that the post-questionnaire mean score of the control group C was significantly lower than that of group B. Once again, this suggests that learning the structure and function of proteins from textbooks alone cannot do a lot to improve students' knowledge or their ability to transfer between the symbol and process levels of chemistry understanding.

Another question students were asked in the pre- and post-questionnaires was:

Amino acids are known for their amphoteric properties (they act as both acid and basis). Provide explanations for this phenomenon.

This question was classified as an **understanding level** question, as students had to demonstrate understanding of acids and base properties and amino acids structure.

Examples of the students' answers and their assigned scores by levels of chemistry understanding are presented in Table 17.4.

Students' mean scores on the question that indicated their understanding of protein structure and function in the pre- and post-questionnaires are presented in Fig. 17.3.

Paired t-test showed statistically significant difference between pre- and post-scores for students who studied biochemistry via CMM either by hands-on manipulation ($t_{(50)} = 5.76$, $p < 0.01$) or teacher's demonstration ($t_{(62)} = 5.76$, $p < 0.05$). This suggests that hands-on CMM and demonstrations of CMM helped students better understand the subject matter; however, the first method (hands-on) is much more efficient. ANCOVA test comparison indicated a statistically significant difference between groups ($F_{(2,170)} = 10.40$, $p < 0.01$). Supporting this claim, Post Hoc Sidak test showed that the post-questionnaire mean score of experimental group A—learning via hands-on CMM, was statistically higher than that of the other two research groups.

The last question in both the pre- and post-questionnaires was: *Although proteins are constructed from the same 20 amino acids, your body includes thousands of different protein molecules, explain how this can be.*

This question was classified as an **application level** question, since students need to apply their knowledge of amino acids to protein structure and function in the human body.

Table 17.4 Examples of students' answers to the question on the amphoteric properties of amino acids

Students' answers Amino acids are amphoteric substances because:	Score	Levels of chemistry understanding
They *can react both as acids and basis*. Amino acids have an amino group ($-NH_2$) that may attract a proton (H^+) due to a pair of non-bounding electrons on the nitrogen atom, thus acting as a base. They also have a carboxylic group ($-COOH$) that may release a proton, thus acting as an acid	10	*Macro*, micro, symbol and process
They consist of both an amino group ($-NH_2$) and a carboxylic group ($-COOH$)	5	Micro, symbol
They consist of a hydrophilic group and a hydrophobic group	0	–

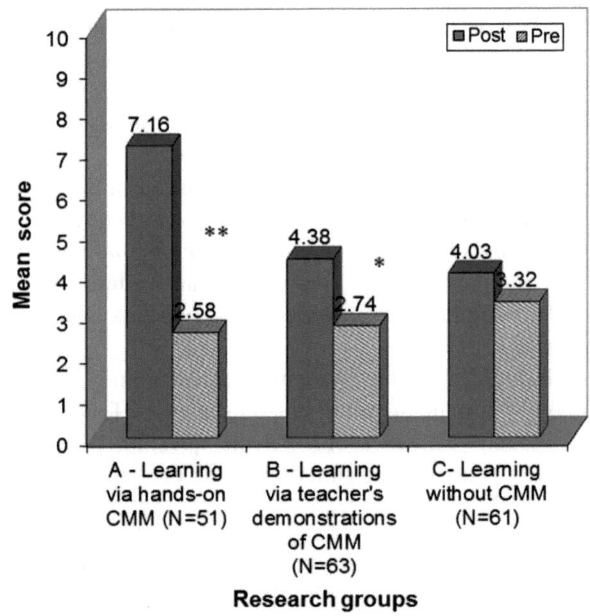

Fig. 17.3 Students' scores in understanding protein structure and function in the pre- and post-questionnaires, comparison between groups, * $p < 0.05$ ** $p < 0.0$

Students' mean scores for the question that indicated their comprehension of protein structure and function in the pre- and post-questionnaires are presented in Fig. 17.4.

ANCOVA test comparison indicated statistically significant difference between groups ($F_{(2,170)} = 4.31$, $p < 0.05$). Post hoc Sidak test showed that the post-questionnaire mean score of experimental group A was higher than that of the other two research groups. This result suggests that learning proteins via hands-on CMM activities improved students' ability to apply knowledge.

The results of the research questionnaires indicate that learning protein structure and function with hands-on CMM activities may improve students' chemistry

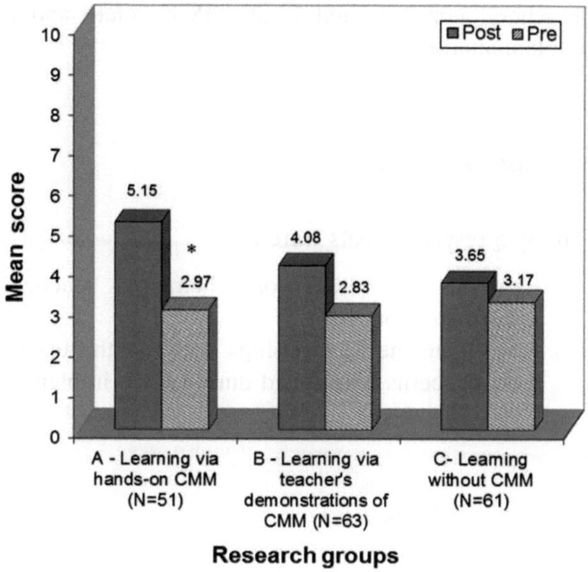

Fig. 17.4 Students' scores in application of protein structure and function in the pre- and post-questionnaires, comparison between groups, * $p < 0.05$

understanding, their ability to transfer across both the four levels of chemistry understanding, and across different types of molecular representation.

Study III: The Effect of CCL Active Learning Environment on Science Teachers' Beliefs and Concerns

The objectives of this study were to examine how different factors, such as teachers' beliefs and concerns, in addition to in-service training and support, relate, and contribute to the complex process of implementing the CCL curriculum.

Research Setting

The objective of integrating the CCL curriculum into high school chemistry was to enhance honors students' higher order thinking skills, teach in a technology-rich environment, and motivate the students by showing the relevance of chemistry to everyday life phenomena (Dori et al. 2004). The CCL study unit integrates computerized (MBL style) desktop experiments and emphasizes scientific inquiry and critical reading of case studies. Implementing this unit in their classes gave teachers an opportunity to be engaged in active learning of both content and skills.

Teachers' concerns and former PCK relate and contribute to the process of implementation.

Research Goals

The two research goals were to:

(1) Examine the PCK of chemistry teachers during CCL classroom implementation.
(2) Investigate the relationships between the teachers' PCK and the change of their concerns before and during CCL implementation.

Assessment Method

The research questions were examined using the following instruments: PCK and SoCQ questionnaires, documentation of support meetings, classroom observations, and teachers' interviews. The PCK questionnaire was administered to a group of 17 chemistry teachers who had implemented CCL for periods of 1–3 years. Aimed at making different PCK aspects explicit, this questionnaire included a list of activities, skills, and topics in the CCL curriculum. The teachers were asked to identify and explain topics or approaches that were most difficult or easy for teachers to teach and for students to understand. Teachers were also asked how they coped with these types of difficulties. Monthly and year-end support meetings during the implementation process were documented, and teachers' remarks and responses were recorded and later analyzed. These meetings gave teachers the opportunity to develop free and spontaneous exchange of ideas which fostered their own active learning of content and skills. In addition, teachers were observed during a sequence of six lessons, in which they taught a case study and a computerized inquiry experiment was performed. The case study and computerized experiment were two unique elements of the CCL curriculum. In this study we report on interviews with two teachers during the CCL implementation. The interviews focused on their CCL teaching methods and reasons for specific actions in the classroom.

Teachers' concerns were captured at different stages of implementation using SoCQ and individual profiles of concerns were devised for each teacher. Findings reported earlier (Dori et al. 2005), showed that teachers' concerns changed during their implementation of the CCL curriculum. Chemistry teachers' concerns regarding self issues and task concerns decreased, while their concerns regarding the impact stage (consequences, collaboration, and refocusing) increased.

Findings

Analyzing the PCK questionnaire, we found that about half (53 %) of the teachers indicated that the approaches which they found to be most difficult to implement in class were also the ones which they thought were most difficult for their students. Teachers expressed their personal difficulties to understand and apply new teaching approaches introduced in the curriculum for the first time. Two examples of such difficulties are the concept of the ***four levels of chemistry understanding*** *(macroscopic, microscopic, process, and symbol)* and *posing* ***inquiry questions***.

Following are a couple of responses of the teachers to the question: *Describe an example of a difficulty you encountered during CCL implementation and how you coped with it.*

- Teacher C. indicated in the interview that her students designed open-ended experiments in which they did not take into account the variables that they had identified as the ones they would like to investigate. In this event teacher C. told the students: *Please decompose your* ***inquiry question*** *into parts, then identify the variable you would like to measure (the dependent variable) and the variable you are going to vary incrementally (the independent variable).*
- Teacher F. reported in the year-end meeting: *I designed a written assignment that contained a list of inquiry questions students had posed and asked them to analyze the questions according to the* ***four chemistry understanding levels***.

At the preliminary stage, some of the teachers raised concerns regarding the technology. For example Teacher I. talked about the interaction between the software and the students: *The first difficulty I'm worried about is managing the software with the students. We have to teach them step by step in order to let them get accustom to the CMM.* However, Teacher V. was more concerned with her need to invest time in order to learn how to use the software: *I anticipate some technical difficulties which will need extra time of preparation.*

Interestingly, and in spite of these concerns, over three quarters (77 %) of the teachers reported in their PCK questionnaire that the easiest topic for their students was adjusting to the new hardware and software of the laboratory, which included temperature, pH, and conductivity data collecting sensors. The explanation teachers provided was that their students nowadays are highly computer literate and that they prepared themselves well before the beginning of the school year since for them it was a new environment. Some even came to the summer workshops accompanied by their laboratory technicians whom they wanted to help with the technical aspects of the CCL throughout the school year. In what follows we present two cases of CCL teachers, their PCK, and their concerns before and during the CCL implementation changed.

R. is a chemistry teacher with 15 years of teaching experience after a short career as chemical engineer. She began to implement CCL with her 11th grade honors students and continued teaching them according to the curriculum during their 12th grade. She showed enthusiasm and willingness to share her experience with her peers during the monthly meetings. The first difficulty she encountered was introducing the CCL hardware and software to her students. She prepared herself extensively to the first computerized experiment and wrote detailed instructions to her class. The second challenge was teaching her students to pose inquiry questions, a topic which she had anticipated would be difficult for them to deal with. To meet this challenge, she applied software for posing questions in general in order to present the characteristics of inquiry questions in chemistry. Students had to draw a star-like map that shows the variables which must be taken into consideration and then suggest different inquiry questions identifying dependent and independent variables. R. taught her students to use this software and later presented it to other CCL teachers in one of the monthly meetings. R. experienced full CCL implementation while teaching the same group of students in 11th and 12th grades.

Using SoCQ, we examined the concerns of R. three times: at the beginning of 11th grade (Practice 0 in Fig. 17.5), at the end of that year (Practice 1), and at the end of 12th grade (Practice 2). Analyzing R.'s concerns in Fig. 17.1 during this 2-year period, and comparing it to her PCK development, shows that concerns and PCK are related. R.'s concerns regarding consequences ('Con' in Fig. 17.5) decreased after 1 year of implementation, while her refocusing ('Ref') concerns, which decreased after a year, increased once again at the end of 12th grade.

This pattern can be explained by the confidence the teacher gained as she solved problems she had encountered in the first year of the CCL implementation. The consequences ('Con') concerns after 1 year decreased, but after another year they increased again, but did not reach the original level. This can be explained by the fact that in the second year, new and alternative assessment standards suitable for CCL were introduced and this presented R. with another challenge. In an interview conducted at the end of second year of CCL implementation she said:

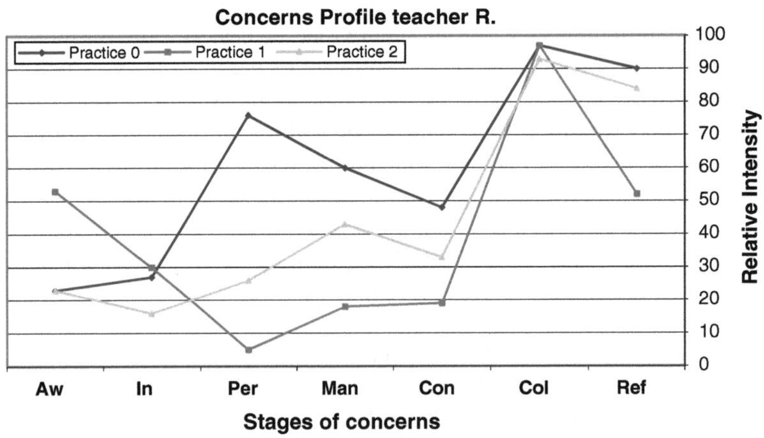

Fig. 17.5 Teacher R.'s individual profile of concerns

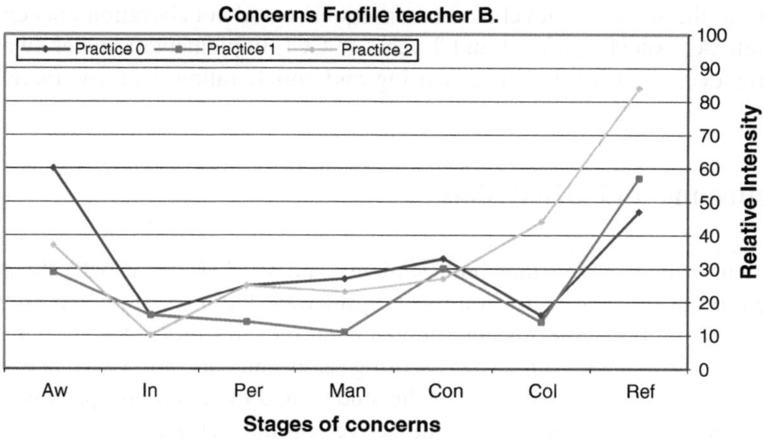

Fig. 17.6 Teacher B.'s individual profile of concerns

> I had hard time checking the students' reports. They were of low quality and I did not know how to assess them. I had to build a rubric and I did not know how to do it. It took me long time to develop a hierarchy for scoring in order to evaluate my students' reports.

Figure 17.1 also shows that R. had high concerns regarding collaboration and refocusing, in line with her willingness to share knowledge and new ideas regarding CCL with her colleagues.

The second teacher, B., a highly experienced chemistry teacher with 30 years of practice and a managerial position at her school, was exposed to the CCL curriculum in a summer workshop. Being excited with the new program, she immediately decided to implement it in the academic year immediately following that summer.

In the annual meetings, B. expressed her confidence in the CCL curriculum and in her students' anticipated outcomes. She was the first teacher to plan and implement the curriculum in her school. Her main concern regarding the curriculum was the difficulties students had in formulating their questions and reports. She worked hard with her students in order for them to attain high achievements. Interviews showed that new knowledge emerged during B.'s practice of the CCL curriculum. For example, she decided to teach a specific sub-unit of visual modeling at the beginning of the program rather than at the end, as suggested by the program developers, because it fits her schedule of teaching organic chemistry. She strengthened the links between the experiments and the subjects taught in class and also tutored a teacher who was new to the CCL curriculum.

Figure 17.6 shows B's concerns, which were examined at three occasions like those of R.

The concerns profile of B at the beginning of the first year (Practice 0) is not typical of a new user. While her awareness concerns are high (as expected of a teacher who is new to CCL), her refocusing concerns, which are expected to be initially at the same low level as the consequence and collaboration concerns, are also high. As expected, after 1 and 2 years of CCL implementation, her awareness concerns decreased, while her refocusing and collaboration concerns increased.

Conclusions and Discussion

Consistent with the idea that both students and teachers must actively process information in order to learn in a meaningful way, the three studies presented in this chapter outlined the design, implementation, and outcomes of two active learning environments. The two learning environments are CCL and CMM. Figure 17.7 presents an overview of the studies and the relationships between the research groups in each study and the active learning settings.

Table 17.5, which presents a summary of the research objectives, learning environment, thinking skills, participants, and findings indicated that our cumulative findings can serve as a basis for well-founded conclusions.

The following implications arise from these studies.

1. Active learning can serve both students and teachers by offering suitable opportunities to exchange ideas, raise questions, and suggest solutions.
2. The combination of technology—the learning environment—and pedagogy—the teaching and learning methods—promotes higher order thinking skills.
3. Technology-enhanced instruction in an active learning setting, properly adapted by teachers, enhances students' meaningful learning.

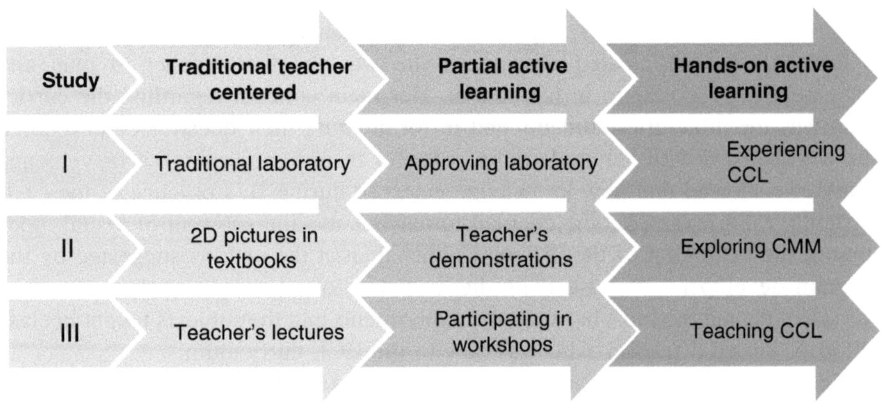

Fig. 17.7 An overview of the three studies presented as a spectrum from traditional teaching to hands-on active learning

Table 17.5 Summary of the three studies

Factor	I	II	III
Research objectives	Investigating the effect of the CCL active learning environment on students' perceptions of science laboratory	Examining whether, and to what extent, learning via CMM affect students' understanding of protein structure and function	Examining teachers' beliefs, concerns, and PCK through implementation
Learning environment	CCL learning unit integrating computerized experiments and inquiry in an active learning setting	CMM assignments including web-based applications and biochemistry learning unit	CCL learning unit integrating computerized experiments and inquiry in an active learning setting
Participants	The experimental group included 383 CCL chemistry students and the control group included 176 chemistry students	175 12th grade students divided into three learning groups: (a) hands-on CMM, (b) teachers' demonstrations of CMM, (c) learning without CMM	17 chemistry teachers who implemented CCL 1–3 years
Thinking skills	Inquiry skills and transferring among the four levels of chemistry understanding	Modeling skills, and transferring among the four levels of chemistry understanding	Inquiry skills and transferring among the four levels of chemistry understanding
Findings	Students using CLL perceived the learning environment more positively in the categories of involvement, integration, and organization in both actual and preferred situations.	Learning via hands-on CMM improved students' conceptual understanding and their ability to transfer across the four levels of chemistry understanding, and molecular representation	Alternating teachers' concerns (toward collaboration); Developing PCK
Conclusions	CLL serves as an active learning environment which creates positive atmosphere and enhances students' higher order thinking skills	CMM is a technology-oriented environment for enhancing chemical conceptual understanding and is preferable than books and model demonstrations	Teachers need an active learning setting in order to develop their own PCK and alter concern to a new curriculum

Study I

Since the 1970s the importance of chemistry laboratory in the clarification and conceptualizing of abstract ideas and processes, as well as, improving students' motivation to study science as a result of working in the laboratory in pairs or teams (Hofstein 2004; Hofstein and Lunetta 2004; Lazarowitz and Tamir 1994; Dori et al. 2005; Dori and Sasson 2008).

The findings indicate a positive attitude of all chemistry students to the laboratory. They all ranked the "Teacher support" category as the highest. "Cohesion among students was also ranked high. This is in line with the findings of Hofstein and Lunetta (1982, 2004), who noted that the goal of the laboratory is to create a learning environment that enables students to interact physically and mentally with each other and the learning materials.

Study II

Successful technology-enhanced instruction often takes advantage of models, simulations, or visualizations to introduce new ideas (Barak and Dori 2005; Hsi et al. 1997). Based on this idea, the novice biochemistry learning unit introduced technology-enhanced learning via CMM manipulation. In accordance to previous studies, our study strengthens the claim that CMM is an effective tool for representing complex molecular structures and enhancing students' conceptual understanding and learning achievements (Barak and Dori 2005; Williamson and Abraham 1995).

The integration of CMM as part of the students' learning environment enhanced students' ability to transfer across the four levels of chemistry understanding, and improved their understanding of biochemical molecules, their spatial structure, and function. It is important to note that although the results of the students who were exposed to CMM via teacher's demonstration improved (relative to their peers in the traditional classrooms), their scores were lower than those of the students who actively manipulated CMM. Our findings suggest that an effort should be made to allow individual use of CMM, even if the school lacks resources.

The presented research is innovative in embedding knowledge representation tools, i.e., CMM enabling 3D manipulation, within technology-enhanced instruction of biochemistry. The new curriculum, as oppose to traditional teacher-centered curriculum, adopts the 'investigative-approach' to biochemistry studies. In addition, the research is unique in its interdisciplinary nature, focusing on students' conceptual understanding, both from the chemical and biological aspects. Our study contributes to the body of knowledge on CMM usage as knowledge representation tools (Jackson et al. 2000) for teaching and learning.

Study III

As students need active settings which foster their active learning, so do teachers need opportunities and active learning environments for building their own knowledge. This knowledge is constructed while teachers built self-understanding and unique ways of demonstrating the new components to their students. Science teachers, who implemented the CCL curriculum, experienced new environments which fostered active learning, both for them and for their students. Support peer meetings enabled them to develop and construct their own PCK and express concerns and solutions to problems they had encountered during the implementation. Teachers who became confident regarding their newly acquired knowledge presented their "products" to their colleagues and were capable of dealing with peer criticism. This openness and readiness to share knowledge and exchange ideas is one of the characteristics of active learning based on social constructivism (Dori and Belcher 2005). In addition it also enabled us as researchers to monitor and document teachers' developing PCK, concerns and changes they went through during the actual process of their first time implementation of CCL syllabus.

The finding that indicates a relation between teachers' PCK and a change in their concerns during the CCL curriculum implementation implies that PCK and change in teachers' concerns may be interdependent. A similar notion was raised by Henze et al. (2009) who connected teachers' type of learners to the teachers' different stages of "concern" development. As teachers gained more experience and as they were exposed to new teaching approaches, their concerns regarding the consequences of introducing the innovation decreased since they felt more competent and accomplished in overcoming pedagogical problems. At the same time, these teachers' collaboration and refocusing concerns increased, as they were thinking more of how to better utilize their newly acquired PCK and share it with peers.

If this pattern will persist with a larger teacher population, then SoCQ might serve as a surrogate tool to monitor growth in teachers' PCK, meaning teachers' growth of knowledge.

In addition to the theoretical importance of this potential finding, it may have practical use for educators and curriculum developers, as they will be able to use this tool to monitor teachers' concerns and recommend special interventions to decrease them by supporting them while they acquire new PCK.

Chemistry understanding relies on making sense of the invisible and untouchable. A good understanding of chemistry requires the ability to navigate properly between four levels of chemistry understanding: macroscopic, microscopic, symbolic, and process levels (Dori and Hameiri 2003; Dori et al. 2003; Kaberman and Dori 2009b). However, research has shown that many students find it difficult to properly link between the different levels of understanding (Gable 1998; Dori and Barak 2001; Chandrasegaran et al. 2008). These difficulties, combined with difficulties in understanding the spatial structures of molecules, obstruct students' ability to solve questions and problems in chemistry (Barnea and

Dori 2000; Gabel 1998; Coll and Treagust 2003). The use of CCL and computerized molecular models (CMM) provides a solution to these problems. Indeed, research has shown that hands-on experiments and visual aids may enhance students' chemical understanding and spatial ability (Barak and Dori 2005; Williamson and Abraham 1995). The CCL and CMM learning environments gave a suitable opportunity for both students and teachers to be engaged in an active mode of learning. Analysis and discussion of students' and teachers' perceptions toward the active learning environment in a laboratory setting contributes to the expansion of the theoretical knowledge about technology-rich learning environments and about cultural and gender diversity.

Researching the perceptions of students and teachers in different laboratory environments contributes to: (1) better implementation of innovative laboratory units by teachers, (2) improved learning by students who study in CCL or CMM environments, (3) better decision making at the Ministry of Education, (4) encouragement of expanding and assimilation of the CCL- or CMM-like units, especially among honors chemistry students, (5) enhancement of higher order thinking skills among learners, and (6) teachers' professional development.

Last but not least, in recent years there is a growing migration from Middle-Eastern countries to Europe and America. The immigrants come from traditional countries with traditional teaching and learning methods and are exposed to variety of teaching and learning approaches. Translating advanced chemistry learning unites to diverse languages of minorities is an issue that concerns many science educators in Israel (Abed and Dori 2007) and around the world (Yagi 2000). Our study suggests that exposing teachers and students who were part of traditional cultures to innovative learning environments, such as CCL and CMM, has a good chance to promote a cultural change from teacher-centered to student-centered active learning. This change might enhance meaningful learning and the promotion of higher order thinking skills among students and teachers alike.

References

Abed, A. & Dori, Y. J. (2007). Fostering question posing and inquiry skills of high school Israeli Arab students in a bilingual chemistry learning environment. *Proceedings of the Annual Meeting of the National Association for Research in Science Teaching (NARST)*, New Orleans, LA, USA.

Abed, A., & Dori, Y. J. (2013). Inquiry, chemistry understanding levels, and bilingual learning. *Educación Química—Emergent topics on chemistry education*, *24*(1), 37–43.

Barak, M., & Dori, Y. J. (2005). Enhancing undergraduate students' chemistry understanding through project-based learning in an IT environment. *Science Education*, *89*(1), 117–139.

Barak, M., & Hussein-Farraj, R. (2013). Integrating model-based learning and animations for enhancing students' understanding of proteins' structure and function. *Research in Science Education, 43*(2), 619–636.

Barnea, N., & Dori, Y. J. (1999). High-school chemistry student's performance and gender differences in a computerized molecular modeling learning environment. *Journal of Science Education and Technology, 8*(4), 257–271.

Barnea, N., & Dori, Y. J. (2000). Computerized molecular modeling the new technology for enhancing model perception among chemistry educators and learners. *Chemistry Education: Research and Practice in Europe, 1*(1), 109–120.

Barnea, N., Dori, Y. J., & Hofstein, A. (2010). Development andimplementation of Inquiry-based and computerized-based laboratories: Reforming high school chemistry in Israel. *Chemistry Education Research and Practice (CERP), 11*, 218–228. http://www.rsc.org/Education/CERP

Bloom, B. S. (Ed.). (1956). *Taxonomy of educational objectives: The classification of educational goals; Handbook I, cognitive domain*. New York: David McKay.

Bonwell, C. C., & Eison, J. A. (1991). *Active learning: Creating excitement in the classroom*. ASHE-ERIC Higher Education Report No. 1. Washington, D.C.: The George Washington University.

Bruner, J. S. (1990). *Acts of meaning*. Cambridge, M.A.: Harvard University Press.

Chandrasegaran, A. L., Treagust, D. F., & Mocerino, M. (2008). An evaluation of a teacher intervention to promote students' ability to use multiple levels of representation when describing and explaining chemical reactions. *Research in Science Education, 38*(2), 237–248.

Coll, R. K., & Treagust, D. F. (2003). Investigations of secondary school, undergraduate, and graduate learners' mental models of ionic bonds. *Journal of Research in Science Teaching, 40*(5), 464–486.

Cox, A. J., Belloni, M., Dancy, M., & Christian, W. (2003). Physlets in introductory physics. *Physics Education, 38*(5), 433–440.

Donovan, W. J., & Nakhleh, M. B. (2001). Students' use of Web-based tutorial materials and their understanding of chemistry concepts. *Journal of Chemical Education, 78*(5), 975–980.

Dori, Y. J., & Barak, M. (2001). Virtual and physical molecular modeling: Fostering model perception and spatial understanding. *Educational Technology & Society, 4*(1), 61–74.

Dori, Y. J., & Belcher, J. W. (2005). How does technology-enabled active learning affect students' understanding of scientific concepts? *The Journal of the Learning Sciences, 14*(2), 243–279.

Dori, Y. J., & Hameiri, M. (2003). Multidimensional analysis system for quantitative chemistry problems—Symbol, macro, micro and process aspects. *Journal of Research in Science Teaching, 40*(3), 278–302.

Dori, Y. J., & Kaberman, Z. (2012). Assessing high school chemistry students' modeling sub-skills in a computerized molecular modeling learning environment. *Instructional Science, 40*(1), 69–91.

Dori, Y. J., & Sasson, I. (2008). Chemical understanding and graphing skills in an honors case-based computerized chemistry laboratory environment: The value of bidirectional visual and textual representations. *Journal of Research in Science Teaching, 45*(2), 219–250.

Dori, Y. J., Barak, M., & Adir, N. (2003). A web-based chemistry course as a means to foster freshmen learning. *Journal of Chemical Education, 80*(9), 1084–1092.

Dori, Y. J., Sasson, I., Kaberman, Z., & Herscovitz, O. (2004). Integrating case-based computerized laboratories into high school chemistry. *The Chemical Educator, 9*, 1–5.

Dori, Y. J., Barak, M., Herscovitz, O., & Carmi, M. (2005). Prepraring pre- and in-service teachers to teach high school science with technology. In C. Vrasidas & G.V. Glass. (Eds.), *Preparing teachers to teach with technology*, 2nd Volume of the book series: Current perspective on applied information technologies. Charlotte :Information Age Publishing.

Fraser, B. J., McRobbie, C. J., & Giddings, G. J. (1993). Development and cross-national validation of a laboratory classroom environment instrument for senior high school science. *Science Education, 77*(1), 1–24.

Fullan, M. (2002). The change. *Educational Leadership*, 59(8), 16-20.

Fullan, M., & Hargreaves, A. (Eds.). (1992). *Teacher development and educational change*. London: FalmerPress.

Fuller, F. F. (1969). Concerns of teachers: A developmental conceptualization. *American Education Research Journal, 6*, 207–226.

Gabel, D. (1998). The complexity of chemistry and implication for teaching. In B. J Fraser & K. G. Tobin (Eds.), *International handbook of science education*, Part One (pp. 233-248). Boston: Kluwer Academic Publisher.

Henze, I., van Driel, H. J., & Verloop, N. (2009). Experienced science teachers' learning in the context of educational innovation (Report). *Journal of Teacher Education, 60*(2), 184.

Hofstein, A. (2004). The laboratory in chemistry education: thirty years of experience with developments, implementation and evaluation. *Chemistry Education Research and Practice, 5*(3), 247–264.

Hofstein, A., & Lunetta, V. N. (1982). The role of the laboratory in science teaching: neglected aspects of research. *Review of Educational Research, 52*(2), 201–217.

Hofstein, A., & Lunetta, V. N. (2004). The laboratory in science education: foundation for the 21st century. *Science Education, 88*(1), 28–54.

Hofstein, A., Cohen, I., & Lazarowitz, R. (1996). The learning environment of high school students in chemistry and biology laboratories. *Research in Science and Technological Education, 14*(1), 103–116.

Hopson, M. H., Simms, R. L., & Knezek, G. A. (2001). Using a technology-enriched environment to improve higher-order thinking skills. *Journal of Research on Technology in Education, 34*(2), 109–119.

Horsley, D. L., & Loucks-Horsley, S. (1998). Tornado of change. *Journal of Staff Development, 19*(4), 17–20.

Hsi, S., Linn, M. C., & Bell, J. (1997). The role of spatial reasoning in engineering and the design of spatial instruction. *Journal of Engineering Education, 86*(2), 151–158.

Jackson, S., Krajcik, J., & Soloway, E. (2000). Model-It: A Design Retrospective. In M. Jacobson & R. Kozma (Eds.), *Advanced designs for the technologies of learning: Innovations in science and mathematics education*. Hillsdale, NJ: Erlbaum.

Johnson, D. W., Johnson, R. T., & Smith, K. A. (1998). *Active learning: Cooperation in the college classroom*. Edina: Interaction Book Company.

Johnston, R. B., & Onwuegbuzie, A. J. (2004). Mixed methods research: A research paradigm whose time has come. *Educational Researcher, 33*(7), 14–26.

Kaberman, Z., & Dori, Y. J. (2009a). Question posing, inquiry, and modeling skills of high school chemistry students in the case-based computerized laboratory environment. *International Journal of Science and Mathematics Education, 7*(3), 597–625.

Kaberman, Z., & Dori, Y. J. (2009b). Metacognition in chemical education: Question posing in the case-based computerized learning environment. *Instructional Science, 37*(5), 403–436.

Kantardjieff, K. A., Hardinger, S. A., & Van Willis, W. J. (1999). Introducing computers early in the undergraduate chemistry curriculum. *Journal of Chemical Education, 76*(5), 694–697.

Keyser, M. W. (2000). Active learning and cooperative learning: understanding the difference and using both styles effectively. *Research Strategies, 17*(1), 35–44.

Lazarowitz, R., & Tamir, P. (1994). Research on using laboratory instruction in science. In D. L. Gabel (Ed.), *Handbook of research on science teaching and learning* (pp. 94–130). New-York: Macmillan.

Linn, M. C., Layman, J. W., & Nachmias, R. (1987). Cognitive consequences of microcomputer-based laboratories: Graphing skills development. *Contemporary Education Psychology, 12*(3), 244–253.

Lipkowitz, K. B., Jalaie, M., Robertson, D., & Barth, A. (1999). Interdisciplinary learning with computational chemistry: A collaboration between chemistry and geology. *Journal of Chemical Education, 76*(5), 684–687.

Lohman, M. C., & Woolf, N. H. (2001). Self-initiated learning activities of experienced public school teachers: Methods, sources, and relevant organizational influences. *Teachers and Teaching: Theory and Practice, 7*(1), 59–74.

Loughran, J., Mulhall, P., & Berry, A. (2004). In search of pedagogical content knowledge in science: Developing ways of articulating and documenting professional practice. *Journal of Research in Science Teaching, 41*(4), 370–391.

Marjieh, C. (2007). *Assessing the learning environment of computerized case-based laboratory of high school chemistry students*. Master Thesis, Technion, Haifa, Israel (In Hebrew with Abstract in English).

Martínez-Jiménez, P., Pontes-Pedrajas, A., Climent-Bellido, M. S., & Polo, J. (2003). Learning in chemistry with virtual laboratories. *Journal of Chemical Education, 80*(3), 346–347.

Meyers, C., & Jones, T. B. (1993). *Promoting active learning strategies for the college classroom*. San Francisco, CA: Jossey-Bass Inc., Publishers.

Niemi, H. (2002). Active learning – a cultural change needed in teacher education and schools. *Teaching and Teacher Education, 18*(7), 763–780.

Russell, D. W., Lucas, K. B., & McRobbie, C. J. (2004). Role of the microcomputer-based laboratory display in supporting the construction of new understanding in thermal physics. *Journal of Research in Science Education, 41*(2), 165–185.

Shulman, L. (1986). Those who understand: Knowledge growth in teaching. *Educational Researcher, 15*(2), 4–14.

Shulman, L. (1987). Knowledge and teaching: Foundations of the new reform. *Harward Educational Review, 57*(1), 1–22.

Stratford, S. J., Krajcik, J., & Soloway, E. (1998). Secondary students' dynamic modeling processes: analyzing, reasoning about, synthesizing, and testing models of stream ecosystems. *Journal of Science Education and Technology, 7*(3), 215–234.

Towns, M. H., & Grant, E. R. (1997). "I believe I will go out of this class actually knowing something": Cooperative learning activities in physical chemistry. *Journal of Research in Science Teaching, 34*(8), 819–835.

Van Driel, J.H., & De Jong, O. (2001). Investigating the development of preservice teachers' pedagogical content knowledge. Paper presented during the NARST Annual Meeting, St. Louis, MO, March 25–28, 2001.

Van Driel, J. H., Verloop, N., & de Vos, W. (1998). Developing science teachers' pedagogical content knowledge. *Journal of Research in Science Teaching, 35*(6), 673–695.

von Glasersfeld, E. (1995). *Radical constructivism: A way of knowing and learning*. London: Falmer Press.

Williamson, V. M., & Abraham, M. R. (1995). The effects of computer animation on the particulate mental models of college chemistry students. *Journal of Research in Science Teaching, 32*(5), 521–534.

Wu, H., Krajcik, J. S., & Soloway, E. (2001). Promoting understanding of chemical representations: Students' use of visualization tool in the classroom. *Journal of Research in Science Teaching, 38*(7), 821–842.

Yagi, S. M. (2000). Language labs and translation booths: Simultaneous interpretation as a learner task. *Language, Culture and Curriculum, 13*(2), 154–173.

Chapter 18
Prospective Chemistry Teachers' Use of Student-Centered Learning During Their Teaching Practicum

Vesna Ferk Savec and Katarina S. Wissiak Grm

The understanding of human learning has undergone dramatic change in the past four decades, due to numerous investigations conducted in the educational research field (National Research Council 1999). The findings obtained through educational research necessitate change in the processes of educational systems and school practices (Gelisli 2009). However, educational research is often viewed by practitioners as not being very "user-friendly"; on the other hand, many existing school practices are inconsistent with what is known about effective science learning (National Research Council 1999; Gabel 1999; De Jong 2000). In overcoming the existing gap, both continuous training of in-service teachers and tertiary education of prospective teachers have a crucial role National Research Council (1999).

The National Research Council (1999) recommended that programs for teachers' education need to provide their students with the opportunity to develop a deep understanding of themselves, of the subject matter they will teach and an ability to facilitate students' transfer of knowledge to related areas. Within the framework of education of prospective teachers, practical pedagogical training is viewed as one of the important components of preparation for their future work (Chelimsky 1997; Trevisan 2004).

This chapter presents the teaching practice of prospective chemistry teachers (students of the third and fourth year of the Faculty of Education) during their practical pedagogical training in primary schools from the perspective of their use of student-centered forms of teaching. 'Student–centered learning' (SCL) has been regarded as an alternative approach to a teacher focused transmission of information, such as through lectures, which has been increasingly criticized (O'Neil and McMullin 2005). Bunce (2009) indicated that teaching is more than lecturing if the teacher accepts the challenge of creating an environment that is conducive to student learning. Students are thereby challenged to accept more responsibility for

V. F. Savec (✉) · K. S. Wissiak Grm
Department of Chemical Education and Informatics, University of Ljubljana,
Faculty of Natural Sciences and Engineering, Ljubljana, Slovenia
e-mail: vesna.ferk@ntf.uni-lj.si

I. Devetak and S. A. Glažar (eds.), *Learning with Understanding in the Chemistry Classroom*, DOI: 10.1007/978-94-007-4366-3_18,

their learning and the resulting shift helps them to move from blind memorization to a deeper understanding of chemistry. In such a learning environment, learning is understood as something that students are capable of accomplishing through their own work and with help from the teacher.

The concept of SCL was used as early as 1905 by Hayward and was further developed by Dewey in 1956 (O'Sullivan 2004). Carl Rogers is associated with expanding this approach into a general theory of education (Burnard 1999; Rogoff 1999). Scott et al. (1997) pointed that a SCL approach encourages students to take more responsibility for their own learning during a course. The definition in the area of learning presented by Harden and Crosby (Harden et al. 2000), describes the relationship between teacher-centered and Student-centered learning strategies as teacher-centered learning occurs when the focus is on the teacher transmitting knowledge as from an expert to a novice. Student-centered learning is when the focus is on the students' learning—what *students* do to achieve this, rather than what the *teacher* does. An even broader and a more comprehensive definition is provided by Lea et al. (2003) who summarize that SCL should include the following tenets: (1) a reliance on active rather than passive learning, (2) an emphasis on deep learning and understanding, (3) increased responsibility and accountability on the part of the student, (4) an increased sense of autonomy in the learner, (5) an interdependence between teacher and learner, (6) mutual respect within the learner—teacher relationship, and (7) a reflexive approach to the teaching and learning process on the part of both teacher and learner. Similar concepts are emphasized by Gibbs (1995) when he describes student-centered courses as those that emphasize learner activity rather than passivity; students' experience in the course outside the institution and prior to the course; process and competence, rather than content; where the key decisions about learning are made by the student through negotiation with the teacher. Gibbs also elaborates this in more detail, arguing that key decisions should include: "What is to be learnt, how and when it is to be learnt, with what outcome, what criteria and what standards are to be used, how the judgments are made and by whom these judgments are made" (Gibbs 1995). In a similar vein in earlier literature, the student–teacher relationship is elaborated by Brandes and Ginnis (1986). Their main principles of SCL are defined as: (1) the learner has full responsibility for her/his learning, (2) involvement and participation are necessary for learning, (3) the relationship between learners is more equal, promoting growth and development, (4) the teacher becomes a facilitator and resource person, (5) the learner experiences confluence in his/her education (affective and cognitive domains flow together), and (6) the learner sees himself/herself differently as a result of the learning experience.

More recently, Dee Fink (2008) proposed a "Model of Active Learning" that suggests how teachers can constitute a meaningful set of student-centered learning activities (Fig. 18.1).

Dee Fink's model (2008) suggests that all learning activities involve some kind of experience or dialog. The two main kinds of dialog are "Dialog with Self" and "Dialog with Others." The two main kinds of experience are "Observing" and "Doing." Each of the four modes of learning described in the model above has its own value, and using

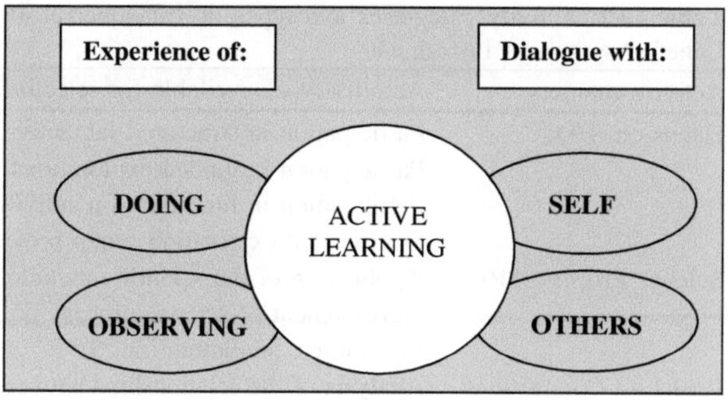

Fig. 18.1 Model of active learning; modified by Dee Fink (2008)

Table 18.1 Classroom observation measure (Ross and Smith 1996)

	Indicators
Overall observation (COM, Part 4)	Cooperative/collaborative learning
	Direct instruction with the entire class
	Math/reading/subject groups
	Independent work
	Independent or group work centers
	Systematic individual instruction
	Individual tutoring
	Teacher provided feedback
	Teacher distributed feedback evenly
	Sustained writing/composition
	Computer as a tool or resource
	Other technologies used as tools or resources
	Integration of subject areas
	Experiential hands-on learning
	Alternative assessment strategies
	Student self-assessment
	Student discussion
	Use of questioning strategies
	Use of a variety of evaluative strategies
	Teacher acted as a coach/facilitator

them can add variety and thereby make learning more interesting for the learner. Furthermore, according to Dee Fink's model (2008) properly connected, the various learning activities can have an impact that is more than additive or cumulative; they can be synergetic and thereby multiply the educational impact.

Various instruments have been developed for the purpose of classroom observations when monitoring the extent to which certain instructional processes or strategies are being used or demonstrated (e.g., Ross and Smith 1996; Cirino et al. 2007; Waxman et al. 2009). For example, Ross and Smith (1996) developed the Classroom Observation Measure (COM), which in Part 4, Overall Observation, includes 20 indicators described in Table 18.1. The COM has been found to be reliable and valid (Ross et al. 1997).

Table 18.2 Activity categories and types in Talanquer et al. (2010) "Teaching and Learning Beliefs Assessment Instrument"

Activity category	Activity types available for selection within a category
Hands-on (HO)	Participation in structured lab activities (no. 1)
	Participation in guided explorations (no. 5)
	Participation in fun hands-on activities (no. 15)
	Open-ended exploration group projects (no. 17)
Science Process (SP)	Application of the scientific method (no. 6)
	Development of science process skills such as observing, making hypotheses, predicting (no. 9)
Real-Life Connections (RLC)	Analysis of the relationships between science and society (no. 2)
	Description of real-life applications of the scientific ideas discussed in class (no. 14)
Personal Ideas (PI)	Discussion of students' personal ideas about scientific concepts (no. 3)
	Discussion of different people's approaches to exploring the natural world (no. 16)
Problem Solving (PS)	Explanation of strategies for solving numerical problems (no. 11)
	Resolution of numerical problems (no. 12)
	Analysis of students' strategies for solving a problem (no. 13)
Knowledge and Understanding (K&U)	Explanation of important scientific facts (no. 4)
	Discussion of central scientific ideas in the discipline (no. 7)
History and Philosophy (H&P)	Reflection about the nature of scientific work (no. 8)
	Discussion of historical events associated with the development of scientific ideas (no. 10)

Talanquer et al. (2010) developed the classification of various activity categories into different activity types of science lesson (Table 18.2) within the framework of the "Teaching and Learning Beliefs Assessment Instrument" (Section C).

Talanquer et al. (2010) report that there was a significant difference in the preferences of entering science teacher candidates for activity types belonging to the categories: Hands-on (HO), Science Process (SP), and Real-Life Connections (RLC) versus activities in the categories Personal Ideas (PI), Problem Solving (PS), Knowledge and Understanding (K&U), and History and Philosophy (H&P). Activity types in the first set of categories (HO/SP/RLC) were selected more frequently than those in the second set (PI/PS/K&U/H&P).

In the light of the above, when planning this study we were eager to get a deeper insight into our prospective chemistry teachers' activities during their teaching within the framework of practical pedagogical training and, based on their reflection about their experiences, better understand their activities. We assumed that we would consequently be able to improve our program of practical pedagogical training in order to provide better support to future generations of prospective chemistry teachers on the part of the university.

Problem Definition and Scope of the Study

The study attempted to discover which student-centered learning methods prospective teachers use in their teaching during their PPT experience, why and to what extent. Based on the results of the study, we would like to improve tertiary education of prospective teachers and thereby influence teaching and learning of chemistry in primary schools.

Research Questions

(1) Which student-centered methods do prospective teachers use in their teaching and to what extent?
(2) Why do prospective teachers apply particular student-centered methods?

Method

Sample

The sample consisted of two groups of prospective teachers:

- Third-year students of the Faculty of Education, University of Ljubljana—prospective teachers of chemistry in primary schools: N = 14 students, average age = 22.3, 2 males and 12 females
- Fourth-year students of the Faculty of Education, University of Ljubljana—prospective teachers of chemistry in primary schools: N = 16 students, average age = 23.1, 1 males and 15 females

Data Collection

Students' practical pedagogical training (PPT) was conducted in April 2009 at 8 primary schools in the Urban Municipality of Ljubljana, Slovenia. Students from the third and fourth year of the Faculty of Education spent one week (5 days) at one of the primary schools, to which they had been previously introduced. Groups of 2–3 of these students conducted PPT at the same time and in the same school.

On average, every student conducted 7.2 lessons and observed 14.3 lessons. Before taking lessons in the classroom, students were acquainted by supervising teachers with the content of current classroom activities, text books, and workbooks that pupils bring to the chemistry lessons. However, the education students were expected to develop the lesson plan by themselves, although teacher supervisors as well as university teachers were available to support them.

Instruments

Classroom Observations

During PPT, third and fourth year students conducted chemistry lessons at a particular primary school. Student peers wrote classroom observations about the teaching conducted by their peers during all PPT lessons. As stated already by Goodwin (1995), a researcher using observational methods must be prepared to counter several problems, such as absence of control, possibilities of observer bias and subject reactivity. In order to reduce biasing effects significantly, the behavior which had to be observed was defined precisely and students were given prior training as classroom observers. Over the course of the last several years of accompanying students to their PPT in primary schools, we have developed and optimized a Science Classroom Observational Protocol (SCOP) and, as part of the students' preparation for PPT, we gave them preliminary training in its use for observation. Students used the SCOP while observing in-service teachers conducting lessons in the classroom throughout the winter semester. In this way, we defined the target behavior of in-service teachers which had to be observed and described by prospective teachers during a selected lesson.

Using the SCOP, the following aspects were systematically followed and described by observers: (1) timeline with current content of the ongoing lesson, (2) indication of teacher-centered (TCA) or student-centered activities (SCA), (3) categorization of student-centered activities into 11 categories of instructional activities: (a) Discussion between pupils and teacher based on pupils' everyday experiences and observations (abbreviation: Everyday experiences); (b) Activities based on games, rebuses, stories, cartoons, movies, etc. (abbreviation: Games and stories); (c) Work with models in groups of pupils accompanied by discussion of results and observations (abbreviation: Work with models); (d) Teacher's presentation of models accompanied by pupils' interpretation of observations (abbreviation: Discussion about models); (e) Presentation of animation of a chemical reaction and its interpretation by pupils' active involvement (abbreviation: Animation of chem. reaction); (f) Experimental work in groups of pupils with explanation of results and observations by students (abbreviation: Pupils' experim. work); (g) Demonstration of a chemical experiment with cooperative explanation of results and observations in small groups (abbreviation: Dem. of chem. experim.); (h) Pupils learning through role playing (abbreviation: Role playing); (i) Preparation of poster presentations by groups of pupils (abbreviation: Poster presentation); (j) Cooperative solving of worksheets (abbreviation: Worksheets); (k) Summarizing discussion between pupils and teacher and synthesis of knowledge gained (abbreviation: Summary and synthesis).

There are several types of observational research as a quantitative method of measuring classroom behavior by direct observation, which specify both events and behaviors that should be observed and how they should be recorded (Waxman 2003). They are designed to obtain information systematically on a student's classroom behavior in the context of an ongoing instructional-learning process (Waxman et al. 2009; Bartholomew et al. 2004; Zohar and Schwartzer 2005; Talanquer et al. 2010). We decided to use SCOP as a systematic observational tool, since it has been designed, developed, and optimized with previous generations of students obtaining practical pedagogical training in our science classrooms (Wissiak Grm and Ferk Savec 2007). In comparison with the categorization of Talanquer et al. (2010), the SCOP categorization is more specific in describing particular activities; it includes the type of learning materials or tools used and an indication of the pupils' role in the learning process.

As suggested by Goodwin (1995), another way of controlling for observer bias, in addition to checklists and observer training, is to use several observers and see whether and to what extent their records agree. In our study, the observers were peer students and a university teacher (one of the authors of the article). For this study, interobserver agreement (Cohen's kappa) was found to be very good, with the interobserver reliability coefficient being 0.92 for the SCOP.

Goodwin (1995) also pointed out that it is possible to reduce bias by audio or video taping lessons and, as a result, the objectivity of the study would increase. In view of the well-known ethical dilemmas associated with issues of invasion of privacy (which inevitably arise when lessons are video or audio taped) and the fact that Slovenian legislation related to primary schools is not favorably inclined, we were unable to use this specific method in our study.

Students' Reflective Essays

In the week after PPT, students were asked to sum up their thoughts about their experiences in a reflective essay. In their writing, students were directed to address the following: (1) describe which student-centered learning methods they used during PPT, (2) explain the purpose of each student-centered learning method used and reflect on how they felt about their success in achieving the goals of these lessons.

Data Analysis

Analysis of Classroom Observations

From the collected 112 lessons accompanied by classroom observations, 20 typical lessons were selected by stratified random selection (10 lessons conducted by third year students and 10 by fourth year students), and analyzed in detail. In order to ensure inter-rater reliability which refers to the consistency of scores that are assigned by two independent raters (Moskal and Leydens 2000), we decided to set criteria to guide the rating process. In response to the risk that two independent raters may not assign the same score to a given lesson, we developed scoring rubrics for the raters and formalized the criteria for each score level. Descriptions of the score levels were additionally used to guide the evaluation process that followed.

In the evaluation process, therefore, bearing in mind the developed scoring rubrics regarding student-centered learning methods, two classroom observations (peer-student, university teacher) for each of the lessons were independently analyzed by both authors of this article to check the overlapping of observations between observers. In order to assess inter-rater reliability, which is defined as the extent to which different coders (two or more), each coding the same content, reach the same coding decisions (Rourke et al. 2000) and the percentage agreement between peers and university teachers, Holsti's coefficient of reliability (C.R) was calculated, which reflects the number of agreements per total number of coding decisions (Rourke et al. 2000; Fahy et al. 2000) and was found to be 96 %.

Analysis of Students' Reflective Essays

In order to develop the appropriate scoring tool to consider the content, construction, and criterion-related evidence that should be examined in students' reflective essays, the large amount of data was first analyzed independently by both authors (Moskal and Leydens 2000). In this context, we first identified the natural units of meaning regarding each of the suggested topics (1. which student-centered learning methods students believed that they had used during PPT, 2. what aims students listed for the use of particular student-centered learning methods, and 3. how successful they felt they had been in achieving the learning goals). Second, the natural units of meaning that had emerged from the described independent analysis were discussed, and units with similar meaning were amalgamated. In this way, codes were ascribed to each of the final natural units of meaning for each of the questions in the reflective essays, enabling a coding table to be established. Inter-rater reliability was 0.95.

Results and Discussion

First research question
Which student-centered methods do prospective teachers use in their teaching and to what extent?

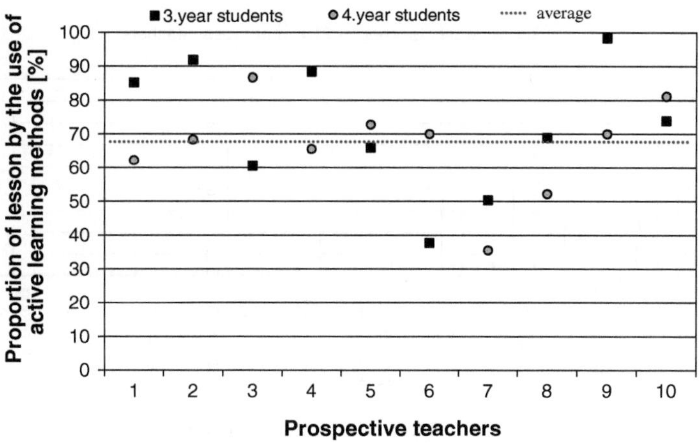

Fig. 18.2 Proportion of time in lesson in which prospective teachers used different student-centered methods

The classroom observations revealed that both groups of prospective chemistry teachers devoted a significant proportion of time during chemistry lessons to pupils' active involvement in the learning process by the use of different student-centered methods. Prospective teachers on average devoted 69.28 % of the lesson; third year students on average 72.11 % of the lesson, and fourth year students 66.44 % of the lesson. It is important to note in this that differences among the teachers within groups are quite large (third year students: min = 37.78, max = 98.33; fourth year students: min = 35.56, max = 86.67) (Fig. 18.2).

Of the 11 student-centered learning methods listed in SCOP, third year students included on average 3.2 different student-centered methods per lesson and fourth year students on average 3.4 different student-centered methods per lesson.

Figure 18.3 shows which student-centered methods were used most among the prospective teachers in our sample. *Discussions about pupils' everyday experiences and observations* and *cooperative solving of worksheets* were used by more than half of the teachers from both groups of prospective teachers. It can be seen from Fig. 18.3 that third-year students considered the active involvement of pupils with models more important than did their fourth-year student peers. It is also possible to see from Fig. 18.3 that prospective teachers highly value pupils' active involvement in experimental work (9/10 third-year students, 5/10 fourth-year students integrated experimental work into the lesson, either as pupils' group work or as guided pupils' observation of a demonstration experiment and collaborative explanation of the results). It is also evident from Fig. 18.3 that many prospective teachers believe that it is worth encouraging pupils to be actively involved in summarizing the discussion and synthesis of the knowledge gained.

Comparison of SCOP categories with the Talanquer et al. (2010) system of eight activity categories (described earlier in this article) revealed that the SCOP

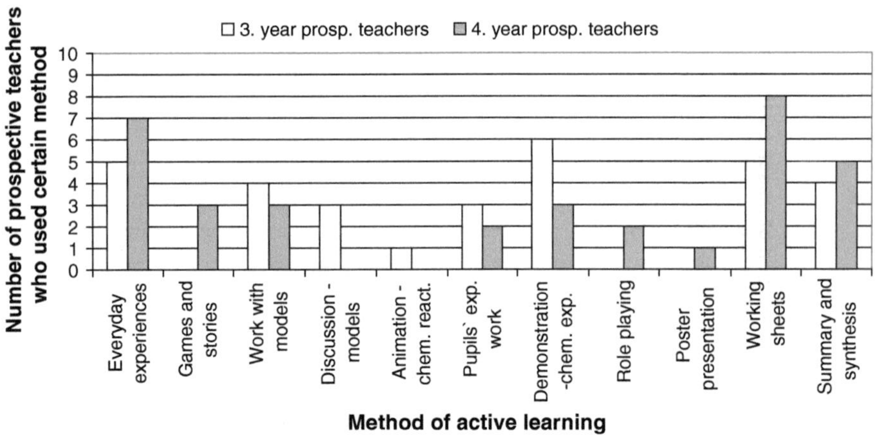

Fig. 18.3 Student-centered learning methods integrated into lessons by prospective teachers

categories that were used most by our prospective teachers can be categorized into three of the Talanquer et al. (2010) activity categories: *Hands-on* (in SCOP: Group work with models, Cooperative solving of work sheets), *Science Process* (in SCOP: Students' experimental work, Discussion–demonstration of chemical experiment) and *Real-Life Connections* (in SCOP: Discussion–everyday experiences). Talanquer et al. (2010) also found in their research that an entering science teacher most preferred the same three activity categories.

Many teachers state in general that they do not use student-centered learning methods in their teaching because they take too much time. We therefore considered the student-centered learning methods in terms of their "time consumption." Figure 18.4 indicates that the average time that an individual group of prospective teachers devoted to a particular student-centered learning method varied greatly. This is probably due to the wide range of purposes and contents in which particular student-centered learning methods can be used. As could be expected, when pupils conduct hands-on activities (e.g., group work with models, students' experimental work) a longer time is required than for similar activities in which they are involved as observers and through discussion about the results (e.g., discussion—models presented, discussion—demonstration of chemical reaction).

In order to illustrate the nature of particular student-centered learning methods found in our investigation, the range of their duration is presented in Fig. 18.5. Although the time used for student-centered learning methods varies greatly for most activities, the widest range was observed with *Group work with models* (max time: 33 min, min time: 4 min), *Cooperative solving of working sheets* (max time: 25 min, min time: 5 min) and, similarly, with *Summary and synthesis of knowledge* (max time: 21 min, min time: 2 min).

It can be seen from the results in Figs. 18.4 and 18.5 that time constraints should not be an obstacle preventing teachers from implementing student-centered

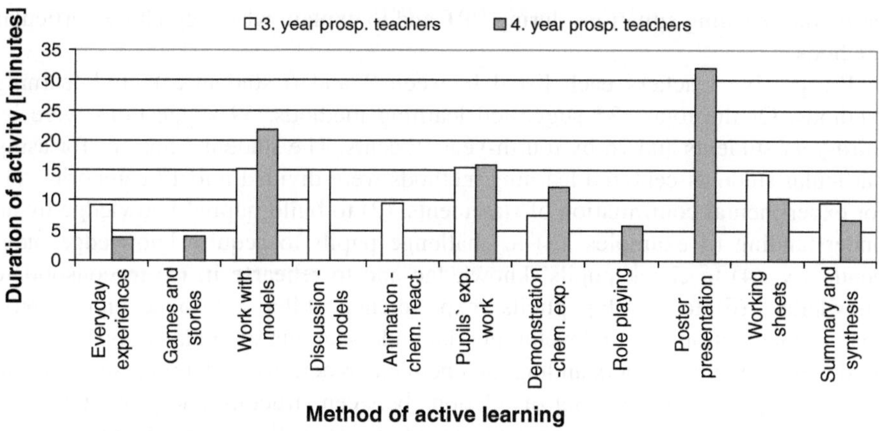

Fig. 18.4 Average duration of student-centered learning methods

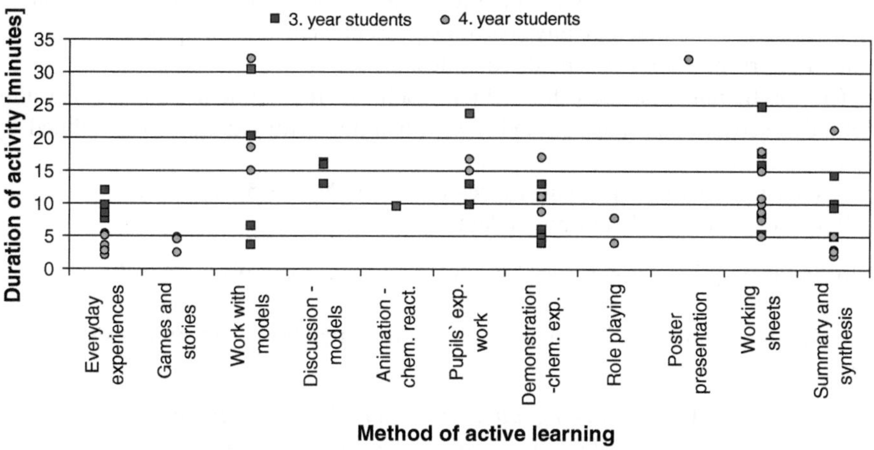

Fig. 18.5 Duration of student-centered learning methods

learning methods in school practice. Namely, prospective teachers use particular student-centered learning methods for a various time periods, which mostly take less then 20 min (with exception of poster presentaion).

Second research question
Why do prospective teachers apply particular student-centered methods?

In the reflective essays that students wrote after they had finished their PPT, they were asked to state which student-centered learning methods they used in particular teaching situations during PPT and to explain why they chose particular methods.

Prospective teachers each listed between 2 and 6 student-centered learning methods. Of the total 135 suggested learning methods, 59 suggestions were by third-year students and 76 by fourth-year students. The students' reasons for using particular students-centered learning methods were divided into 17 categories: (1) for experimental confirmation of statements, (2) to build pupils' knowledge by an understanding of examples, (3) to challenge pupils to acquire knowledge independently, (4) to check pupils' knowledge and to rehearse it, (5) to consolidate knowledge, (6) to develop pupils' experimental skills, (7) to develop pupils' spatial imagination, (8) to develop pupils' understanding of chemical bonding, (9) to develop pupils' understanding of chemical reactions and processes, (10) to develop pupils' understanding of relations between structure and properties, (11) to enable pupils to participate in the learning process actively through a game, (12) to help pupils to remember, (13) to help pupils better understand the submicro world based on macroscopic representations, (14) to help pupils relax, (15) to examine pupils' pre-knowledge, (16) to make lessons more interesting for pupils by their active mental and physical engagement, and (17) to practice chemical stoichiometry based on practical examples.

The particular purposes for which each of the student-centered learning methods used are summarized and then illustrated with examples (as defined by students) as follows:

Discussion—everyday experiences ($N = 11$; $N_{3y} = 5$, $N_{4y} = 6$)[1]

Purposes of use as listed by students:

- to build pupils' knowledge by understanding examples ($N = 6$; $N_{3y} = 3$, $N_{4y} = 3$)
- to develop pupils' understanding of chemical reactions and processes ($N = 3$; $N_{3y} = 1$, $N_{4y} = 2$)
  - *Example: »When learning about polymers in everyday life, pupils had to classify various products in terms of the materials used. Pupils learn that there are many different kinds of polymer materials.«*
- to challenge pupils through independent acquisition of knowledge ($N = 1$; $N_{3y} = 0$, $N_{4y} = 1$)
  - *Example: »I gave articles about carbon monoxide and its effect on people to pair of pupils. I found the articles on the web news-page entitled "24 ur" and I added three short questions in a list below each of the articles to guide pupils during studying.«*
- to discover pupils' pre-knowledge ($N = 1$; $N_{3y} = 1$, $N_{4y} = 0$)

[1] N-number of all students; N_{3y}-number of students from third year, N_{4y}-number of students from fourth year.

– *Example: »Short questions at the beginning of the lesson are a good method of discovering pupils' pre-knowledge. I first asked general questions related to everyday life. When I didn't get the answers that I wanted, I asked additional questions to discover how the pupils think.«*

Games and stories (N = 21; $N_{3y} = 0$, $N_{4y} = 21$)
Purposes of use as listed by students:

- to build pupils' knowledge by understanding examples (N = 6; $N_{3y} = 0$, $N_{4y} = 6$)
  – *Example: »Through playing the connection game, pupils learn which kinds of molecules are present in certain foods, e.g., carbohydrates in pasta, proteins in milk,* etc.*«*
- to develop pupils' understanding of chemical reactions and processes (N = 2; $N_{3y} = 0$, $N_{4y} = 2$)
  – *Example: »As an introduction to new topics, each pupil received a piece of paper that he/she had to put in the right place on a table drawn on the blackboard. In such a way, pupils developed an understanding of the underlying chemical reactions and were actively involved in the learning process.«*
- to check pupils' knowledge and to rehearse it (N = 4; $N_{3y} = 0$, $N_{4y} = 4$)
- to make lessons more interesting for pupils (N = 5; $N_{3y} = 0$, $N_{4y} = 5$)
- to enable pupils to participate actively in the learning process through a game (N = 1; $N_{3y} = 0$, $N_{4y} = 1$).
  – Example: *»At the end of the topic of natural polymers, I prepared 10 questions on the topic in the form of a quiz. Pupils split into smaller groups and were very competitive in answering. I was satisfied because we had rehearsed the knowledge gained in such a dynamic and—for the pupils—interesting way.«*
- to help pupils to relax (N = 1; $N_{3y} = 0$, $N_{4y} = 1$)
- to challenge pupils with independent acquiring of knowledge (N = 2; $N_{3y} = 0$, $N_{4y} = 2$)
  – Example: *»Instead of me explaining what are the components of air, pupils discovered the answers by solving a rebus. It was a complete success. Pupils quickly got the right answers and also relaxed.«*

Group work with models (N = 17; $N_{3y} = 9$, $N_{4y} = 8$)
Purposes of use listed by students:

- to build pupils' knowledge by an understanding of examples (N = 1; $N_{3y} = 0$, $N_{4y} = 1$)

- Example: »*Pupils in groups had to assemble models of molecules that are examples of alcohols with varying numbers of carbon atoms.*«

- to check pupils' knowledge and to rehearse it ($N = 2$; $N_{3y} = 1$, $N_{4y} = 1$)
  - Example: »*By their ability successfully to assemble models, pupils checked their understanding of new topics, which was also feedback for me.*«
- to make the lesson more interesting for pupils ($N = 2$; $N_{3y} = 1$, $N_{4y} = 1$)
- to challenge pupils to acquire new knowledge independently ($N = 1$; $N_{3y} = 1$, $N_{4y} = 0$)
- to discover pupils' pre-knowledge. ($N = 1$; $N_{3y} = 1$, $N_{4y} = 0$)
  - Example: »*First of all, I intended to make the lesson more interesting and challenging for pupils but from pupils' problems in assembling models, I also discovered about their pre-knowledge.*«
- to develop pupils' spatial imagination. ($N = 4$; $N_{3y} = 2$, $N_{4y} = 2$)
- to help pupils better understand the submicro world based on macroscopic representations ($N = 1$; $N_{3y} = 0$, $N_{4y} = 1$)
- to develop pupils' understanding of chemical bonding ($N = 4$; $N_{3y} = 2$, $N_{4y} = 2$)
  - Example: »*The purpose of pupils' group work with models was to develop their understanding of chemical bonding in organic molecules. They also thereby develop spatial imagination.*«
- to help pupils to relax ($N = 1$; $N_{3y} = 1$, $N_{4y} = 0$)
  - Example: »*I believe that assembling models is relaxing for pupils, they like it very much.*«

Discussion—models presented ($N = 4$; $N_{3y} = 3$, $N_{4y} = 1$)
Purposes of use as listed by students:

- to build pupils' knowledge by an understanding of examples ($N = 1$; $N_{3y} = 1$, $N_{4y} = 0$)
- to develop pupils' understanding of chemical bonding ($N = 3$; $N_{3y} = 2$, $N_{4y} = 1$)
  - *Example: "I showed pupils examples of models of monosaccharide, disaccharides, and polysaccharides for easier understanding of their chemical structure based on examples."*

Interpretation of animations of chemical reaction ($N = 1$; $N_{3y} = 1$, $N_{4y} = 0$)
Purposes of use as listed by students:

- to develop pupils' understanding of chemical reactions and processes ($N = 1$; $N_{3y} = 1$, $N_{4y} = 0$).

- *Example: »Pupils saw animations of two chemical reactions on a videotape, we then discussed them in terms of reactants and products. Finally, the pupils entered the results into the worksheets. I believe that the method was appropriate because the pupils were successful in solving further examples of chemical reactions in the worksheet.«*

Pupils' experimental work ($N = 25$; $N_{3y} = 12$, $N_{4y} = 13$)
Purposes for the use as listed by students:

- to facilitate pupils remembering better ($N = 2$; $N_{3y} = 0$, $N_{4y} = 2$)
- to develop pupils' experimental skills ($N = 7$; $N_{3y} = 4$, $N_{4y} = 3$)
- to build pupils' knowledge by understanding of examples ($N = 2$; $N_{3y} = 1$, $N_{4y} = 1$)
  - *Example: »When pupils are conducting chemical experiments by them-selves, they better remember which reagents they used; for example, iodine solution to test for the presence of starch in food. The students also thereby develop experimental skills and they understand—based on the results of experimental work—in which food starch is present.«*
- to develop pupils' understanding of chemical reactions and processes ($N = 9$; $N_{3y} = 4$, $N_{4y} = 5$)
  - *Example: »The purpose of this experiment was for the pupils to discover what happens to bread in the mouth from chemical and biological points of view. Pupils guessed that there is starch in bread, which falls apart into smaller units due to enzymes in the saliva.«*
- to make the lesson more interesting for pupils by their active mental and physical engagement ($N = 1$; $N_{3y} = 1$, $N_{4y} = 0$)
- to challenge pupils to acquire new knowledge independently ($N = 2$; $N_{3y} = 1$, $N_{4y} = 1$)
- to discover pupils' pre-knowledge ($N = 1$; $N_{3y} = 1$, $N_{4y} = 0$)
  - *Example: »I wanted to discover what pre-knowledge pupils had, which they had to use in explaining the experimental results. The pupils had thus to make an effort to gain knowledge.«*
- experimental confirmation of statements, e.g., from books etc. ($N = 1$; $N_{3y} = 0$, $N_{4y} = 1$)
  - *Example: »Based on their experimental work, students came to conclusions that confirmed statements in their chemistry workbook.«*

Discussion–demonstration of chemical experiments ($N = 33$; $N_{3y} = 21$, $N_{4y} = 12$)
Purposes of use as listed by students:

- to facilitate pupils remembering better ($N = 1$; $N_{3y} = 1$, $N_{4y} = 0$)

- *Example: »Students performing the experiments themselves makes them more motivated for learning and improves their remembering.«*

- to build pupils' knowledge by understanding examples ($N = 1$; $N_{3y} = 1$, $N_{4y} = 0$)
- to develop pupils' understanding of chemical reactions and processes ($N = 19$; $N_{3y} = 11$, $N_{4y} = 8$)
- to make the lesson more interesting for pupils by their active mental and physical engagement ($N = 7$; $N_{3y} = 3$, $N_{4y} = 4$)
- to practice chemical stoichiometry based on practical examples ($N = 2$; $N_{3y} = 2$, $N_{4y} = 0$)
  - *Example: »Performing the chemical experiment between HCl and NH_3 solutions motivates pupils to write a chemical equation for the specific chemical reaction observed.«*
- experimental confirmation of statements, e.g., from books etc. ($N = 2$; $N_{3y} = 2$, $N_{4y} = 0$)
  - *Example: »A chemical experiment demonstrating how nitrogen can be proved to be present in human hair can be used as good introductory motivation for pupils.«*
- to develop pupils' understanding of the relation between structure and properties ($N = 1$; $N_{3y} = 1$, $N_{4y} = 0$)
  - *Example: »Performing several experiments connected with determining the properties of matter leads children to a recognition of the structure-properties relationship.«*

Role playing ($N = 4$; $N_{3y} = 1$, $N_{4y} = 3$)
Purposes of use as listed by students:

- to develop pupils' spatial imagination ($N = 1$; $N_{3y} = 1$, $N_{4y} = 0$)
  - *Example:* »Work with models in groups of pupils enables them to practice how many bonds can be formed, as well as to improve their spatial imagination.«
- to help pupils better to understand the submicro world based on macroscopic representations ($N = 1$; $N_{3y} = 0$, $N_{4y} = 1$)
- to develop pupils' understanding of chemical bonding ($N = 1$; $N_{3y} = 0$, $N_{4y} = 1$)
- to enable pupils to participate actively through a game ($N = 1$; $N_{3y} = 0$, $N_{4y} = 1$)
  - *Example: »A didactic game in which pupils, through role playing, acquire new knowledge about the structure of amino acids is a very useful tool on their path towards understanding the structure of polypeptides.«*

Poster presentation (N = 3; $N_{3y} = 0$, $N_{4y} = 3$)
Purposes of use as listed by students:

- to make the lesson more interesting for pupils by their active mental and physical engagement (N = 1; $N_{3y} = 0$, $N_{4y} = 1$)
  - *Example: »A poster presentation was good stimulation for students to learn about polypeptides and properly to present their work and knowledge to others.«*
- to develop pupils' understanding of chemical bonding (N = 1; $N_{3y} = 1$, $N_{4y} = 0$)
- to consolidate knowledge (N = 1; $N_{3y} = 0$, $N_{4y} = 1$)
  - *Example:* »Work with models in groups of pupils enables them to practice knowledge already well gained about the shapes of molecules.«

Cooperative solving of worksheets (N = 7; $N_{3y} = 3$, $N_{4y} = 4$)
Purposes of use as listed by students:

- to develop pupils' understanding of chemical reactions and processes (N = 1; $N_{3y} = 1$, $N_{4y} = 0$)
- to check pupils' knowledge and to revise it (N = 3; $N_{3y} = 1$, $N_{4y} = 2$)
  - *Example: »In order to discover pupils' understanding of the knowledge gained during the lesson, a method was used that offered the opportunity to give pupils enough time to listen to the teacher's explanation as well as to enable them simultaneously to solve the tasks given by the teacher.«*
- to challenge pupils to acquire new knowledge independently (N = 3; $N_{3y} = 1$, $N_{4y} = 2$)
  - *Example: »Pupils need to think individually as well as to share their opinions in a team. This leads them to develop their critical thinking and supports their growing sense of managing and working in a team.«*

Summary and synthesis of knowledge (N = 7; $N_{3y} = 3$, $N_{4y} = 4$)
Purposes of use as listed by students:

- to check pupils' knowledge and to revise it (N = 6; $N_{3y} = 2$, $N_{4y} = 4$)
  - *Example: »At the end of the lesson I asked students to solve a task in their notebook and I walked through the class. I looked into the notebook of each of the students and gave feedback.«*
- to make the lesson more interesting for pupils by their active mental and physical engagement (N = 1; $N_{3y} = 1$, $N_{4y} = 0$)
  - *Example: »Pupils were mentally and physically active because they had to write down formulas on the blackboard in front of the class. They also discovered whether they really understood.«*

In addition to the categories of student-centered learning presented above, which appeared from an analysis of classroom observations of students' sample lessons (equal to those presented in research question 1), a further category appeared from students' descriptions of student-centered learning methods used during their overall PPT (reflective essay)—*Pupils' self-study from books and workbooks.*

Pupils' self-study from books and workbooks ($N = 2$; $N_{3y} = 1$, $N_{4y} = 1$)
Purposes of use as listed by students:

- to make the lesson more interesting for pupils by their active mental and physical engagement ($N = 1$; $N_{3y} = 1$, $N_{4y} = 0$)
- to challenge pupils to acquire new knowledge independently ($N = 1$; $N_{3y} = 0$, $N_{4y} = 1$)
  - *Example: »It is challenging for students to read a text from their workbooks and to discuss it with their peers. In this way, they are physically and mentally involved in learning.«*

It is evident from the analysis of the students' reflective essays that although they generally described methods in which pupils are actively involved in the construction of new knowledge, several examples of learning methods in which pupils were» active" just by listening and writing down the teacher's explanations were also defined as student-centered learning methods.

- *Example: »I presented new topics by PowerPoint, in which I included a lot of visual material. This was an additional explanation of my words, and pupils discovered more easily what is important and must be written down in their notebooks.«*
- *Example: »I showed pupils a photo of a cupola in the form of a fullerene built by the famous architect Mr. Richard Fuller. In this way, they discovered an additional interesting fact that indicates the origin of the name.«*

The final examples indicate that students' understanding of student-centered learning methods is not congruent and is closely related to issues raised by Huet et al. (2009) and Bonwell and Eison (1991), indicating that it is necessary to clearly define what an active role of pupils is in the learning process means.

Conclusions and Implications for Teaching

The main findings of the study and their implications for future practical pedagogical training of prospective chemistry teachers as part of their tertiary education are as follows:

Use of Student-Centered Learning Methods in Teaching Chemistry

The presented research showed that all groups of prospective chemistry teachers devote significant proportions of chemistry lessons to students' active involvement in the learning process by the use of different student-centered learning methods (from 66.44 to 72.11 %). The following student-centered learning methods were found: (1) Discussion between pupils and teacher based on pupils' everyday experiences and observations; (2) Activities based on games, rebuses, stories, cartoons, movies; (3) Work with models in groups of pupils, accompanied by discussion of results and observations; (4) Teacher's presentation of models, accompanied by pupils' interpretation of observations; (5) Presentation of animation of a chemical reaction and its interpretation by pupils' active involvement; (6) Experimental work in groups of pupils, with an explanation of the results and observations; (7) Demonstration of a chemical experiment with cooperative explanation of results and observations; (8) Pupils learning through role playing; (9) Preparation of poster presentations in groups of pupils; (10) Cooperative solving of worksheets; and (11) Summarizing discussion between pupils and teacher and synthesis of knowledge gained.

The above SCOP categories can be categorized into three of the activity categories of Talanquer et al. (2010): *Hands-on* (in SCOP: Group work with models, Cooperative solving of worksheets), *Science Process* (in SCOP: Students' experimental work, Discussion–demonstration of chemical experiment) and *Real-Life Connections* (in SCOP: Discussion–everyday experiences). The findings are congruent with the research of Talanquer et al. (2010), which revealed that entering science teachers in his sample most prefer the same three activity categories as our pre-service teachers.

The above methods were used by different groups of prospective teachers to a different extent, e.g., their duration varied from a minimum time of 2 min observed in *Summary and synthesis of knowledge* to a maximum time of 33 min devoted to *Group work with models*.

Purposes of Use of Student-Centered Learning Methods in Teaching Chemistry

Prospective teachers listed a number of reasons for the use of specific student-centered learning methods, which were divided into 17 categories. The students' aim behind these reasons can be seen as a wish to influence the following three aspects of pupils' learning: (1) *science knowledge and skills*—through the development of new content knowledge with understanding, development of process skills, checking, and consolidation of students' knowledge, (2) *interest in learning*—through pupils' active physical and cognitive engagement, and (3) *pleasure during learning*—through the selection of learning activities that are fun and relaxing.

The reasons students gave for using particular student-centered learning methods show that most prospective teachers are aware of the wide range of possibilities of effective use of student-centered learning methods in teaching chemistry.

It would be worth extending this study through further research, e.g., in the following areas: (1) Evaluation of the quality of knowledge gained by pupils through specific student-centered learning methods in comparison with methods in which pupils are more passive; (2) Study of added value in terms of pupils' competences gained through specific student-centered learning methods; (3) Evaluation of the efficiency of prospective chemistry teachers' use of specific student-centered learning methods in comparison to their use by in-service chemistry teachers; and (4) Investigation of prospective chemistry teachers' capacity to select the most suitable method for a particular classroom situation.

References

Bartholomew, H., Osborne, J., & Ratclife, M. (2004). Teaching students»Ideas-about-science«: Five dimensions of effective practice. *Science Education, 88*(5), 655–682.

Bonwell, C. C., & Eison, J. A. (1991). *Active learning: Creating excitement in the classroom.* Washington, D. C.: The George Washington University.

Brandes, D., & Ginnis, P. (1986). *A guide to student centred learning*. Oxford: Blackwell.

Bunce, D. M. (2009). Teaching is more than lecturing and learning is more than memorizing. *Journal of Chemical Education, 86*(6), 674–680.

Burnard, P. (1999). Carl Rogers and postmodernism: Challenges in nursing and health sciences. *Nursing and Health Sciences, 1*(4), 241–247.

Cirino, P. T., Pollard-Durodola, S. D., Foorman, B. R., Carlson, C. D., & Francis, D. J. (2007). Teacher characteristics, classroom instruction, and student literacy and language outcomes in bilingual kindergartners. *The Elementary School Journal, 107*(4), 341–364.

Chelimsky, E. (1997). The political environment of evaluation and what it means for the development of the field. In E. Chelimsky & W. Shadish (Eds.), *Evaluation for the 21st century* (pp. 53–68). Thousand Oaks: Sage Publications.

Dee Fink, L. (2008). Active learning. Retrieved June 5, 2008, from http://www2.honolulu.hawaii.edu/facdev/guidebk/teachtip/m-files/m-active.htm

De Jong, O. (2000). Crossing the borders: Chemical education research and teaching practice. *University Chemistry Education, 4*(1), 31–34.

Fahy, P. J., Crawford, G., Ally, M., Cookson, P., Keller, V., & Prosner, F. (2000). The development and testing of a tool for analysis of computer mediated conferencing transcripts. *Alberta Journal of Educational Research, 46*(1), 85–88.

Gabel, D. (1999). Improving teaching and learning through chemistry education research: A look to the future. *Journal of Chemical Education, 76*(4), 548–554.

Gelisli, Y. (2009). The effect of student centered instructional approaches on student success. *Procedia Social and Behavioral Sciences, 1*(1), 469–473.

Gibbs, G. (1995). *Assessing Student Centred Courses*. Oxford: Oxford Centre for Staff Learning and Development.

Goodwin, C. J. (1995). Other research methods. In K. Dubno (Ed.), *Research in psychology. Methods and design*. New York: Wiley.

Harden, R. M., & Crosby, J. (2000). AMEE guide no 20: The good teacher is more than a lecturer—the twelve roles of the teacher. *Medical Teacher, 22*(4), 334–347.

Huet, I., Oliveira, J. M., Costa, N., & Estima de Oliveira, J. (2009). The effectiveness of curriculum maps of alignment in higher education. In C. Holtham & C. C. Nygaard (Eds.), *Improving students' learning outcomes* (pp. 275–287). Copenhagen: Copenhagen Business School Press.

Lea, S. J., Stephenson, D., & Troy, J. (2003). Higher Education Students' attitudes to student centred learning: Beyond educational bulimia. *Studies in Higher Education, 28*(3), 321–334.

Moskal, B. M. & Leydens, J. A. (2000). Scoring rubric development: Validity and reliability. *Practical Assessment, Research & Evaluation, 7*(10). Retrieved September 5, 2010, from http://pareonline.net/getvn.asp?v=7&n=10

National Research Council. (1999). *How people learn: Bridging research and practice*. Washington, DC: National Academy Press.

O'Sullivan, M. (2004). The reconceptualisation of learner-centred approaches: A Nambian case study. *International Journal of Educational Development, 24*(6), 585–602.

O'Neill, G., McMullin, B. (2005). Student–centred learning: What does it mean for students and lecturers? In G. O'Neill,.S. Moore, & B. McMullin (Eds.), *Emerging issues in the practice of university learning and teaching*. Dublin: AISHE. Retrieved June 5, 2008, from http://www.aishe.org/readings/2005-1/

Ross, S. M., & Smith, L. J. (1996). *Classroom observation measure observer's manual*. Memphis: University of Memphis, Center for Research in Educational Policy.

Ross, S. M., Troutman, A., Horgan, P., Maxwell, S., Laitinen, R., & Lowther, D. (1997). The success of schools in implementing eight restructuring designs: A synthesis of first-year evaluation outcomes. *School Effectiveness and School Improvement, 8*(1), 95–124.

Rogoff, B. (1999). Cognitive development through social interaction: Vygotsky and Piaget. In P. Murphy (Ed.), *Learners, learning and assessment* (pp. 69–82). London: Open University Press.

Rourke, L., Anderson, T., Garrison, D. R., & Archer, W. (2000). Methodological issues in the content analysis of computer conference transcripts. *International Journal of Artificial Intelligence in Education.* Retrieved July 5, 2009, from http://iaied.org/pub/951/file/951_paper.pdf

Scott, J., Buchanan, J., & Haigh, N. (1997). Reflections on student-centered learning in a large class setting. *British Journal of Educational Technology., 28*(1), 19–30.

Talanquer, V., Novodvorsky, I., & Tomanek, D. (2010). Factors influencing entering teacher candidates' preferences for instructional activities: A glimpse into their orientations towards teaching. *International Journal of Science Education, 32*(10), 1389–1406.

Trevisan, M. S. (2004). Practical training in evaluation: A review of the literature. *American Journal of Evaluation, 25*(2), 255–272.

Waxman, H. C. (2003). Systematic classroom observation. In J. W. Gurthrie (Ed.), *Encyclopedia of education* (pp. 303–310). New York: Macmillan.

Waxman, H. C., Padron, Y. N., Franco-Fuenmayor, S. E., & Huang, S.-Y. L. (2009). Observing classroom instruction for ELLs from student, teacher, and classroom perspectives. *The TABE Journal, 11*(1), 63–95.

Wissiak Grm, K., & Ferk Savec, V. (2007). Spremljanje in ocenjevanje bodočih učiteljev kemije med praktičnim pedagoškim usposabljanjem. In M. Juriševič, M. Lipec-Stopar, Z. Magajna, & M. Krajnčan (Eds.), *Praktično pedagoško usposabljanje : izhodišča—model—izkušnje* (pp. 192–213). Ljubljana: University of Ljubljana, Faculty of Education.

Zohar, A., & Schwartzer, N. (2005). Assessing teacher's pedagogical knowledge in the context of teaching higher-order thinking. *International Journal of Science Education, 27*(13), 1595–1620.

About the Authors

Sevil Akaygun finished her Ph.D. at the University of Northern Colorado, USA. She is an Instructor at Bogazici University. She has conducted research in both chemistry education and nanotechnology. Her current research focuses on learning chemistry from simulations and animations and employs novel cognitive science-based research methods that she has developed. She taught secondary-school chemistry for 12 years and college-level methods courses in the teaching of chemistry. She organized a national conference on science education and has coordinated a variety of international education projects.

Miri Barak is a Senior Researcher and Lecturer in the Department of Education in Technology and Science, Technion—Israel Institute of Technology. Her activities in the past 15 years have focused on developing, integrating, and assessing science education curricula at the high school and higher education levels. Her research concerns theories on constructivist learning environments and applications of educational technologies for enhancing such environments. Her studies involve the use of information and communication technologies, with emphasis on emerging Web capabilities, to foster science teaching and meaningful learning.

Katharina Beck studied chemistry, food-chemistry, geography, and education at the Universities of Hamburg and Bremen (both Germany) for becoming a teacher in secondary schools. Part of her examination was a thesis based on a Participatory Action Research study in cooperation with an action research group of teachers. The study dealt with cooperative learning under inclusion of ICT and open lab instruction in introductory chemistry classes. She is now completing her teacher training by the mandatory in-service training program as a young teacher at the KGS Kirchweyhe (comprehensive school) and within the teacher training center Verden.

George Bodner is the Arthur E. Kelly Distinguished Professor of Chemistry, Education and Engineering at Purdue University. After receiving a Ph.D. in both Inorganic and Organic Chemistry from Indiana University, he began his academic career teaching general chemistry at the University of Illinois. In 1977, he took a

I. Devetak and S. A. Glažar (eds.), *Learning with Understanding in the Chemistry Classroom*, DOI: 10.1007/978-94-007-4366-3,

faculty position in chemical education at Purdue, where he has worked with more than 65 graduate students pursuing M.S. and/or Ph.D. degrees. He has received the Nyholm Medal in Chemical Education from the Royal Society of Chemistry and the Pimentel Award in Chemical Education from the American Chemical Society.

Diane M. Bunce is a Professor of Chemistry at The Catholic University of America, Washington, DC where she directs the Ph.D. program in Chemical Education. She serves as the Associate Editor of the *Journal of Chemical Education for Chemical Education Research* and is a co-author of the book Nuts and Bolts of Chemical Education Research. Diane has also served as one of the original authors for the American Chemical Society's curricula projects (ChemCom, Chemistry in Context and Chemistry). Dr. Bunce's research focuses on how students learn chemistry and the mismatch between how students learn and the way chemistry is traditionally taught.

Liberato Cardellini is an Associate Professor in Chemistry at the Polytechnical University of Marche (Italy). He has more than 150 research publications in chemical education and on study of charge-transfer complexes and chemistry of free radicals. He is the author of two books, one on how to solve chemical problems and another on the use of computers in chemical education. He has spent three periods of study leave doing radical chemistry research at the Chemical Centre of the University of St. Andrews in Scotland, and he worked with Prof. Alex Johnstone, Director of the Centre for Science Education at the University of Glasgow.

Miriam Carmi is a chemistry teacher at the Gymnasia Herzeliyya High School in Tel-Aviv, Israel. Her Ph.D. research on chemistry teachers' leadership was advised by Prof. Avi Hofstien and Dr. Ruth Ben-Zvi and completed at the Department of Science Education at Weizmann Institute of Science. Her postdoctoral research on the assessment of teachers' pedagogical content knowledge during the implementation of a new curriculum was completed at the Department of Education in Science and Technology, Technion, Israel Institute of Technology, advised by Prof. Yehudit Judy Dori. Miriam Carmi is the leading author of a high-school chemistry textbook on thermodynamics. During the last decade, she has been a mentor and an instructor for chemistry teachers in numerous workshops.

Kirsten Lowrey Casey is a Professor of Chemistry and chair of the science department at Anne Arundel Community College. Having decided to become a Chemistry Professor at the age of 11, she eventually left her home in Greenbelt, MD to journey from coast to coast to earn a B.S. in Chemistry from the University of California San Diego. As a chemistry student, tutor, and teaching assistant at UCSD she concluded that there had to be a better way to teach chemistry. She therefore came to Purdue, where she received her M.S and Ph.D. degrees in

Chemical Education. She then moved back to Maryland and took a position at Anne Arundel Community College, which is roughly 30 miles from where her journey started, and where she received the Rookie Professor of the Year award.

Sanoe Chairam holds a Ph.D. in Science Education from Mahidol University studying the materials science based on nanostructure synthesis and applications, and experimental design based on inquiry-based experiments for enhancing student learning in chemistry. Presently, he is a Lecturer at the Department of Chemistry, Ubon Ratchathani University, Thailand. His current research interests include the polymer-assisted synthesis of nanostructures and catalytic applications, and the development of new inquiry-based experiments in chemistry for science students.

A. L. Chandrasegaran has been a postdoctoral researcher at the Science and Mathematics Education Centre, Curtin University of Technology for the past 5 years. He was previously a science and chemistry teacher for 38 years in Malaysia, Singapore, and Western Australia, including 12 years as a secondary school principal in Malaysia.

Gail Chittleborough is a Senior Lecturer in Science Education at Deakin University, Victoria Australia, working with preservice primary, secondary science, and chemistry teachers. Gail studied Science, majoring in Chemistry and Education at Macquarie University in NSW and worked as a Science and Chemistry teacher in Australia, PNG, and UK for over a decade. Gail studied Computer Education at Melbourne University before working on an Australian Research Council study into the use of models and modeling in secondary science education. Gail completed her doctoral studies into the role of teaching models and chemical representations in developing students' mental models of chemical phenomena at Curtin University, Perth Western Australia.

Richard K. Coll is an Associate Professor of Science Education at the University of Waikato, New Zealand. Richard holds a Ph.D. in Chemistry from Canterbury University and an EdD in Science Education from Curtin University of Technology. He worked as a Lecturer in chemistry in the Pacific and Caribbean before taking up his current appointment at Waikato. Richard is currently Director of Cooperative Education, and Deputy Dean of the School of Science and Engineering. His research interests are concerned with mental models of science concepts, and a variety of aspects of work-integrated learning.

Chanyah Dahsah holds a B.Sc. in Chemistry from Mahidol University, Thailand and a Ph.D. in Science Education, from Kasetsart University, Thailand. She is a Lecturer in the Institute for Innovative Learning, at Mahidol University, Thailand, and her research interests are in the areas of conceptual change in science, students' conceptual understanding and problem-solving in quantitative chemistry, professional development in enhancing pedagogical content knowledge in chemistry, active learning, and formative assessment.

Iztok Devetak is an Assistant Professor of Chemical Education at University of Ljubljana, Faculty of Education, Slovenia. His research focuses on how students from elementary school to university learn chemistry at all three levels of chemical concepts (i.e., macro, submicro, and symbolic). Dr. Devetak also studies the influence of motivation and ICT and active learning approaches on science learning. He is also involved in research projects about implementing liquid crystals to schools and PROFILES project (7th Framework Program). He is a member of ESERA (**E**uropean **S**cience **E**ducation **R**esearch **A**ssociation), Slovenian Chemical Society, and editorial boards of International Journal of Environmental and Science Education, Cypriot Journal of Educational Sciences, Eurasian Journal of Physics and Chemistry Education and CEPS Journal.

Yehudit Judy Dori is the Dean of Continuing Education and External Studies. She is also a Professor at the Department of Education in Technology and Science, Technion—Israel Institute of Technology. She was also a visiting Professor at MIT. Her research interests are scientific visualizations, higher-order thinking skills, and educational assessment at both high school and university levels. Her studies include development and implementation of teaching approaches and curricula, and assessment of their educational value. She advised 35 graduate students and received numerous grants from the Israeli Ministry of Education and the European Union. She was the guest editor of *Journal of Science Education and Technology* special issues on the educational reform at MIT.

Nina Dunker has got a Ph.D. in Education and has carried out research in the area of primary science education. She investigated effects of conceptual understanding based on laboratory courses and the use of concept maps. She is also the coordinator of the science laboratory CHEMOL for primary school students and teacher education at the University of Oldenburg.

Ingo Eilks is a full-time Professor for didactics of chemistry at the University of Bremen (Germany). In 1997, he received his Ph.D. in Chemistry Education at the University of Oldenburg. Following a period as grammar school teacher, he started working in the field of chemistry teacher education and commenced research chemistry education. His actual research interests encompass Participatory Action Research in chemistry education, cooperative learning, sociocritical approaches to science education, or studies about science teachers' beliefs and knowledge. Among several other activities, Ingo Eilks is an editor of the textbook series Chemie Interaktiv for German middle and comprehensive schools.

Wiebke Endres has studied Chemistry and Mathematics for secondary teaching and is currently working as a teacher at a local Gymnasium. She is also involved in university teacher education as an assistant teacher and has started a Ph.D. project on transfer in context-based education.

Saša Aleksij Glažar is a Professor for Chemical Education at University of Ljubljana, Faculty of Education, Slovenia. He is a Lecturer at the undergraduate and postgraduate level of chemical education. He has been a member of International Centre for Chemical Studies (ICCS) University of Ljubljana since its foundation. Dr. Glažar has been involved in defining, classifying, and categorization of science concepts in building relational systems for various levels of education, structuring information into knowledge maps, i.e., hierarchically built relational systems in science education. These systems have been tested through evaluation of knowledge of students at various levels, thus identifying misconceptions. The same approach was also applied in designing study programs for all levels of education. Through the Tempus project, which is focused on early science learning, the author is concerned with the development of comprehension for concrete and abstract concepts to learning rules and solving problems. As expert is taking part in development of new science curricula (Kosovo, Montenegro, Serbia) and in international research dealing with evaluation of science knowledge of pupils in primary and secondary school (TIMSS and PISA).

Katarina Senta Wissiak Grm is an Assistant Professor and Researcher in Chemical Education at The University of Ljubljana, Slovenia. She took a B.Sc., M.Sc., and a Ph.D. in Chemistry, all at the University of Ljubljana. She taught chemistry in secondary (high) schools, before moving into both chemical and chemical education research. Her chemical research has been on the biodegradation and photolysis of organic pollutants in the atmosphere. Her current chemical education research interest is in evaluation of methods for teaching and learning science, including associated teaching material.

Ninna Jansoon is an Assistant Professor in the Department of Chemistry, Faculty of Science, at Thaksin University, Songkhla, Thailand. She holds a Ph.D. in Science and Technology Education from Mahidol University, Bangkok, Thailand. The subjects she teaches include General Chemistry, Analytical Chemistry (lectures and Practical), and Integrative Science. Her main research interests are in designing and implementing intervention strategies in chemistry education, especially for undergraduate students. She is currently also interested in using mental models and analogies to develop students' understanding of chemistry concepts.

Loretta Jones finished her Ph.D. and D.A. at the University of Illinois at Chicago, USA and is an Emeritus Professor of Chemistry at the University of Northern Colorado. She conducts research on chemical education, particularly the active involvement of students in their learning, the applications of advanced technologies, and visual learning. She is a principal developer of two award-winning multimedia chemistry courses and helped to lead a large undergraduate chemistry reform project. She has taught chemistry at the college level since 1979. She was a Consulting Scholar for IBM and the 2006 Chair of the American Chemical Society Division of Chemical Education.

Mojca Juriševič is an Associate Professor in Educational Psychology at University of Ljubljana, Faculty of Education. Her research interests lie in personality factors which influence learning processes, primarily academic self-concept and motivation for learning. She works closely with teachers in preservice and in-service teacher training, especially through pedagogical practice, teacher's portfolios, and building of professional self-concept. She is a member of the National Expert Team for Gifted Pupils and Head of the Centre for Research and Stimulation of Giftedness at University of Ljubljana, Faculty of Education.

Stephan Kienast is a teacher for Chemistry and Physics at the Ursulinengymnasium (grammar school) in Werl (Germany) since 1998. In 1996, he received his Ph.D. for a study on students' misconceptions concerning chemical equilibria under supervision of H.-J. Schmidt at the University of Dortmund. He is a member of a Participatory Action Research group accompanied by the University of Bremen focusing innovative methods in chemistry teaching, i.e., cooperative learning or sociocritical approaches to science teaching. He is also a part of an implementation project focusing context-based chemistry curricula and co-author of the textbook series Chemie Interaktiv for German middle and comprehensive schools.

Paweł Kościelniak is a Professor at the Jagiellonian University (JU, Kraków, Poland) and the Institute of Forensic Research (Kraków). Since 1997 he is a Head of Department of Analytical Chemistry (JU). The main areas of his educational and scientific interest are flow analysis, environmental analysis, and forensic chemistry. On his own initiative, the regular courses in the field of forensic chemistry have been established at the Faculty of Chemistry (JU) as at first educational center in Poland. He led the TEMPUS—EDUFORMAK project and is an author of hundreds of research papers and several educational contributions (e.g., presented on the European Variety in Chemistry conferences in 2005 and 2007).

Iwona Maciejowska is an Assistant Professor at the Department of Chemical Education, Jagiellonian University (Kraków, Poland), with 20 years university and high-school teaching experience. Currently she teaches students of chemistry and environmental sciences, as well as future teachers. She is a member of International Advisory Board "Chemical Education Research and Practice," "*Journal of Science Education*," and 'GU/NSE' and a vice-president of Division on Chemical Education EuCheMS. Her research interests focused mostly on use of active methods in chemistry education and academic teachers training. She contributed over 90 publications in those and other fields of science education.

Patricia Metz is a program director with the Johns Hopkins University Center for Talented Youth where she works with both the academic summer programs and the international programs. She earned her Ph.D. in Chemistry Education from Purdue University. Before coming to CTY, she taught chemistry at the university level for 25 years.

Bettina Most is teacher for Chemistry and Biology at the Wilhelm-Röntgen-Realschule (middle school) in Dortmund (Germany) since 2001. From 2004, she also started working as a teacher trainer within the mandatory in-service chemistry teacher training program at the teacher training center Dortmund. As a practitioner researcher, she is part of a Participatory Action Research group accompanied by the University of Bremen dealing with innovative curricula and methods for chemistry teaching, i.e., cooperative learning, sociocritical approaches to science teaching, and context-based curricula. Bettina Most is co-author of the textbook series Chemie Interaktiv, a textbook series for German middle and comprehensive schools.

Zuzi Mthembu completed her doctoral studies at Curtin University of Technology while seconded from the Centre for Advancement of Science and Mathematics Education (CASME) in Durban South Africa where she conducted professional development programs for teachers.

Ilka Parchmann became Director of the department of chemistry education at the Leibniz Institute for Science Education (IPN) and a Professor for chemistry education at the University of Kiel in October 2009. Before, she worked as a Professor for chemistry education at the University of Oldenburg, where she carried out research and developmental projects in the areas of context-based learning, teacher education, and science laboratories for school students. She is currently the Chair of the division of chemistry education of EuCheMS and the editor of two German chemistry education journals.

Renata Wietecha-Posłuszny is an Assistant Professor at the Laboratory for Forensic Chemistry, Jagiellonian University Kraków, Poland, researcher and doctoral students supervisor, teaching students of forensic and analytical chemistry. She cooperates with the Institute of Forensic Research (Kraków), and other European forensic centers (e.g., NTU, England), participates in international projects, e.g., TEMPUS (EDUFORMAK), Socrates-Comenius (CITIES). She is an author of laboratory exercises, educational materials, and courses for students of forensic chemistry, and advanced spectroscopy in chemistry. Author of 22 publications and over 30 conference presentations (e.g., "Is this last will fraudulent?—An experiment for the forensic chemistry laboratory course," Third European Academy of Forensic Science Meeting, 2003).

Norman Reid has taught chemistry at school and university and has also been a headteacher of a large secondary school. He directed a large international research center exploring learning in the sciences and mathematics. Born in Scotland, he travels and lectures widely throughout the world and has published extensively. Currently, he is Emeritus Professor of Science Education, University of Glasgow; Honorary Professor of Science Education, University of Dundee; Honorary Senior Research Fellow, Department of Physics, University of Glasgow; and is the President of the Education Division, the Royal Society of Chemistry.

Vesna Ferk Savec is an Assistant Professor and researcher in chemical education at the University of Ljubljana, Slovenia. She took a B.Sc. in Chemistry and Chemical Education, and an M.Sc. and Ph.D. in Chemical Education, all at the University of Ljubljana. Her current research interests are in evaluation of methods for teaching and learning science, including the usefulness of help-tools and associated teaching materials.

Kristina Sodja was preparing her diploma thesis as an undergraduate student of the Faculty of Education—program chemistry and biology at the time that the research on hands-on approach in teaching chemistry was conducted. She was awarded for her diploma work at the Faculty of natural sciences and engineering, with a prestigious Prešern's Prize for the best students' achievement. She is now a respected teacher at one of the primary school in Slovenia.

Keith S. Taber is a Senior Lecturer in Science Education at the University of Cambridge (UK). His doctoral research concerned students' developing understanding of chemical bonding. He is a chemistry graduate who spent 17 years teaching in schools and further education. He has been the Royal Society of Chemistry's Teaching Fellow, and chair of its Chemical Education Research Group. His books include "Chemical Misconceptions: prevention, diagnosis and cure" (RSC: 2002) and "Progressing Science Education: Constructing the scientific research programme into the contingent nature of learning science" (Springer: 2009). His research interests relate to various aspects of student learning in science.

David Treagust is a Professor of Science Education in the Science and Mathematics Education Centre at Curtin University of Technology in Perth, Western Australia and teaches courses in campus-based and international programs related to teaching and learning science. He taught secondary science for 10 years in schools in England and in Australia prior to working in universities in the USA and Australia. His research interests are related to understanding students' ideas about science concepts, and how these ideas contribute to conceptual change and can be used to enhance the design of curricula and teachers' classroom practice.

Georgios Tsaparlis is Professor of Science Education at the Department of Chemistry of University of Ioannina, Greece. His research interests are on *science/ chemistry education, and has* authored or co-authored: 47 original publications in international peer-reviewed journals; 7 chapters in international collective volumes; 12 chemistry and physical chemistry books in Greek for schools and for universities. He is the founding editor and since 2005 joint editor of the *Electronic Journal of Chemistry Education Research and Practice*, published by the Royal Society of Chemistry. He has been invited plenary or keynote speaker at many international conferences, and he is a member of the editorial boards of most leading science education journals.

Metka Vrtačnik has been full-time Professor of Chemical Education at the University of Ljubljana since 1998. She took an M.Sc. in Chemical Education in 1980 and a Ph.D. in Chemistry in 1986. She is a member of the Slovenian Chemical Society, and chairperson of the section of Chemical education. She is also a chairperson of the Department of Chemical Education and Informatics, at the University of Ljubljana, Faculty of Natural Sciences and Engineering. She teaches several courses, such as Chemical didactics, Experimental methods for chemistry teachers, Information methods in chemistry, Information-communication technology for chemists, etc. Her research interest is application of QSAR studies in biodegradation of organic pollutants, university-industry cooperation, pattern recognition, ICT in chemical education, active teaching/ learning strategies and motivation for learning science.

Vickie M. Williamson received her Ph.D. from the University of Oklahoma. She has experience teaching at secondary, community college, and university levels. Since 1997 she has been a faculty member at Texas A&M University, where she teaches freshman chemistry and graduate courses in chemical education. Dr. Williamson has written inquiry-based curriculum materials for middle school through university and served on editorial boards. Her research and teaching articles have appeared in a number of journals and books. Her research interests include visualization to aid student understanding of the particulate nature of matter and inquiry-based teaching in lecture and laboratory.

Torsten Witteck is teacher of Chemistry and Mathematics since 1999, currently working at the Engelbert-Kaempfer-Gymnasium (grammar school) in Lemgo (Germany). In 2006, he received a Ph.D. in Chemistry Education under supervision of I. Eilks at the University of Bremen dealing with the development and evaluation of chemistry lesson plans based on cooperative learning, ICT, and open lab instruction. Actually, he is part of a Participatory Action Research group focusing alternative methods and curricula in chemistry education accompanied by the University of Bremen and co-author of the textbook series Chemie Interaktiv for German middle and comprehensive schools.

Michał Woźniakiewicz is a Research Assistant at the Laboratory for Forensic Chemistry at the Jagiellonian University (JU, Kraków, Poland); a researcher, an academic teacher, and a lecturer at the TEMPUS—EDUFORMAK project. He focuses his scientific interest on forensic chemistry but also persistently develops new exercises for students, especially using the "teaching-in-context" approach. He is the author of 21 scientific papers and several conference contributions (e.g., "Bank robbery?—example of forensic chemistry training," Variety in Chemistry Education, 2004). Currently, he conducts a new project concerning the implementation of blended learning in the forensic chemistry module at the JU.

Author Index

I. Devetak and S. A. Glažar (eds.), *Learning with Understanding in the Chemistry Classroom*, DOI: 10.1007/978-94-007-4366-3,

Subject Index

I. Devetak and S. A. Glažar (eds.), *Learning with Understanding in the Chemistry Classroom*, DOI: 10.1007/978-94-007-4366-3,

MIX
Papier aus verantwortungsvollen Quellen
Paper from responsible sources
FSC® C105338

If you have any concerns about our products,
you can contact us on
ProductSafety@springernature.com

In case Publisher is established outside the EU,
the EU authorized representative is:
Springer Nature Customer Service Center GmbH
Europaplatz 3, 69115 Heidelberg, Germany

Printed by Libri Plureos GmbH
in Hamburg, Germany